Strong Coulomb Correlations in Electronic Structure Calculations

Advances in Condensed Matter Science
A series edited by ***D.D. Sarma, G. Kotliar and Y. Tokura***

Volume 1
Strong Coulomb Correlations in Electronic Structure Calculations: Beyond the Local Density Approximation, edited by Vladimir I. Anisimov

Forthcoming titles in the series
Colossal Magnetoresistive Oxides, edited by Y. Tokura
Introduction to DV-Xα Quantum Chemistry, by J. Kawai and H. Adachi
Advances in Amorphous Semiconductors, by J. Singh and K. Shimakawa
Electronic Structure of Non-periodic Solids: Alloys, Surfaces and Clusters, by A. Mookerjee and D.D. Sarma

This book is part of a series. The publisher will accept continuation orders which may be cancelled at any time and which provide for automatic billing and shipping of each title in the series upon publication. Please write for details.

Strong Coulomb Correlations in Electronic Structure Calculations

Beyond the Local Density Approximation

Edited by

Vladimir I. Anisimov

Institute of Metal Physics
Ekaterinburg, Russia

CRC Press is an imprint of the
Taylor & Francis Group, an **informa** business

CRC Press
Taylor & Francis Group
6000 Broken Sound Parkway NW, Suite 300
Boca Raton, FL 33487-2742

Visit the Taylor & Francis Web site at
http://www.taylorandfrancis.com

and the CRC Press Web site at
http://www.crcpress.com

ISBN: 9789056991319 (hbk)
ISBN: 9780367578961 (pbk)
ISBN: 9780429081613 (ebk)

Contents

Chapter 2. LDA+U Method: Screened Coulomb Interaction in the Mean-field Approximation

Vladimir I. Anisimov and A.I. Lichtenstein

Chapter 3. LSDA and Self-Interaction Correction

Takeo Fujiwara, Masao Arai and Yasushi Ishii

Chapter 4. Orbital Functionals in Density Functional Theory: The Optimized Effective Potential Method

T. Grabo, T. Kreibich, S. Kurth and E.K.U. Gross

Contributors

Vladimir I. Anisimov
Institute of Metal Physics, Ekaterinburg, GSP-170, Russia

Masao Arai
Department of Applied Physics, University of Tokyo, Bunkyo-ku, Tokyo 113, Japan

Ferdi Aryasetiawan
Department of Theoretical Physics, University of Lund, Sölvegatan 14A, S-223 62 Lund, Sweden

Takeo Fujiwara
Department of Applied Physics, University of Tokyo, Bunkyo-ku, Tokyo 113, Japan

T. Grabo
Institut für Theoretische Physik, Universität Würzburg, Am Hubland, 97074 Würzburg, Germany

E.K.U. Gross
Institut für Theoretische Physik, Universität Würzburg, Am Hubland, 97074 Würzburg, Germany

Yasushi Ishii
Department of Applied Physics, University of Tokyo, Bunkyo-ku, Tokyo 113, Japan

T. Kreibich
Institut für Theoretische Physik, Universität Würzburg, Am Hubland, 97074 Würzburg, Germany

S. Kurth
Department of Physics and Quantum Theory Group, Tulane University, New Orleans, LA 70118, USA

A.I. Lichtenstein
Institut für Festkörperforschung des Forschungszentrum Jülich, D-52425 Jülich, Germany

Introduction

In the last 30 years, numerical electronic structure calculations have become a well established branch of solid state physics. While for finite systems, such as atoms and molecules, more sophisticated and rigorous calculation methods exist, for the extended systems studied in condensed matter physics the only widely used practical tool up to now is Density Functional Theory (DFT) in the Local Density Approximation (LDA). It has such a great predictive power, that not only are charge and spin density, one-electron and total energies obtained in LDA in general in good agreement with experimental data, but it was possible to develop *ab initio* molecular dynamics methods, based on LDA. Such methods achieved the level of numerical experiment, because even such complicated effects as reconstruction of the crystal surface can be correctly described by them.

However, all this is true only as long as one is treating electronic states extended over the whole system, so that electrons can be considered completely delocalized. If there are reasons to believe that the electrons partially preserve their localized atomic-like nature, as is the case for 4f-states in rare-earth atoms and sometimes for 3d-states of transition metal atoms, then the LDA is in serious trouble. The typical examples are rare-earth metals, where LDA predicts a strong peak of 4f-orbital origin in density of states on the Fermi level, which is in total disagreement with experiment, and some insulating transition metal oxides, which LDA predicts to be metallic. The problem of LDA here could be traced to the Coulomb correlations between electrons, because the LDA equations describe non-interacting electrons moving in an effective self-consistent mean field.

Such kinds of systems are not an exception, quite the opposite, the majority of the new promising materials with anomalous electronic and magnetic properties belong to this class. Examples are heavy-fermion compounds, copper-oxides based superconductors, colossal magnetoresistance manganites etc. There is a serious demand for adequate information about electronic structure of such compounds and there were many attempts to develop calculation methods, which would be able to treat those systems. The present book describes the approaches which were most successful in solving this problem. All these approaches are based on LDA as a starting point and introduce additional terms intended to treat Coulomb correlations between electrons.

The best justified is the so-called *GW* approximation, which is formally the first term in the perturbation expansion of self-energy operator in powers of the screened Coulomb interaction. In contrast to DFT, which gives by construction the properties of the ground state of the system and where an excitation spectrum is not a well defined property, the Green function formalism used in the derivation of the *GW* approximation is a proper way to describe the propagation of a hole or an added electron, which is essential for describing spectral experiments, such as photoemission and inverse photoemission. Another way to look at the *GW* approximation is to regard it as a Hartree-Fock approximation with a frequency dependent screened interaction. The crucial part of the *GW* approximation method is the calculation of the response (or polarization) function to describe screening. In existing calculation schemes it is done by using the eigenvalue spectrum and corresponding eigenfunctions obtained by solving some simpler approximate Hamiltonian, usually LDA or Hartree-Fock. If this spectrum is a good approximation to the exact one, then the response function calculated with it gives a good self-energy operator and an excellent agreement with experimental spectra, as is the case for semiconductors like Si. However, if the correlation effects are strong and the LDA spectrum is a rather bad approximation for exact spectra, as it is for rare-earth and transition metal oxides, the response function must be calculated with corrected spectra or using the Green function with included correlations (the so called self-consistent *GW* approximation). It is not only a much more complicated procedure calculationally, but also taking into account only the first order of screened potential (which is the *GW* approximation) is not well justified. It is possible (although very complicated in realization) to go beyond *GW* including vertex corrections and some attempts in this direction have been made (cumulant expansion, T-matrix approximation).

In spite of these problems the *GW* approximation was used for strongly correlated systems and the results are described in the *GW* contribution to the present book. As self-energy is a non-local and energy dependent operator, and the screened Coulomb interaction requires calculation of full response function, the *GW* approximation is computationally very heavy and some more simple methods were proposed.

One of them is the so called LDA+U method, where instead of the energy-dependent interaction the averaged static Coulomb interaction between localized orbitals is used. This approach uses the ideology of the Hubbard and Anderson models and can be viewed as a promising way of connecting the theory of strongly correlated systems, which is based on such models, with the *ab initio* electronic structure calculation methods. In contrast to the *GW* approximation, where the screened Coulomb interaction is used in the form of a non-local energy dependent operator, in the Hubbard model a set of site-centered atomic-type orbitals is introduced and the Coulomb interaction is present only between electrons on such orbitals. If a particular LDA method uses an atomic-type orbitals basis set for realization of its calculation scheme, then the resulting hamiltonian matrix in this basis can be interpreted as a mean-field solution of the extended Hubbard model. This can be called the 'orbital-restricted' solution because LDA-potential is the same for all orbitals (for example for all five 3d-orbitals of a transition metal ion).

The corresponding extended Hubbard model can be solved in a spin-orbital unrestricted mean-field approximation, which is able to treat Mott insulators with long-range magnetic order. The difference between those two solutions gives an orbital-dependent correction to the LDA potential which has the form of a projection operator.

The resulting LDA+U method can reproduce the splitting of d- (or f-)band into occupied lower and unoccupied upper Hubbard bands, which was the main problem of LDA. While taking into account the strongest part of Coulomb correlations, the LDA+U method preserves the main advantage of LDA, its *ab initio* nature, where all potentials, wave functions and parameters are calculated inside and no empirical parameters are used. That leads to the possibility of investigating the influence of correlation effects on structural properties, such as Jahn-Teller distortions. Orbital dependence of the LDA+U potential allows one to treat orbital and charge ordering in transition metal compounds, which is very important for modern materials with anomalous magnetic and electronic properties. The deficiency of the method is the necessity to define explicitly the set of localized orbitals for interacting electrons. While for rare-earth and transition metal ions the good approximation is to use atomic-type f- or d-orbitals, for more extended systems like, for example, semiconductors, some more complicated Wannier-type orbitals are needed.

Another method described in the present book is the Self Interaction Correction (SIC) method, where the major part of the LDA error in treating localized states, the Coulomb interaction of the electron with itself, is explicitly subtracted. In this method, in contrast to LDA+U, the natural set of the localized orbitals can be determined minimizing the total energy. Ideologically it means restoring to the LDA functional the properties of the Hartree-Fock method, where the self-interaction is absent. As is known, the main problem of Hartree-Fock is the unscreened nature of the Coulomb interaction leading to overestimation of energy gap values by 2–3 times as compared with experimental data. In the SIC-method the screening is partially taken into account via the exchange-correlation part of the SIC-potential. However, the resulting potential splitting between occupied and unoccupied orbitals is still significantly overestimated.

The appeal of SIC is that the potential correction is still the 'LDA-style' one and the method has an ideological consistency. It was applied to various systems from solid hydrogen and transition metal oxides to rare-earth compounds and the results were reasonably good. Even the above mentioned overestimation of potential splitting is reduced in one of the versions of SIC-method described in the corresponding chapter of this book.

The most close in spirit to Density Functional Theory is the Optimized Effective Potential (OEP) method, where one uses a functional of spin orbitals. It can be regarded as an important step on the way from LDA to the Exact Density Functional. While the Hohenberg-Kohn theorem proves that there exists a functional of density giving the exact ground state electron density and exact ground state energy, the theorem gives no hints on the possible form of this functional. In OEP method the exact local exchange-correlation potential can be found by solving an integral equation if the explicit form of the exchange-correlation energy as a functional of one-electron orbitals is given. The simplest form is the exchange-only functional, but the correlation effects can also be included.

All these methods rather satisfactorily describe Mott insulators with long range magnetic order, which were the major problem of LDA. However the most physically interesting problem is strongly correlated metals, where the metal-insulator transition happens due to the doping of a Mott insulator, or due to the critical value of ratio of U/W (U–Coulomb interaction parameter, W–effective band width). For such systems there are still no adequate methods of electronic structure calculation and the development of them is a great challenge of our times. To study this problem a lot was done by solving model hamiltonians. The first attempts to combine these achievements with *ab initio* calculation schemes are described in the LDA+U chapter of this book.

1. The *GW* Approximation and Vertex Corrections

FERDI ARYASETIAWAN

1 Introduction

The most widely used method for calculating groundstate properties and electronic structures of solids is the density functional theory (DFT) (Hohenberg and Kohn 1964, Kohn and Sham 1965) within the local density approximation (LDA) (Slater 1951a,b, 1953, Kohn and Sham 1965). For recent reviews of DFT we refer to Jones and Gunnarsson (1989) and Dreizler and Gross (1990). In this theory, one solves the set of single-particle Kohn-Sham equations with an exchange-correlation potential V^{xc} containing the effects of many-body correlations. It is customary to interpret the Kohn-Sham eigenvalues as quasiparticle energies measured in photoemission experiments although there is no clear theoretical justification, except for the highest occupied states (Almbladh and von Barth 1985a,b). For sp systems, it is found that the agreement is often rather good. However, there are serious discrepancies: the band gaps in semiconductors (Si, GaAs, Ge, etc.) are systematically underestimated (see table 2). The bandwidth in Na is about 15 % too large (see section 4). The discrepancies become worse in strongly correlated 3d and 4f systems. The bandwidth in Ni, for instance, is 30 % too large in the LDA (Hüfner *et al.* 1972, Himpsel, Knapp, and Eastman 1979). In f systems, the LDA density of states is in strong disagreement with experiment (see e.g. Anisimov, Aryasetiawan, and Liechtenstein 1996). In the Mott-Hubbard insulators of transition metal oxides, the LDA band gap is much too small compared with experiment. In some cases, the LDA gives qualitatively wrong results. For example, the Mott-Hubbard insulator CoO and the undoped parent compound of the high T_c material La_2CuO_4 are predicted to be metals whereas experimentally they are insulators (see e.g. Pickett 1989). In alkali-metal clusters, it is known that the ionization energies calculated within the LDA are too low compared to experiment (Ishii, Ohnishi, and Sugano 1986, Saito and Cohen 1988) and the discrepancy becomes worse the smaller the cluster. The band gap of solid C_{60} is about 1 eV in the LDA (Erwin and Pickett 1991, Troullier and Martins 1992, Satpathy *et al.*. 1992), which is substantially smaller than the experimental value, 2.3 eV (Lof *et al.*. 1992), obtained from photoemission and inverse photoemission. In the LDA, the image potential seen by an electron in the vacuum far from a surface decays exponentially instead of the expected $-1/4(z - z_0)$ decay where z is the coordinate normal to the surface and z_0 is the position of the image plane (Lang and Kohn 1973).

A number of attempts have been made to go beyond the LDA. Among the most successful of these are the LDA+U (Anisimov, Zaanen, and Andersen 1991, Anisimov *et al.* 1993, Lichtenstein, Zaanen, and Anisimov 1995) and the self-interaction correction (SIC) method (Cowan 1967, Lindgren 1971, Zunger, Perdew, and Oliver 1980, Perdew and Zunger 1981, Svane and Gunnarsson 1990, Szotek, Temmerman, and Winter 1993, Arai and Fujiwara 1995) described in the other contributions to this book. Another popular approach is the generalized gradient approximation (GGA) (see e.g. Perdew, Burke, and Ernzerhof 1997 and references therein for a recent development, Becke 1988, 1992, 1996, Svendsen and von Barth 1996, Springer, Svendsen, and von Barth 1996) which was initiated by the work of Langreth and Mehl (1983). While in general gradient corrections give a significant improvement in the total energy (Causa and Zupan 1994, Philipsen and Baerends 1996, Dal Corso 1996) there is almost no major improvement for quasiparticle energies (Dufek *et al.* 1994). A recent approach of improving the LDA is the optimized effective potential method (Kotani 1995, Bylander and Kleinman 1995a,b, Kotani and Akai 1996) where the exchange energy is calculated exactly and a local exchange potential is obtained by taking a functional derivative of the exchange energy with respect to the density. The correlation energy can be approximated by the LDA value. The original idea of this method is due to Talman and Shadwick (1976) in their work on atoms as a restricted minimum search of the Hartree-Fock total energy within local potentials, i.e. the orbitals are restricted to be solutions to single-particle Hamiltonians with local potentials. Another recent attempt is a generalized Kohn-Sham scheme (Seidl *et al.* 1996) with a non-local exchange-correlation potential.

These methods, however, are meant to address ground-state rather than excitation properties. A time-dependent extension of DFT (Runge and Gross 1984, Gross, Dobson, and Petersilka 1996) can treat excitation properties but so far the method is limited to finite systems. Excitation properties have also been studied using the configuration interaction (CI) method (Boys 1950, Dykstra 1988) but this method is very much restricted to small systems such as small molecules or clusters (Bonacic-Koutecky, Fantucci, and Koutecky 1990a,b, Bonacic-Koutecky *et al.* 1992).

A proper way of calculating quasiparticle energies is provided by the Green function theory (Galitskii and Migdal 1958, Galitskii 1958). In this approach, the many-body effects are contained in the self-energy operator which is non-local and energy dependent. We can think of the V^{xc} in DFT as a local and energy independent approximation to the self-energy and which gives the correct ground state density. Unfortunately, for extended systems the self-energy is rather hard to calculate even for the electron gas. Approximations must be used and the simplest theory beyond the Hartree-Fock approximation (HFA) (Hartree 1928, Fock 1930) that takes into account screening effects is the GW approximation (GWA) (Hedin 1965, Hedin and Lundqvist 1969). The earliest attempt to include correlations beyond the HFA in the form of GW theory was probably the work of Quinn and Ferrel (1958) for the electron gas. Their calculations, however, were limited to states around the Fermi energy and several approximations were made. DuBois (1959a, 1959b) also calculated the self-energy of the electron gas within a GW type theory but his calculations were only for high densities or for small values of the electron gas parameter $r_s < 1$ since $(4\pi/3)r_s^3 = \rho$, where ρ is the electron density. These results are not directly relevant to real metals which have $r_s \sim 2-5$ (Al $r_s \sim 2$, Cs $r_s \sim 5$). The first full calculation of the self-energy within the GWA for the electron gas was performed by Hedin (1965).

The GWA is formally the first term in a perturbation expansion of the self-energy in powers of the screened interaction W (Hedin 1965). In fact, many-body perturbation theory for calculating the self-energy is rather ill-defined for extended systems. The problem arises from the long-range nature of the Coulomb interaction which causes the perturbation expansion divergent or at best conditionally convergent. A straightforward evaluation of the self-energy in the perturbation expansion can yield unphysical quantities such as negative density of states (Minnhagen 1974) and this difficult problem is far from being solved. It is therefore more useful and physical to regard the GWA as a generalization of the Hartree-Fock approximation but with a dynamically screened interaction.

The GWA was applied to the electron gas by Hedin (1965) and more extensively by Lundqvist (1967a,b, 1968, 1969) and Lundqvist and Samathiyakanit (1969). For a review of the GWA and its application to the electron gas we refer to Hedin and Lundqvist (1969). Many theoretical results in many-body theory were also derived around the same period. Since then the field has been dormant since calculations for real systems were hampered by the size of the computation. It was not until the mid eighties that such calculations were possible for semiconductors (Hybertsen and Louie 1985a,b, 1986, 1987a,b, Godby, Schlüter, and Sham 1986, 1987a,b, 1988). We should also mention an earlier calculation for diamond using the tight-binding approach by Strinati, Mattausch, and Hanke (1982). The GWA has now been applied with success to many systems ranging from alkali metals, semiconductors, transition metals, surfaces to clusters.

One purpose of this article is to present a detailed description for performing GW calculations, in particular for applications to systems containing 3d or 4f states. Conventional methods for calculating the self-energy use plane waves as basis functions which are suitable for $s-p$ systems. However, applications to 3d and 4f systems are not feasible due to a large number of plane waves needed to describe the localized 3d or 4f states. In this contribution, a method for treating systems with localized states is described. The method is based on the linear muffin-tin orbital (LMTO) method (Andersen 1975) and basis functions needed to describe the response function and the self-energy are products of the LMTO's. Although the method is described in terms of the LMTO, it can be easily generalized to be used in conjunction with other band structure schemes such as the linear augmented plane-wave (LAPW) (Andersen 1975) method or the linear combinations of atomic orbitals (LCAO) method. We describe applications of the GWA to more strongly correlated systems of 3d materials. For a review of applications to other systems and the present status of the GWA we refer to a recent article (Aryasetiawan and Gunnarsson 1998). Finally, we discuss the question of self-consistency and a new development beyond the GWA.

2 Theory

2.1 The Green function and the self-energy

In a photoemission experiment, photons are used to excite electrons out of a crystal leaving holes behind and giving information about the occupied states whereas in an inverse photoemission (BIS) electrons are sent into a solid to probe the unoccupied states. The propagation of a hole or an added electron can be described by means of the one-particle Green function. For completeness, the main results of Green function theory are presented below. More details may be found in a review by Hedin and Lundqvist (1969) or in text books on many-body theory, e.g. Nozières (1964), Fetter and Walecka (1971), Inkson (1984), or Mahan (1990).

The Green function is defined as:

$$iG(x,x') = \langle N|T[\widehat{\psi}(x)\widehat{\psi}^{\dagger}(x')]|N\rangle \tag{1}$$

$$= \begin{cases} \langle N|\widehat{\psi}(x)\widehat{\psi}^{\dagger}(x')|N\rangle & \text{for } t > t'(\text{electron}) \\ -\langle N|\widehat{\psi}^{\dagger}(x')\widehat{\psi}(x)|N\rangle & \text{for } t < t'(\text{hole}) \end{cases} \tag{2}$$

We have used a notation $x \equiv (\mathbf{r}, t)$. T is the time ordering operator, and $|N\rangle$ is the groundstate of N Fermions. Thus, for $t > t'$, the Green function is the probability amplitude that an electron added at x' will propagate to x, and for $t' > t$, the probability amplitude that a hole created at x will propagate to x'. A possible spin flip can be incorporated in the definition of G. The time-ordering operator arises naturally from the time-development operator defined later in equation (12). The above definition of the time-ordered Green function treats particles and holes on equal footing which is convenient for a diagrammatic expansion.

From the Green function we can obtain:

- The expectation value of any single-particle operator in the ground state.
- The ground state energy
- The excitation spectrum

The field operators are defined in the Heisenberg representation and they satisfy the equation of motion

$$i\frac{\partial}{\partial t}\hat{\psi}(x) = [\hat{\psi}(x), \hat{H}] \tag{3}$$

where the Hamiltonian is given by

$$\hat{H} = \int dx\ \hat{\psi}^{\dagger}(x)h_0(x)\hat{\psi}(x) + \frac{1}{2}\int dxdx'\ \hat{\psi}^{\dagger}(x)\hat{\psi}^{\dagger}(x')v(x-x')\hat{\psi}(x')\hat{\psi}(x) \tag{4}$$

with

$$v(x-x') = v(|\mathbf{r}-\mathbf{r}'|)\delta(t-t') \tag{5}$$

h_0 is the kinetic energy plus any local external field which may be time-dependent. By evaluating the commutator $[\hat{\psi}(x), \hat{H}]$, we obtain the equation of motion for the Green function

$$\left[i\frac{\partial}{\partial t} - h_0(x)\right]G(x,x') + i\int d^3r_1\ v(\mathbf{r}-\mathbf{r}_1)\ \langle N|T[\hat{\psi}^{\dagger}(\mathbf{r}_1,t)\hat{\psi}(\mathbf{r}_1,t)\hat{\psi}(\mathbf{r},t)\hat{\psi}^{\dagger}(\mathbf{r}',t')]|N\rangle = \delta(x-x') \tag{6}$$

The quantity

$$\langle N|T[\hat{\psi}^{\dagger}(\mathbf{r}_1,t)\hat{\psi}(\mathbf{r}_1,t)\hat{\psi}(\mathbf{r},t)\hat{\psi}^{\dagger}(\mathbf{r}',t')]|N\rangle \tag{7}$$

is a special case of the two-particle Green function defined by

$$G_2(1,2,3,4) = (i)^2\langle N|T[\hat{\psi}(1)\hat{\psi}(3)\hat{\psi}^{\dagger}(4)\hat{\psi}^{\dagger}(2)|N\rangle \tag{8}$$

describing the propagation of two particles from $2,4$ to $1,3$ where we use a short-hand notation $1 \equiv x_1$. The two-particle Green function, in turn, satisfies an equation of motion involving a three-particle Green function and so on, forming a hierarchy of equations. To break the hierarchy, the two-particle Green function is expressed in terms of the one-particle Green functions by introducing the mass operator (Hartree potential + self-energy) $M(x,x')$ such that

$$\int dx_1 M(x,x_1)G(x_1,x') = -i\int d^3r_1\ v(\mathbf{r}-\mathbf{r}_1)\ \langle N|T[\hat{\psi}^{\dagger}(\mathbf{r}_1,t)\hat{\psi}(\mathbf{r}_1,t)\hat{\psi}(\mathbf{r},t)\hat{\psi}^{\dagger}(\mathbf{r}',t')]|N\rangle \tag{9}$$

so that from equation (6)

$$\left[i\frac{\partial}{\partial t} - h_0(x)\right] G(x,x') - \int dx_1 M(x,x_1)G(x_1,x') = \delta(x-x') \tag{10}$$

First-principles calculation of the self-energy for extended systems is a major challenge. For molecules or finite systems, the self-energy can be calculated using a finite order perturbation theory. The perturbation expansion in powers of the Coulomb potential converges since the system is finite. In contrast, finite order perturbation expansion does not work in extended systems except to first order only which corresponds to the HFA. This is due to the long-range nature of the Coulomb interaction. It is essential that the perturbation series for a particular class of diagrams is summed to infinite order in order to have physically sensible results. There is, however, no systematic way of choosing which classes of diagrams should be summed. The choice is usually dictated by physical intuition and by the requirement that the resulting self-energy has the correct general properties.

There are at least two ways of evaluating the perturbation expansion of the self-energy, either by using Wick's theorem (see, e.g. Fetter and Walecka 1971) or by the functional derivative technique (Hedin 1965, Hedin and Lundqvist 1969, Inkson 1984). We follow the latter and give a summary of the steps. This is done by introducing a time varying field $\phi(\mathbf{r},t)$ which is used as a mathematical tool for evaluating the self-energy and it will be set to zero once the self-energy is obtained. With a time-varying field, it is convenient to work in the interaction (Dirac) picture:

$$|\psi_D(\mathbf{r},t)\rangle = \hat{U}(t,t_0)\,|\psi_D(\mathbf{r},t_0)\rangle \tag{11}$$

The time development operator $\hat{U}$ is given by

$$\hat{U}(t,t_0) = T\,\exp\left[-i\int_{t_0}^{t} d\tau\hat{\phi}(\tau)\right] \tag{12}$$

$$\hat{\phi}(t) = \int d^3r\,\phi(\mathbf{r},t)\,\hat{\psi}_D^{\dagger}(\mathbf{r},t)\hat{\psi}_D(\mathbf{r},t) \tag{13}$$

The relationship between operators in the Heisenberg and the interaction representation is

$$\hat{\psi}(x) = \hat{U}^{\dagger}(t,0)\hat{\psi}_D(x)\hat{U}(t,0) \tag{14}$$

The field operator $\hat{\psi}_D(x)$ satisfies

$$i\frac{\partial}{\partial t}\hat{\psi}_D(x) = [\hat{\psi}_D(x), \hat{H}(\hat{\phi}=0)] \tag{15}$$

so it is the same as the unperturbed ($\phi = 0$) Heisenberg operator. The Green function can now be written as

$$iG(1,2) = \frac{\langle N^0|\,T[\hat{S}\hat{\psi}_D(1)\hat{\psi}_D^{\dagger}(2)|N^0\rangle}{\langle N^0|\hat{S}|N^0\rangle} \tag{16}$$

where $\hat{S} = \hat{U}(\infty, -\infty)$ and $|N^0\rangle$ is the unperturbed ground state (an interacting ground state but with $\phi = 0$). Similarly for G_2

$$G_2(1,2,3,4) = \frac{(i)^2 \langle N^0 | T[\hat{S}\hat{\psi}_D(1)\hat{\psi}_D(3)\hat{\psi}_D^{\dagger}(4)\hat{\psi}_D^{\dagger}(2)|N^0\rangle}{\langle N^0|\hat{S}|N^0\rangle} \tag{17}$$

By taking a functional derivative of $\hat{S}$ respect to ϕ it follows from the above definition of $\hat{S}$ that

$$\frac{\delta \hat{S}}{\delta \phi(3)} = T\left[\hat{S}\hat{\psi}_D^{\dagger}(3)\hat{\psi}_D(3)\right] \tag{18}$$

Therefore

$$\frac{\delta G(1,2)}{\delta \phi(3)} = G(1,2)G(3,3^+) - G_2(1,2,3,3^+) \tag{19}$$

Using this result in the definition of the self-energy in equation (9) we get

$$\int d3\; M(1,3)G(3,2) = V_H(1)G(1,2) + i\int d3\; v(1,3)\frac{\delta G(1,2)}{\delta \phi(3)} \tag{20}$$

where V_H is the Hartree potential

$$V_H(1) = \int d2\; v(1-2)\rho(2) \tag{21}$$

Defining the self-energy Σ as

$$\Sigma = M - V_H \tag{22}$$

and using the identity

$$\frac{\delta}{\delta\phi}(G^{-1}G) = G^{-1}\frac{\delta G}{\delta \phi} + \frac{\delta G^{-1}}{\delta \phi}G = 0 \;\rightarrow\; \frac{\delta G}{\delta \phi} = -G\frac{\delta G^{-1}}{\delta \phi}G \tag{23}$$

we have

$$\begin{aligned}\Sigma(1,2) &= -i\int d3\; d4\; v(1,4)G(1,3)\frac{\delta G^{-1}(3,2)}{\delta \phi(4)} \\ &= -i\int d3\; d4\; d5\; v(1,4)G(1,3)\frac{\delta G^{-1}(3,2)}{\delta V(5)}\frac{\delta V(5)}{\delta \phi(4)}\end{aligned} \tag{24}$$

where

$$V = V_H + \phi, \tag{25}$$

Defining

$$\Lambda(1,2,3) = -\frac{\delta G^{-1}(1,2)}{\delta V(3)} \tag{26}$$

$$\epsilon^{-1}(1,2) = \frac{\delta V(1)}{\delta \phi(2)} \tag{27}$$

$$W(1,2) = \int d3\ v(1,3)\epsilon^{-1}(2,3) \tag{28}$$

we have

$$\Sigma(1,2) = i\int d3\ d4\ G(1,3)W(1,4)\Lambda(3,2,4) \tag{29}$$

W is a screened Coulomb interaction and Λ is known as the vertex function which can be written

$$\begin{aligned}\Lambda(1,2,3) &= -\frac{\delta G^{-1}(1,2)}{\delta V(3)} \\ &= \delta(1-2)\delta(2-3) + \frac{\delta\Sigma(1,2)}{\delta V(3)} \\ &= \delta(1-2)\delta(2-3) \\ &\quad + \int d(4567)\frac{\delta\Sigma(1,2)}{\delta G(4,5)}G(4,6)G(7,5)\Lambda(6,7,3)\end{aligned} \tag{30}$$

The second line is obtained from

$$G^{-1} = i\frac{\partial}{\partial t} - H_0 - \Sigma, \tag{31}$$

which follows from equation (10) where we have defined

$$H_0 = h_0 + V_H \tag{32}$$

The last line is obtained by using the chain rule $\delta\Sigma/\delta V = (\delta\Sigma/\delta G)(\delta G/\delta V)$ and by using the identity in equation (23) and the definition of Λ.

Fourier transformation of equation (10) gives

$$[\omega - H_0(\mathbf{r})]\,G(\mathbf{r},\mathbf{r}',\omega) - \int d\mathbf{r}_1\,\Sigma(\mathbf{r},\mathbf{r}_1,\omega)G(\mathbf{r}_1,\mathbf{r}',\omega) = \delta(\mathbf{r}-\mathbf{r}') \tag{33}$$

We define a quasiparticle wave function Ψ with energy E as a solution to

$$H_0(\mathbf{r})\Psi(\mathbf{r},E) + \int d\mathbf{r}_1\,\Sigma(\mathbf{r},\mathbf{r}_1,E)\Psi(\mathbf{r}_1,E) = E\Psi(\mathbf{r},E) \tag{34}$$

If we define a zeroth order Green function G_0

$$\left[i\frac{\partial}{\partial t} - H_0(x)\right] G_0(x, x') = \delta(x - x') \tag{35}$$

then we have the Dyson equation

$$G(x, x') = G_0(x, x') + \int dx_1 dx_2 G_0(x, x_1)\Sigma(x_1, x_2)G(x_2, x') \tag{36}$$

which may be verified by applying $i\partial/\partial t - H_0(x)$ on both sides of the above equation. Treating G and Σ as matrices we have

$$G = G_0 + G_0\Sigma G \Longrightarrow G^{-1} = G_0^{-1} - \Sigma \tag{37}$$

2.2 The response and polarization function

It follows from equations (21), (25) and (27) that the inverse dielectric function is

$$\begin{aligned} \epsilon^{-1} &= 1 + v\frac{\delta\rho}{\delta\phi} \\ &= 1 + v\frac{\delta\rho}{\delta V}\frac{\delta V}{\delta\phi} \end{aligned} \tag{38}$$

The response and the polarization function are defined respectively by

$$R(1, 2) = \frac{\delta\rho(1)}{\delta\phi(2)} \tag{39}$$

$$P(1, 2) = \frac{\delta\rho(1)}{\delta V(2)} \tag{40}$$

R is defined to be the change in the density with respect to the *external* potential ϕ whereas P with respect to the *total* potential V. The above response function is defined to be time-ordered and it is related to the physical (retarded) one by

$$\text{Re}R(\omega) = \text{Re}R^R(\omega), \quad \text{Im}R(\omega)\text{sgn}\omega = \text{Im}R^R(\omega) \tag{41}$$

Similarly for P.

Noting that

$$\rho(1) = -iG(1, 1^+) \tag{42}$$

we can write using the identity (23)

$$P(1, 2) = -i\int d3\, d4\, G(1, 3)\Lambda(3, 4, 2)G(4, 1^+) \tag{43}$$

In summary, we have

$$\epsilon^{-1} = 1 + vR \tag{44}$$

$$\epsilon = 1 - vP \tag{45}$$

$$R = P + PvR \tag{46}$$

$$\begin{aligned} W &= v + vPW \\ &= v + vRv \end{aligned} \tag{47}$$

The response function and the polarization function play an important role in the calculation of the self-energy.

2.3 The Hedin equations

Summarizing the results in the previous sections, we arrive at the well-known set of coupled integral equations (Hedin 1965, Hedin and Lundqvist 1969). From equations (29), (36), (30), (43, and (47) we have

$$\Sigma(1,2) = i \int d(34) G(1,3^+) W(1,4) \Lambda(3,2,4) \tag{48}$$

$$G(1,2) = G_0(1,2) + \int d(34) G_0(1,3) \Sigma(3,4) G(4,2) \tag{49}$$

$$\Lambda(1,2,3) = \delta(1-2)\delta(2-3) + \int d(4567) \frac{\delta\Sigma(1,2)}{\delta G(4,5)} G(4,6) G(7,5) \Lambda(6,7,3) \tag{50}$$

$$P(1,2) = -i \int d3\, d4\; G(1,3) \Lambda(3,4,2) G(4,1^+) \tag{51}$$

$$W(1,2) = v(1,2) + \int d(34) v(1,3) P(3,4) W(4,2) \tag{52}$$

Like G, Λ and W satisfy Dyson-like equations. Starting from a given approximation for Σ the above set of equations can be used to generate higher order approximations. Although the equations are exact, a straightforward expansion for the self-energy in powers of the screened interaction may yield unphysical results such as negative spectral functions. In fact, the expansion itself is only conditionally convergent due to the long-range nature of the Coulomb potential. So far there is no systematic way of choosing which diagrams to sum. The choice is usually dictated by physical intuition.

2.4 The self-energy expansion

Starting from $G = G_0$ or $\Sigma = 0$ we get from equation (50)

$$\Lambda(1,2,3) = \delta(1-2)\,\delta(2-3) \tag{53}$$

which yields the GW approximation:

$$\Sigma(1,2) = iG(1,2)W(1,2) \tag{54}$$

We can continue evaluating Σ to higher order in W by using $\Sigma = iGW$ in the vertex function and solving the integral equation (50). Alternatively, we can take the functional derivative of Σ with respect to V as in equation (30) which involves derivatives with respect to G and W:

$$\frac{\delta G(1,2)}{\delta V(3)} = \int d4\, d5\, G(1,4)\Lambda(4,5,3)G(5,2) \tag{55}$$

$$\frac{\delta W(1,2)}{\delta V(3)} = \int d4\, d5\, W(1,4)\frac{\delta P(4,5)}{\delta V(3)}W(5,2) \tag{56}$$

obtained by taking a functional derivative of $G^{-1}G = 1$ and W in equation (52) with respect to V. By inserting the previous vertex function in the above expressions and using the expression for P in equation (43), we generate the second order self-energy diagram in W and so on. One advantage of the functional derivative technique is that we do not need to worry about double-counting diagrams.

It is useful to interpret qualitatively the Dyson equation (36) in the spirit of Feynman's path integral. The electron Green function is the probability amplitude that an electron introduced at x' is to be found at x. This is then the sum over all possible paths for an electron to arrive at x starting from x'. Thus, the first path is represented by G_0 corresponding to a straight propagation to x without interaction. The self-energy represents all possible events (interactions) that can happen to the electron on its way to x. These events may be represented by diagrams and it is qualitatively clear from the form of the Dyson equation, and it can be proven rigorously, that the self-energy contains only irreducible diagrams, i.e. diagrams which cannot be decomposed into several diagrams by cutting Green function lines. Reducible diagrams would lead to double-counting because they are already generated from the Dyson equation using irreducible self-energy diagrams. Thus, for a given order in the Coulomb interaction, we generate all topologically distinct irreducible diagrams and the results are equivalent to the functional derivative technique.

Evaluation of the self-energy diagrams is facilitated by Feynman's rules which may be found in e.g. Fetter and Walecka (1971).

2.5 Spectral representations

In general, the Fourier transforms of G, P, R, and Σ^c may be expressed in terms of their spectral representations. For the Green function it is given by

$$G(\mathbf{r},\mathbf{r}',\omega) = \int_{-\infty}^{\mu} d\omega' \frac{A(\mathbf{r},\mathbf{r}',\omega')}{\omega - \omega' - i\delta} + \int_{\mu}^{\infty} d\omega' \frac{A(\mathbf{r},\mathbf{r}',\omega')}{\omega - \omega' + i\delta} \tag{57}$$

The spectral function A (density of states) is proportional to the imaginary part of G

$$A(\mathbf{r}, \mathbf{r}', \omega) = -\frac{1}{\pi}\text{Im } G(\mathbf{r}, \mathbf{r}', \omega) sgn(\omega - \mu) \tag{58}$$

and it is given by

$$\begin{aligned} A(\mathbf{r}, \mathbf{r}', \omega) = & \sum_i h_i(\mathbf{r}) h_i^*(\mathbf{r}') \delta[\omega - \mu + e(N-1, i)] \\ & + \sum_i p_i(\mathbf{r}) p_i^*(\mathbf{r}') \delta[\omega - \mu - e(N+1, i)] \end{aligned} \tag{59}$$

where

$$h_i(\mathbf{r}) = \langle N-1, i|\hat{\psi}(\mathbf{r}, 0)|N\rangle \quad \text{(hole)} \tag{60}$$

$$p_i(\mathbf{r}) = \langle N|\hat{\psi}(\mathbf{r}, 0)|N+1, i\rangle \quad \text{(particle)} \tag{61}$$

$|N \pm 1, i\rangle$ is the ith eigenstate of the $N \pm 1$ electrons with excitation energy

$$e(N \pm 1, i) = E(N \pm 1, i) - E(N \pm 1) \tag{62}$$

which is positive and $E(N \pm 1)$ is the groundstate energy of the $N \pm 1$ electrons. The quantity μ is the chemical potential

$$\mu = E(N+1) - E(N) = E(N) - E(N-1) + O(1/N) \tag{63}$$

The poles of the Green function are therefore the exact excitation energies of the $N \pm 1$ electrons. For a non-interacting system h_i and p_i become the single-particle wave functions and the poles of G correspond to the single-particle eigenvalues. The spectral representation has been obtained from the definition of the Green function in equation (1) by noting that

$$\hat{\psi}(\mathbf{r}, t) = \exp(i\hat{H}t)\ \hat{\psi}(\mathbf{r}, 0)\ \exp(-i\hat{H}t) \tag{64}$$

and inserting a complete set of $N \pm 1$-particle eigenstates of $\hat{H}$. In time space the spectral function is given by

$$A(x, x'; t) = \frac{1}{2\pi}[\langle N|\hat{\psi}(x, t)\hat{\psi}^\dagger(x', 0)|N\rangle + \langle N|\hat{\psi}^\dagger(x', 0)\hat{\psi}(x, t)|N\rangle] \tag{65}$$

which when Fourier transformed $A(\omega) = \int dt exp(i\omega t) A(t)$ yields equation (59).

From equation (37), the spectral function A is schematically given by

$$\begin{aligned} A(\omega) &= \frac{1}{\pi}\sum_i |\text{Im } G_i(\omega)| \\ &= \frac{1}{\pi}\sum_i \frac{|\text{Im } \Sigma_i^c(\omega)|}{|\omega - \varepsilon_i - \text{Re } \Sigma_i(\omega)|^2 + |\text{Im } \Sigma_i^c(\omega)|^2} \end{aligned} \tag{66}$$

where G_i is the matrix element of G in an eigenstate ψ_i of the non-interacting system H_0. The above expression assumes that Σ is diagonal in the basis $\{\psi_i\}$. A is usually peaked at each energy $E_i = \varepsilon_i - \text{Re}\ \Sigma_i(E_i)$ (quasiparticle peak) with a life-time given by $|\text{Im}\ \Sigma_i^c(E_i)|$ and a renormalization factor (weight of the Lorentzian)

$$Z_i = \left[1 - \frac{\partial \text{Re}\ \Sigma_i^c(\varepsilon_i)}{\partial \omega}\right]^{-1} < 1 \tag{67}$$

At some other energies ω_p, the denominator may be small and $A(\omega_p)$ could also show peaks or satellite structure which can be due to plasmon excitations or other collective phenomena.

By using the spectral representation for G, the polarization function can also be expressed in terms of its spectral representation:

$$P(\mathbf{r}, \mathbf{r}', \omega) = \int_{-\infty}^{0} d\omega' \frac{S(\mathbf{r}, \mathbf{r}', \omega')}{\omega - \omega' - i\delta} + \int_{0}^{\infty} d\omega' \frac{S(\mathbf{r}, \mathbf{r}', \omega')}{\omega - \omega' + i\delta} \tag{68}$$

where S is proportional to the imaginary part of P and defined to be anti-symmetric in ω:

$$S(\mathbf{r}, \mathbf{r}', \omega) = \frac{1}{\pi}\text{Im}\ P(\mathbf{r}, \mathbf{r}', \omega)\text{sgn}(\omega) \tag{69}$$

$$S(\mathbf{r}, \mathbf{r}', -\omega) = -S(\mathbf{r}, \mathbf{r}', \omega) \tag{70}$$

From equation (46), the response function is given by

$$R = [1 - Pv]^{-1} P \tag{71}$$

If we assume that $[1 - Pv]^{-1}$ does not have poles, then the analytic structure of R is determined by P and therefore the spectral representation has the same form as that of P. In the case of the electron gas within the random phase approximation (RPA) (Pines and Bohm 1952, Bohm and Pines 1953, Pines 1961, Gell-Mann and Brueckner 1957), $[1 - Pv]^{-1}$ does have a pole at the plasmon energy but this pole approaches the positive real axis from the lower half plane so that it does not destroy the analytic structure of P.

Since

$$W = \epsilon^{-1} v = v + vRv = v + W^c \tag{72}$$

the frequency dependent part of W can be written

$$W^c(\mathbf{r}, \mathbf{r}', \omega) = \int_{-\infty}^{0} d\omega' \frac{D(\mathbf{r}, \mathbf{r}', \omega')}{\omega - \omega' - i\delta} + \int_{0}^{\infty} d\omega' \frac{D(\mathbf{r}, \mathbf{r}', \omega')}{\omega - \omega' + i\delta} \tag{73}$$

D is proportional to the imaginary part of W and defined to be anti-symmetric in ω:

$$D(\mathbf{r}, \mathbf{r}', \omega) = -\frac{1}{\pi} \text{Im } W(\mathbf{r}, \mathbf{r}', \omega) \text{sgn}(\omega) \tag{74}$$

$$D(\mathbf{r}, \mathbf{r}', -\omega) = -D(\mathbf{r}, \mathbf{r}', \omega) \tag{75}$$

Similarly, the correlation part of the self-energy can be written

$$\Sigma^c(\mathbf{r}, \mathbf{r}', \omega) = \int_{-\infty}^{\mu} d\omega' \frac{\Gamma(\mathbf{r}, \mathbf{r}', \omega')}{\omega - \omega' - i\delta} + \int_{\mu}^{\infty} d\omega' \frac{\Gamma(\mathbf{r}, \mathbf{r}', \omega')}{\omega - \omega' + i\delta} \tag{76}$$

where the spectral function Γ is given by

$$\Gamma(\mathbf{r}, r', \omega) = -\frac{1}{\pi} \text{Im } \Sigma^c(\mathbf{r}, r', \omega) \text{sgn}(\omega - \mu) \tag{77}$$

2.6 The *GW* approximation

As discussed above, the GWA is formally obtained as the first term in the expansion of the self-energy in the screened interaction W. This is, however, not very useful from the physical point of view. If we do include higher order diagrams, we do not necessarily get better results (in the sense of closer agreement with experiment). Straightforward higher order diagrams can give a self-energy with wrong analytic properties which in turns results in an unphysically negative spectral function (density of states) for some energies (Minnhagen 1974; Schindlmayr and Godby, 1998). It is more physical and useful to regard the GWA as a Hartree-Fock approximation with a frequency dependent screened interaction which is essential in metals. The lack of dynamical screening in the HFA leads to unphysical results such as zero densities of states at the Fermi level (see, e.g., Ashcroft and Mermin 1976). The GWA is physically sound because it is qualitatively correct in some limiting cases (Hedin 1995) which allows applications to a large class of materials, metals or insulators.

- In atoms, screening is small and the GWA approaches the HFA which is known to work well for atoms. Taking into account polarization improves the HF results (Shirley and Martin 1993).
- In the electron gas, screening is very important which is taken into account in the GWA and for semiconductors it can be shown that the GWA reduces the Hartree-Fock gaps.
- For a core-electron excitation, the GWA corresponds to the classical Coulomb relaxation energy of the other electrons due to the adiabatic switching-on of the core-hole potential, which is just what is to be expected physically.

- For a Rydberg electron in an atom, the GWA gives the classical Coulomb energy of the Rydberg electron due to the adiabatic switching-on of an induced dipole in the ion core.
- For the decay rate and the energy loss per unit time of a fast electron in an electron gas, the GWA gives the correct formula.

2.6.1 Explicit expressions for Σ_{GW} and the polarization P

It is useful to split up the screened interaction W into the bare Coulomb potential v and the induced potential W^c as in equation (72). The self-energy in the GWA is a product of G and W in space-time. It therefore becomes a convolution in frequency space when it is Fourier transformed:

$$\Sigma(\mathbf{r}, \mathbf{r}', E) = \frac{i}{2\pi} \int_{\infty}^{\infty} d\omega \, e^{i\eta\omega} G(\mathbf{r}, \mathbf{r}', E + \omega) W(\mathbf{r}, \mathbf{r}', \omega) \tag{78}$$

The Fourier transform is defined as $f(\omega) = \int dt \, \exp(i\omega t) f(t)$. The infinitesimal η is needed for the bare Coulomb (exchange) term. It arises from the bare Coulomb interaction which is instantaneous ($t = t'$) so that according to Feynman's rule $G(\mathbf{r}t, \mathbf{r}'t' = t) = G(\mathbf{r}t, \mathbf{r}'t^{+})$. It is understood that the self-energy may have spin dependence which is carried by G only because W has no spin dependence (within the RPA). The bare Coulomb term gives the exchange (Fock) self-energy:

$$\begin{aligned} \Sigma^{x}\left(\mathbf{r}, \mathbf{r}'\right) &= \frac{i}{2\pi} \int_{\infty}^{\infty} d\omega \, e^{i\eta\omega} G(\mathbf{r}, \mathbf{r}', E + \omega) v(\mathbf{r} - \mathbf{r}') \\ &= \frac{i}{2\pi} v\left(\mathbf{r} - \mathbf{r}'\right) \int_{-\infty}^{\mu} d\omega' \int_{\infty}^{\infty} d\omega \, e^{i\eta\omega} \frac{A\left(\mathbf{r}, \mathbf{r}', \omega'\right)}{E + \omega - \omega' - i\delta} \\ &= -v\left(\mathbf{r} - \mathbf{r}'\right) \int_{-\infty}^{\mu} d\omega' A\left(\mathbf{r}, \mathbf{r}', \omega'\right) \end{aligned} \tag{79}$$

where we have closed the contour of the integration on the upper plane.

The correlation part W^c gives

$$\Sigma^{c}(E) = \frac{i}{2\pi} \int_{-\infty}^{\infty} d\omega \, G(E + \omega) W^{c}(\omega) \tag{80}$$

The space variables have been dropped out for clarity. Using the spectral representations of G and W^c in equations (57) and (73) we get

$$\begin{aligned} \Sigma^{c}(E) = \frac{i}{2\pi} \int_{-\infty}^{\infty} d\omega & \left\{ \int_{-\infty}^{\mu} d\omega_1 \, \frac{A(\omega_1)}{E + \omega - \omega_1 - i\delta} \right. \\ & \left. + \int_{\mu}^{\infty} d\omega_1 \, \frac{A(\omega_1)}{E + \omega - \omega_1 + i\delta} \right\} \\ & \times \int_{0}^{\infty} d\omega_2 \, D(\omega_2) \left\{ \frac{1}{\omega - \omega_2 + i\eta} - \frac{1}{\omega + \omega_2 - i\eta} \right\} \end{aligned} \tag{81}$$

Performing the contour integration in ω yields

$$\Sigma^c(E) = \int_{-\infty}^{\mu} d\omega_1 \int_0^{\infty} d\omega_2 \frac{A(\omega_1)D(\omega_2)}{E + \omega_2 - \omega_1 - i\delta} + \int_{\mu}^{\infty} d\omega_1 \int_0^{\infty} d\omega_2 \frac{A(\omega_1)D(\omega_2)}{E - \omega_2 - \omega_1 + i\delta} \quad (82)$$

The spectral function of Σ^c in equation (77) is then

$$\begin{aligned} \Gamma(E) = & -\operatorname{sgn}(E-\mu) \int_{-\infty}^{\mu} d\omega_1 \int_0^{\infty} d\omega_2 A(\omega_1)D(\omega_2)\delta(E + \omega_2 - \omega_1) \\ & + \operatorname{sgn}(E-\mu) \int_{\mu}^{\infty} d\omega_1 \int_0^{\infty} d\omega_2 A(\omega_1)D(\omega_2)\delta(-E + \omega_2 + \omega_1) \\ = & -\operatorname{sgn}(E-\mu) \int_0^{\infty} d\omega_2 \theta(\mu - E - \omega_2) A(E + \omega_2) D(\omega_2) \\ & + \operatorname{sgn}(E-\mu) \int_0^{\infty} d\omega_2 \theta(-\mu + E - \omega_2) A(E - \omega_2) D(\omega_2) \end{aligned} \quad (83)$$

For a non-interacting G, the spectral function is given by

$$A^0(\mathbf{r}, \mathbf{r}', \omega) = \sum_i \psi_i(\mathbf{r})\psi_i^*(\mathbf{r}')\delta(\omega - \varepsilon_i) \quad (84)$$

as follows from equation (59). For a non-interacting system, the spectral function consists of delta function peaks at the single-particle eigenvalues whereas in the interacting system, apart from the quasiparticle peaks, the spectral function may have a satellite structure. The exchange self-energy in equation (79) becomes the familiar expression

$$\Sigma^x\left(\mathbf{r}, \mathbf{r}'\right) = -v\left(\mathbf{r} - \mathbf{r}'\right) \sum_i^{\text{occ}} \psi_i(\mathbf{r})\psi_i^*(\mathbf{r}') \quad (85)$$

and the correlation part of the self-energy becomes

$$\Gamma(\mathbf{r}, \mathbf{r}', E) = \begin{cases} \sum_i^{\text{occ}} \psi_i(\mathbf{r})\psi_i^*(\mathbf{r}')D(\mathbf{r}, \mathbf{r}', \varepsilon_i - E)\theta(\varepsilon_i - E) & \text{for } E \le \mu \\ \sum_i^{\text{unocc}} \psi_i(\mathbf{r})\psi_i^*(\mathbf{r}')D(\mathbf{r}, \mathbf{r}', E - \varepsilon_i)\theta(E - \varepsilon_i) & \text{for } E > \mu \end{cases}.$$

The real part of the self-energy can be obtained from the Hilbert transform in equation (76).

In the GWA, the polarization function becomes

$$P(1,2) = -iG(1,2)G(2,1^+) \quad (87)$$

Since it is a product of two Green functions in space-time, the Fourier transformation into frequency space becomes a convolution:

$$P(\mathbf{r},\mathbf{r}',\omega) = -\frac{i}{2\pi}\int_{\infty}^{\infty} d\omega'\, G(\mathbf{r},\mathbf{r}',\omega'+\omega)G(\mathbf{r},\mathbf{r}',\omega') \tag{88}$$

Using the spectral representation of a non-interacting G, the polarization can then be written as

$$P^0(\mathbf{r},\mathbf{r}',\omega) = \sum_{\text{spin}}\sum_{i}^{\text{occ}}\sum_{j}^{\text{unocc}} \psi_i(\mathbf{r})\psi_i^*(\mathbf{r}')\psi_j^*(\mathbf{r})\psi_j(\mathbf{r}') \times\left\{\frac{1}{\omega-\varepsilon_j+\varepsilon_i+i\delta} - \frac{1}{\omega+\varepsilon_j-\varepsilon_i-i\delta}\right\} \tag{89}$$

The spectral function of P is then

$$S^0(\mathbf{r},r',\omega) = \sum_{\text{spin}}\sum_{i}^{\text{occ}}\sum_{j}^{\text{unocc}} \psi_i(\mathbf{r})\psi_i^*(\mathbf{r}')\psi_j^*(\mathbf{r})\psi_j(\mathbf{r}')\,\delta[\omega-(\varepsilon_j-\varepsilon_i)] \tag{90}$$

2.6.2 The Coulomb hole and screened exchange approximation (COHSEX)

A physically appealing way of expressing the self-energy is to divide it into a screened exchange term Σ_{SEX} and a Coulomb hole term Σ_{COH} (COHSEX) (Hedin 1965, Hedin and Lundqvist 1969). It is straightforward to verify that the real part of the self-energy can be written as

$$\text{Re}\Sigma_{\text{SEX}}\left(\mathbf{r},\mathbf{r}',E\right) = -\sum_{i}^{\text{occ}}\psi_i(\mathbf{r})\psi_i^*(\mathbf{r}')\text{Re}W\left(\mathbf{r},r',E-\varepsilon_i\right) \tag{91}$$

$$\text{Re}\Sigma_{\text{COH}}\left(\mathbf{r},\mathbf{r}',E\right) = \sum_{i}\psi_i(\mathbf{r})\psi_i^*(\mathbf{r}')\text{P}\int_0^{\infty} d\omega\frac{D\left(\mathbf{r},\mathbf{r}',\omega\right)}{E-\varepsilon_i-\omega} \tag{92}$$

The first term is simply the exchange term but with a frequency-dependent screened interaction. The physical interpretation of Σ_{COH} becomes clear in the static approximation due to Hedin. If we are interested in states close to the Fermi level, the matrix element $\langle\psi|\text{Re}\Sigma_{\text{COH}}|\psi\rangle$ picks up most of its weight from states close to ψ in energy. We may then assume that $E-\varepsilon_i$ is small compared to the main excitation energy of D, which is about the plasmon energy. If we set $E-\varepsilon_i=0$, we get

$$\text{Re}\Sigma_{\text{COH}}\left(\mathbf{r},\mathbf{r}'\right) = \frac{1}{2}\delta\left(\mathbf{r}-\mathbf{r}'\right)W^c\left(\mathbf{r},\mathbf{r}',0\right) \tag{93}$$

This is simply the interaction energy of the quasiparticle with the induced potential due to the screening of the electrons around the quasiparticle. The factor of 1/2 arises from the adiabatic growth of the interaction. In this static COHSEX approximation, Σ_{COH} becomes local.

A number of simplifications to the GWA have been proposed. One approach is based on the short-range property of the electron gas self-energy leading to a local-density-like approximation (Sham and Kohn 1966, Hedin and Lundqvist 1971, Watson *et al.* 1976, Wang and Pickett 1983, Pickett and Wang 1984, Hybertsen and Louie 1988a). An extension of this approach to semiconductors (Wang and Pickett 1983, Pickett and Wang 1984) is made by using a model dielectric function corresponding to a model semiconducting homogeneous electron gas (Penn 1962, Levine and Louie 1982). Another simplified approach uses the static COHSEX approximation for $\delta\Sigma = \Sigma - V^{xc}_{\mathrm{LDA}}$ rather than on the full Σ (Gygi and Baldereschi 1989). Bechstedt *et al.* (1992) and Palummo *et al.* 1995 included the dynamical renormalization factor into a simplified *GW* scheme. There are also a number of works based on an extreme tight-binding approach (Ortuno and Inkson 1978, Sterne and Inkson 1984, Hanke and Sham 1988, Bechstedt and Del Sole 1988, Gu and Ching 1994) suitable to describe directionally bonded materials such as Si.

2.6.3 Generalization of the HFA

The GWA may be regarded as a generalization of the Hartree-Fock approximation (HFA) but with a dynamically screened Coulomb interaction. The non-local exchange potential in the HFA is given by equation (85). In the Green function theory, the exchange potential is written as

$$\Sigma^x(\mathbf{r}, \mathbf{r}', t - t') = iG(\mathbf{r}, \mathbf{r}', t - t')v(\mathbf{r} - \mathbf{r}')\delta(t - t') \tag{94}$$

which when Fourier transformed yields equation (85). The GWA corresponds to replacing the bare Coulomb interaction v by a screened interaction W:

$$\Sigma(1, 2) = iG(1, 2)W(1, 2) \tag{95}$$

which is formally the first-order term in the expansion of the self-energy in the screened interaction as in equation (54).

From equation (78) and using the spectral representation of G_0, the correlation part of the self-energy can be written as

$$\Sigma^c(\mathbf{r}, \mathbf{r}'; \omega) = \sum_i^{\mathrm{occ}} \psi_i(\mathbf{r})\psi_i^*(\mathbf{r}')W^c_-(\mathbf{r}, \mathbf{r}'; \omega - \epsilon_i) + \sum_i^{\mathrm{unocc}} \psi_i(\mathbf{r})\psi_i^*(\mathbf{r}')W^c_+(\mathbf{r}, \mathbf{r}'; \omega - \epsilon_i) \tag{96}$$

where

$$W_{\pm}^{c}(\mathbf{r},\mathbf{r}';\omega)=\frac{i}{2\pi}\int_{-\infty}^{\infty}d\omega'\frac{W^{c}(\mathbf{r},\mathbf{r}';\omega')}{\omega+\omega'\pm i\delta} \tag{97}$$

In short we can write

$$\Sigma(\mathbf{r},\mathbf{r}';\omega)=-\sum_{i}\psi_{i}(\mathbf{r})\psi_{i}^{*}(\mathbf{r}')W_{0}(\mathbf{r},\mathbf{r}';\omega-\epsilon_{i}) \tag{98}$$

where

$$\begin{aligned}W_{0}(\mathbf{r},\mathbf{r}';\omega-\epsilon_{i})\equiv&\left\{v(\mathbf{r}-\mathbf{r}')-W_{-}^{c}(\mathbf{r},\mathbf{r}';\omega-\epsilon_{i})\right\}\,\theta(\mu-\epsilon_{i})\\&-W_{+}^{c}(\mathbf{r},\mathbf{r}';\omega-\epsilon_{i})\,\theta(\epsilon_{i}-\mu)\end{aligned} \tag{99}$$

Thus, the self-energy in the GWA has the same form as in the HFA except that it depends on energy and contains a term which depends on unoccupied states as a consequence of correlation effects. The GWA can be interpreted as a generalization of the Hartree-Fock approximation (HFA) with a potential W_0 which contains dynamical screening of the Coulomb potential. We note, however, that W_0 is not the same as the dynamically screened potential W. Using a static W_0 is equivalent to the static COHSEX approximation.

As previously mentioned, the HFA does not take into account the effects of screening which, for insulators, results in too large band gaps. It can be shown that the GWA gives the expected band gap, at least for localized states which are well isolated from the other states. Consider the self-energy for an occupied core-like state ψ_d.

$$\begin{aligned}\langle\psi_{d}|\Sigma^{c}(\epsilon_{d})|\psi_{d}\rangle=&\langle\psi_{d}\psi_{d}|W_{-}^{c}(0)|\psi_{d}\psi_{d}\rangle\\&+\sum_{i\neq d}^{\text{occ}}\langle\psi_{d}\psi_{i}|W_{-}^{c}(\epsilon_{d}-\epsilon_{i})|\psi_{i}\psi_{d}\rangle\\&+\sum_{i}^{\text{unocc}}\langle\psi_{d}\psi_{i}|W_{+}^{c}(\epsilon_{d}-\epsilon_{i})|\psi_{i}\psi_{d}\rangle\end{aligned} \tag{100}$$

Strictly speaking, the self-energy should be evaluated at the new energy $E_d = \epsilon_d + \langle\psi_d|\Sigma\,(E_d)\,|\psi_d\rangle$ and it is understood to be the case here. If ψ_d is localized and well separated in energy from other states, then the first term is evidently much larger than the rest. Thus we may make the following approximation:

$$\begin{aligned}\langle\psi_{d}|\Sigma^{c}(\epsilon_{d})|\psi_{d}\rangle&\approx\langle\psi_{d}\psi_{d}|W_{-}^{c}(0)|\psi_{d}\psi_{d}\rangle\\&=-\frac{1}{2}\langle\psi_{d}\psi_{d}|W^{c}(0)|\psi_{d}\psi_{d}\rangle\end{aligned} \tag{101}$$

We show below that $W^c_-(0) = -W^c(0)/2$.

From the spectral representation of W we have

$$W^c(0) = \int_{-\infty}^{0} d\omega' \frac{D(\omega')}{-\omega' - i\delta} + \int_{0}^{\infty} d\omega' \frac{D(\omega')}{-\omega' + i\delta}$$
$$= -2\int_{0}^{\infty} d\omega' \frac{D(\omega')}{\omega' - i\delta} \tag{102}$$

using the fact that $D(\omega)$ is odd.

$$W^c_-(0) = \frac{i}{2\pi}\int_{-\infty}^{\infty} d\omega' \frac{W^c(\omega')}{\omega' - i\delta}$$
$$= \frac{i}{2\pi}\int_{-\infty}^{\infty} d\omega' \frac{1}{\omega' - i\delta}$$
$$\times \left\{ \int_{-\infty}^{0} d\omega'' \frac{D(\omega'')}{\omega' - \omega'' - i\delta} + \int_{0}^{\infty} d\omega'' \frac{D(\omega'')}{\omega' - \omega'' + i\delta} \right\}$$
$$= \int_{0}^{\infty} d\omega'' \frac{D(\omega'')}{\omega'' - i\delta}$$
$$= -\frac{1}{2} W^c(0) \tag{103}$$

This is a correction due to the work done on the electron by the polarization field from zero to $W^c(0)$ which is just the negative of Σ_{COH} (Hedin 1965) described in the next section. A similar result,

$$+\frac{1}{2}\langle \psi_d \psi_d | W^c(0) | \psi_d \psi_d \rangle \tag{104}$$

is obtained for an unoccupied core-like state of the same character so that the energy separation of the states is

$$\Delta = \epsilon_2^{HF} - \epsilon_1^{HF} + \langle \psi_d \psi_d | W^c(0) | \psi_d \psi_d \rangle$$
$$= \langle \psi_d \psi_d | v | \psi_d \psi_d \rangle + \langle \psi_d \psi_d | W^c(0) | \psi_d \psi_d \rangle$$
$$= \langle \psi_d \psi_d | W(0) | \psi_d \psi_d \rangle \tag{105}$$

which agrees with the intuitive result that the "gap" is given by the screened Coulomb interaction: $\Delta = U \approx W(0)$. Since $W_c(0)$ is negative, we see that the self-energy correction to the HFA raises an occupied state and lowers an unoccupied state. This is still true also for states which are not so localized.

3 Computational method

The calculation of the self-energy involves the calculations of the following quantities:

1. A self-consistent bandstructure which is the input to the self-energy calculation. In principle we may use any non-interacting Hamiltonian but we use the LDA in practice.
2. The bare Coulomb matrix v.
3. The polarization function P in equations (90) and (68), which is the most time consuming part.
4. The response function R in equation (71) and the screened Coulomb matrix W in equation (72).
5. The self-energy Σ in equations (86) and (76).

3.1 Basis functions

To construct a minimal basis, let us consider the polarization P. Due to the symmetry of P with respect to a lattice translation $\mathbf{T}$,

$$P(\mathbf{r}+\mathbf{T}, \mathbf{r}'+\mathbf{T}, \omega) = P(\mathbf{r}, \mathbf{r}', \omega), \tag{106}$$

it follows that P may be expanded as follows:

$$P(\mathbf{r}, \mathbf{r}', \omega) = \sum_{\mathbf{k}ij} B_{\mathbf{k}i}(\mathbf{r}) P_{ij}(\mathbf{k}, \omega) B^*_{\mathbf{k}j}(\mathbf{r}') \tag{107}$$

where $\{B_{\mathbf{k}i}\}$ are any basis functions satisfying Bloch's theorem and they are normalized to unity in the unit cell with volume Ω. All other quantities depending on two space variables such as G, W, and Σ can be expanded in a similar fashion. From equation (90) we get

$$S^0_{ij}(\mathbf{q}, \omega) = \sum_{\text{spin}} \sum_{\mathbf{k}} \sum_{n \leq \mu} \sum_{n' > \mu} \langle B_{\mathbf{q}i} \psi_{\mathbf{k}n} | \psi_{\mathbf{k}+\mathbf{q},n'} \rangle \langle \psi_{\mathbf{k}+\mathbf{q},n'} | \psi_{\mathbf{k}n} B_{\mathbf{q}j} \rangle \times \delta[\omega - (\varepsilon_{\mathbf{k}+\mathbf{q},n'} - \varepsilon_{\mathbf{k}n})] \tag{108}$$

In the above expression, the wave function $\psi_{\mathbf{k}n}$ is normalized to unity in the unit cell. The matrix elements are given by

$$\langle \psi_{\mathbf{k}+\mathbf{q}n'} | \psi_{\mathbf{k}n} B_{\mathbf{q}j} \rangle = \int_{\Omega} d^3r \; \psi^*_{\mathbf{k}+\mathbf{q}n'} \psi_{\mathbf{k}n} B_{\mathbf{q}j} \tag{109}$$

The problem is to choose a minimal number of basis functions needed to describe the response function. It is clear from the above expression that the basis functions must span the space formed by products of the wave functions.

If the wave functions are expanded in plane waves, as in the pseudopotential theory, then the basis functions will be products of plane waves which are also plane waves, in which case

$$B_{\mathbf{k}j} \Longrightarrow \frac{\exp[i(\mathbf{k}+\mathbf{G})\cdot\mathbf{r}]}{\sqrt{\Omega}}, \quad S^0_{ij} \Longrightarrow S^0_{\mathbf{GG}'}$$

One simply needs to have a large enough **G** vector in order to have a complete basis. The advantages of the plane-wave basis are that matrix elements can be easily calculated and the Coulomb matrix is simple since it is diagonal, $v_{\mathbf{GG}'}(\mathbf{k}) = 4\pi\delta_{\mathbf{GG}'}/|\mathbf{k}+\mathbf{G}|^2$. Other advantages are a good control over convergence and programming ease. There are, however, serious drawbacks:

1. It is not feasible to do all electron calculations. In many cases, it is essential to include core electrons. For example, the exchange of a $3d$ valence state with the $3s - 3p$ core states in the late $3d$ metals is overestimated by the LDA by as much as 1 eV which would lead to an error of the same order in the pseudopotential method.
2. The size of the response matrix becomes prohibitively large for narrow band systems due to a large number of plane waves.

Moreover, the plane waves have no direct physical interpretation.

To overcome these drawbacks, we use the LMTO method which allows us to treat any system. The LMTO method uses a minimal number of basis functions and we carry over the concept of minimal basis in bandstructures to the dielectric matrix ϵ. Instead of a planewave basis, we use a "product basis" which consists of products of LMTO's. As will be clear later, the product basis constitutes a minimal basis for ϵ within the LMTO formalism. A method for inverting the dielectric matrix using localized Wannier orbitals instead of planewaves has also been used in the context of local field and excitonic effects in the optical spectrum of covalent crystals (Hanke and Sham 1975).

In the LMTO method within the atomic sphere approximation (ASA), the basis functions are given by

$$\chi_{RL\nu}(\mathbf{r}) = \phi_{RL\nu}(\mathbf{r}) + \sum_{R'L'\nu'} \dot{\phi}_{R'L'\nu'}(\mathbf{r})h_{R'L'\nu',RL\nu} \tag{110}$$

The orbital ϕ is a solution to the Schrödinger equation for a given energy ε_ν and $\dot{\phi}$ is the energy derivative at ε_ν. The label R denotes atom type, $L = lm$ denotes angular momentum, and ν denotes the principal quantum number when there are more than one orbital per L channel (Aryasetiawan and Gunnarsson 1994a). Therefore a product of wave functions consists of products of the orbitals ϕ and $\dot{\phi}$:

$$\phi_{RL\nu}\phi_{RL'\nu'}, \quad \phi_{RL\nu}\dot{\phi}_{RL'\nu'}, \quad \dot{\phi}_{RL\nu}\phi_{RL'\nu'} \tag{111}$$

This means that these product orbitals form a complete set of basis functions for the polarization function and also for the response function and the self-energy as discussed below (Aryasetiawan and Gunnarsson 1994b). From equation (46) we have

$$R = P + PvP + PvPvP + \ldots \tag{112}$$

Writing $P = \sum_{ij} |i\rangle P_{ij} \langle j|$, the second term can be written

$$PvP = \sum_{ij,kl} |i\rangle P_{ij} \langle j|v|k\rangle P_{kl} \langle l| \tag{113}$$

Similar expressions can be written down for the other terms and we can therefore write $R = \sum_{ij} |i\rangle R_{ij} \langle j|$, i.e. P and R span the same space.

From equation (47) the self-energy of a given state $\psi_{\mathbf{k}n}$ in the GWA schematically has the form

$$\begin{aligned}\Sigma(\mathbf{k}n, \omega) &= < \psi_{\mathbf{k}n}|GW|\psi_{\mathbf{k}n} > \\ &= < \psi_{\mathbf{k}n}\psi|v|\psi\psi_{\mathbf{k}n} > + < \psi_{\mathbf{k}n}\psi|vRv|\psi\psi_{\mathbf{k}n} >\end{aligned} \tag{114}$$

The two $\psi\psi$ come from the G. We see that Σ is sandwiched by products of two wave functions and it is therefore sufficient to have v expanded in the product orbitals. Thus, the product orbitals in equation (111) form a complete basis for a GW calculation.

The number of product functions can still be quite large. For example, with spdf orbitals we have 1 ss, 3 sp, 5 sd, 7 sf, 6 pp, 15 pd, 15 dd, 35 df, and 28 ff which amounts to $115 \times 3 = 345$ product functions. Inclusion of core states would make the basis even larger. For instance, including a s core state doubles the number of basis functions. We can reduce the number of product orbitals substantially in three steps:

1. We neglect terms containing $\dot{\phi}$ since they are small ($\dot{\phi}$ is typically 10% of ϕ). This reduces the number of product functions by a factor of 3. In some cases, it may be necessary to include $\dot{\phi}$.
2. If we are only interested in valence states, then there are no products between conduction states. Therefore, in sp systems, products of $\phi_d\phi_d$ can be neglected and similarly in d systems, products of $\phi_f\phi_f$ may be neglected.
3. The remaining product orbitals turn out to have a large number of linear dependencies, typically $30 - 50\%$. These linear dependencies can be eliminated systematically, which is described below.

A product orbital is defined by

$$\begin{aligned}b_i(\mathbf{r}) &\equiv \phi_{\mathbf{R}L\nu}(\mathbf{r})\phi_{\mathbf{R}L'\nu'}(\mathbf{r}) \\ &= \varphi_{\mathbf{R}l\nu}(r)\varphi_{\mathbf{R}l'\nu'}(r)y_L(\hat{\mathbf{r}})y_{L'}(\hat{\mathbf{r}})\end{aligned} \tag{115}$$

where $i = (\mathbf{R}, L\nu, L'\nu')$. Due to the ASA, this function is only non-zero inside a sphere centered on atom **R**. Thus, there are no products between orbitals centered on different spheres. For a periodic system we need a Bloch basis and perform a Bloch sum

$$b_{\mathbf{k}i}(\mathbf{r}) = \sum_{\mathbf{T}} e^{i\mathbf{k}\cdot\mathbf{T}} \phi_{\mathbf{R}L\nu}(\mathbf{r}-\mathbf{R}-\mathbf{T})\phi_{\mathbf{R}L'\nu'}(\mathbf{r}-\mathbf{R}-\mathbf{T}) \tag{116}$$

The **k** dependence is in some sense artificial because the function has no overlap with neighbouring spheres, similar to core states. After leaving out unnecessary products (step 2), we optimize the basis by eliminating linear dependencies (step 3). This is done by orthogonalizing the overlap matrix

$$O_{ij} = < b_i | b_j > \tag{117}$$

$$Oz = ez \tag{118}$$

and neglecting eigenvectors z with eigenvalues ϵ < tolerance. The resulting orthonormal basis is a linear combination of the product functions:

$$B_i = \sum_j b_j z_{ji}, \tag{119}$$

and typically we have $\approx 70 - 100$ functions per atom with $spdf$ orbitals. The above procedure ensures that we have the smallest number of basis functions. Further approximations may be introduced to reduce the basis.

In Table 1 we show a completeness test for the basis. The slight discrepancy for high lying states is due to the neglect of $\dot{\phi}$ in the product basis which become more important for the broad high lying conduction states, and also because the optimization procedure puts less weight on those products which have smaller overlap. This is not crucial for two reasons: the matrix elements become smaller for the higher states, and in relation to GW calculations, there is a factor of $1/\omega$ which makes the higher states less important.

3.2 Special directions

When calculating matrix elements using the product basis, we encounter angular integrals of the form

$$\int d\Omega \; y_{L_1} y_{L_2} y_{L_3} y_{L_4} \tag{120}$$

Analytic evaluation of these integrals are computationally expensive. Instead, they are calculated by using special directions which are analogous to Gaussian quadratures in one dimension. In general, Gaussian integration over a unit sphere

Table 1: *A completeness test of the optimal product basis for Nickel. A product of two wavefunctions is expanded in the basis:* $\psi^*_{\mathbf{k}n}\psi_{\mathbf{k}'n'} = \sum_i B_i c_i$ *with* $\mathbf{k} = (0\ 0\ 0)$, $\epsilon_{\mathbf{k}n} = -1.22\ eV$ *(the highest valence state at the* Γ *point) and* $\mathbf{k}' = (.5\ .5\ .5)2\pi/a$. *The basis is complete if the third column is equal to the second column. The number of optimal product basis functions is* 101 *and* 82 *with and without* $3s$, $3p$ *core states respectively. After Aryasetiawan and Gunnarsson 1994a.*

Core	$\int \lvert\psi_{\mathbf{k}n}\psi_{core}\rvert^2$	$\sum_i \lvert c_i\rvert^2$	error
3s	.158114	.158113	.000001
3p	.066833	.066832	.000001
3p	.209174	.209172	.000002
3p	.168184	.168183	.000001
$\epsilon_{\mathbf{k}'n'}(eV)$	$\int \lvert\psi_{\mathbf{k}n}\psi_{\mathbf{k}'n'}\rvert^2$	$\sum_i \lvert c_i\rvert^2$	error
−9.09	.052222	.052220	.000002
−2.23	.181724	.181722	.000002
−2.23	.101124	.101124	.000000
−2.23	.167753	.167743	.000010
−1.22	.119586	.119583	.000003
−1.22	.014565	.014537	.000028
24.39	.060638	.060634	.000004
28.19	.013188	.013074	.000114
28.19	.008852	.008736	.000116
28.19	.001883	.001738	.000145
42.29	.015563	.015400	.000163
42.29	.018353	.018167	.000186
42.29	.006292	.006040	.000252
73.89	.017300	.016660	.000640
73.89	.011847	.011409	.000438
73.89	.013877	.012911	.000966

means that we try to find M directions Ω_i and weights w_i such that (von Barth and Aryasetiawan 1990; Aryasetiawan *et al.* 1990)

$$\sum_{i=1}^{M} w_i y_L(\Omega_i) = \frac{\delta_{l,0}}{\sqrt{4\pi}} \quad \text{for } 0 \le l \le l_{\max}, \ -l \le m \le l \tag{121}$$

Gaussian accuracy is achieved when the number of correctly integrated spherical harmonics is equal to the number of free parameters which is $3M - 2$, since the spherical harmonics transform among themselves under rotation so that one direction can thus be taken to be the z direction and the sum of the weights is one. In one dimension, Gaussian accuracy is always achieved and the mesh points are uniquely determined. In two dimensions, it is rarely achieved although one usually comes rather close. There are often several sets of directions that yield the same accuracy. We have found by minimization of

$$\sum_L \left| \sum_{i=1}^{M} w_i y_L(\Omega_i) - \frac{\delta_{l,0}}{\sqrt{4\pi}} \right|^2 \tag{122}$$

a set of 62 cubic directions which correctly integrates all spherical harmonics up to and including $l = 11$ plus a few more for the $l = 12$ harmonics, a total of 168 functions. Full Gaussian accuracy would mean $3 \times 62 - 2 = 184$ functions and the cubic constraint only leads to a 9% less effective integration formula. The directions and the corresponding weights are

Direction	Weight/4π
(1, 0, 0)	0.130 612 244 897 931 /6
(1, 1, 1)/$\sqrt{3}$	0.128 571 428 571 554/8
(0.846 433 804 070 399, 0.497 257 813 599 068, 0.190 484 860 662 438)	0.740 816 326 530 515 /48

(123)

plus all possible cubic variations of these (sign changes and permutations). We have also found a larger set with 114 cubic directions which integrate up to $l = 15$. The directions and weights are

Direction	Weight/4π
(1, 0, 0)	0.076 190 476 192 774 /6
(1, 1, 0)/$\sqrt{2}$	0.137 357 478 197 258 /12
(0.733 519 276 107 007, 0.570 839 829 704 020, 0.368 905 625 333 822)	0.344 086 737 167 612 /48
(0.909 395 474 471 327, 0.385 850 474 128 732, 0.155 303 839 700 451)	0.442 365 308 442 356 /48

(124)

Using these directions, any angular function can be integrated easily

$$\int d\Omega\, f(\Omega) = \sum_i w_i f(\Omega_i) \tag{125}$$

The error in $\int d\Omega y_L(\Omega)$ is $\approx 10^{-10}$ for $L \neq 0$ and zero for $L = 0$ by construction.

Similar integration formulas for higher L have also been worked out by Lebedev (1975, 1976, 1977, 1978).

3.3 Evaluation of the Coulomb Matrix

We consider one atom per unit cell for simplicity. Extension to several atoms is straightforward. The Coulomb matrix is given by

$$v_{ij}(\mathbf{q}) = \frac{1}{N}\int d^3r \int d^3r' \, \frac{B^*_{\mathbf{q}i}(\mathbf{r})B_{\mathbf{q}j}(\mathbf{r}')}{|\mathbf{r}-\mathbf{r}'|} \tag{126}$$

where $B_{\mathbf{q}i}$ is normalized to unity in the unit cell. The integrations over the whole space may be reduced to integrations over a unit cell Ω by using the property

$$B_{\mathbf{q}i}(\mathbf{r}+\mathbf{T}) = e^{i\mathbf{q}\cdot\mathbf{T}}B_{\mathbf{q}i}(\mathbf{r}) \tag{127}$$

and noting that the integration over $\mathbf{r}'$ is independent of the origin of $\mathbf{r}$. This gives

$$v_{ij}(\mathbf{q}) = \int_\Omega d^3r \int_\Omega d^3s \; B^*_{\mathbf{q}i}(\mathbf{s})E_{\mathbf{q}}(\mathbf{s},\mathbf{r})B_{\mathbf{q}j}(\mathbf{r}) \tag{128}$$

where

$$\begin{aligned} E_{\mathbf{q}}(\mathbf{s},\mathbf{r}) &= \sum_{\mathbf{T}} \frac{e^{i\mathbf{q}\cdot\mathbf{T}}}{|\mathbf{s}-\mathbf{r}-\mathbf{T}|} \\ &= \frac{4\pi}{\Omega}\sum_{\mathbf{G}} \frac{e^{-(\mathbf{q}+\mathbf{G})^2/4\alpha^2}}{(\mathbf{q}+\mathbf{G})^2} e^{i(\mathbf{q}+\mathbf{G})\cdot(\mathbf{s}-\mathbf{r})} \\ &\quad + \alpha\sum_{\mathbf{T}} e^{i\mathbf{q}\cdot\mathbf{T}}\frac{\operatorname{erfc}(\alpha|\mathbf{s}-\mathbf{r}-\mathbf{T}|)}{\alpha|\mathbf{s}-\mathbf{r}-\mathbf{T}|} \end{aligned} \tag{129}$$

The Ewald method (see e.g. Ziman 1972) has been used to obtain the above decomposition into summations in the reciprocal and real space. erfc is the complementary error function equal to (1-erf), and α is an arbitrary constant whose value is chosen to give a fast convergence in the number of reciprocal lattice vectors and the number of neighbours. The essence of the Ewald method is to add and subtract a Gaussian charge distribution which breaks the Coulomb potential from a point charge into a short- and long-range part. The short-range part is done in real space and the long-range part is done in reciprocal space. The main task is to calculate the potential

$$\begin{aligned} \Phi_{\mathbf{q}j}(\mathbf{s}) &= \int_\Omega d^3r \; E_{\mathbf{q}}(\mathbf{s},\mathbf{r})B_{\mathbf{q}j}(\mathbf{r}) \\ &= \sum_i p_{\mathbf{q}i}(\mathbf{s})z_{ij} \end{aligned} \tag{130}$$

with z given by equation (118) and

$$p_{\mathbf{q}i}(\mathbf{s}) = \int_\Omega d^3r\, b_{\mathbf{q}i}(\mathbf{r}) E_\mathbf{q}(\mathbf{s}, \mathbf{r}) = \sum_\mathbf{G} p_{\mathbf{q}i}(\mathbf{s}, \mathbf{G}) + \sum_\mathbf{T} p_{\mathbf{q}i}(\mathbf{s}, \mathbf{T}) \tag{131}$$

where

$$p_{\mathbf{q}i}(\mathbf{s}, \mathbf{G}) = \frac{4\pi}{\Omega} \frac{e^{-(\mathbf{q}+\mathbf{G})^2/4\alpha^2}}{(\mathbf{q}+\mathbf{G})^2} e^{i(\mathbf{q}+\mathbf{G})\cdot\mathbf{s}} \int_\Omega d^3r\, e^{-i(\mathbf{q}+\mathbf{G})\cdot\mathbf{r}} b_{\mathbf{q}i}(\mathbf{r}) \tag{132}$$

and

$$p_{\mathbf{q}i}(\mathbf{s}, T) = \alpha e^{i\mathbf{q}\cdot\mathbf{T}} \int_\Omega d^3r\, \frac{\mathrm{erfc}(\alpha|\mathbf{s}-\mathbf{r}-\mathbf{T}|)}{\alpha|\mathbf{s}-\mathbf{r}-\mathbf{T}|} b_{\mathbf{q}i}(\mathbf{r}) \tag{133}$$

It is straightforward to calculate $p_{\mathbf{q}j}(\mathbf{s}, \mathbf{G})$, since it is a Fourier transform of $b_{\mathbf{q}j}(\mathbf{r})$. To calculate $p_{\mathbf{q}i}(\mathbf{s}, \mathbf{T})$, we use the following expansion formulas

$$\frac{1}{|\mathbf{s}-\mathbf{r}|} = \sum_L \frac{4\pi}{2l+1} \frac{r_<^l}{r_>^{l+1}} y_L(\hat{\mathbf{s}}) y_L(\hat{\mathbf{r}}) \tag{134}$$

and

$$\frac{\mathrm{erf}(\alpha|\mathbf{s}-\mathbf{r}|)}{\alpha|\mathbf{s}-\mathbf{r}|} = \sum_L \frac{4\pi}{2l+1} g_l(r, s) y_L(\hat{\mathbf{s}}) y_L(\hat{\mathbf{r}}) \tag{135}$$

The coefficients $g_l(r, s)$, which depend on α, are determined by numerical integrations using special directions in equation (123).

At the central sphere, $b_{\mathbf{q}i}(\mathbf{r})$ has no $\mathbf{q}$ dependence and it is given by equation (115), so that

$$p_{\mathbf{q}i}(\mathbf{s}, \mathbf{T}) = \alpha e^{i\mathbf{q}\cdot\mathbf{T}} \sum_L w_{li}(s_\mathbf{T}) y_L(\hat{\mathbf{s}}_\mathbf{T}) \int d\Omega\, y_{L_1} y_{L_2} y_L \tag{136}$$

where $\mathbf{s}_\mathbf{T} = \mathbf{s} - \mathbf{T}$ and

$$w_{li}(s_\mathbf{T}) = \frac{4\pi}{2l+1} \int_0^R dr\, r^2 \left\{ \frac{r_<^l}{r_>^{l+1}} - g_l(r, s_\mathbf{T}) \right\} b_i(r) \tag{137}$$

We note that the sum over l is limited to $l_1 + l_2$ as follows from the product function b_i in equation (115). Finally,

$$v_{ij}(\mathbf{q}) = \int_\Omega d^3s\, B^*_{\mathbf{q}i}(\mathbf{s}) \Phi_{\mathbf{q}j}(\mathbf{s}) \tag{138}$$

which is easily done with special directions in equation (123).

3.4 Evaluation of the polarization P

Calculations of P are the most time consuming due to a large number of matrix elements, a summation over the Brillouin zone, which is not restricted to the irreducible zone, and a sum over occupied and unoccupied states as may be seen from the following expression:

$$P_{ij}(\mathbf{q},\omega) = \sum_{\text{spin}} \sum_{\mathbf{k}} \sum_{n \leq \mu} \sum_{n' > \mu} \langle B_{\mathbf{q}i} \psi_{\mathbf{k}n} | \psi_{\mathbf{k+q},n'} \rangle \langle \psi_{\mathbf{k+q},n'} | \psi_{\mathbf{k}n} B_{\mathbf{q}j} \rangle \times \left\{ \frac{1}{\omega - \varepsilon_{\mathbf{k+q},n'} + \varepsilon_{\mathbf{k}n} + i\delta} - \frac{1}{\omega + \varepsilon_{\mathbf{k+q},n'} - \varepsilon_{\mathbf{k}n} - i\delta} \right\} \quad (139)$$

For real frequencies, we calculate S^0 in equation (90). The δ function is replaced by a Gaussian

$$\delta(x) \Longrightarrow \frac{\exp -(x/\sigma)^2}{\sigma\sqrt{\pi}} \quad (140)$$

The self-energy is not sensitive to the choice of σ.

The matrix elements reduce into integrals of four orbitals

$$\int d^3r\, \phi_{RL_1\nu_1}\phi_{RL_2\nu_2}\phi_{RL_3\nu_3}\phi_{RL_4\nu_4} = \int dr\, r^2 \varphi_{Rl_1\nu_1}\varphi_{Rl_2\nu_2}\varphi_{Rl_3\nu_3}\varphi_{Rl_4\nu_4} \times \int d\Omega\, y_{L_1} y_{L_2} y_{L_3} y_{L_4} \quad (141)$$

The angular integral is calculated by using special directions in equation (123).

To obtain the real part of P we calculate the Hilbert transform in equation (68) using the anti-symmetry of S:

$$\text{Re } P(\omega) = \text{P} \int_0^\infty d\omega'\, S^0(\omega') \left\{ \frac{1}{\omega - \omega'} - \frac{1}{\omega + \omega'} \right\} \quad (142)$$

Although the integrand diverges when $\omega' = \omega$, the integral is well-defined because it is a principal value integral. In practice, S^0 is expanded in Taylor series around ω within an interval $\omega - h$ and $\omega + h$.

For imaginary frequency, we calculate P directly from equation (139) by setting $\omega \to i\omega$:

$$P_{ij}(\mathbf{q}, i\omega) = \sum_{\text{spin}} \sum_{\mathbf{k}} \sum_{n \leq \mu} \sum_{n' > \mu} \langle B_{\mathbf{q}i} \psi_{\mathbf{k}n} | \psi_{\mathbf{k+q},n'} \rangle \langle \psi_{\mathbf{k+q},n'} | \psi_{\mathbf{k}n} B_{\mathbf{q}j} \rangle \times \frac{-2(\varepsilon_{\mathbf{k+q},n'} - \varepsilon_{\mathbf{k}n})}{\omega^2 + (\varepsilon_{\mathbf{k+q},n'} - \varepsilon_{\mathbf{k}n})^2} \quad (143)$$

Thus $P(\mathbf{r}, \mathbf{r}', i\omega)$ is real along the imaginary axis although the matrix representation P_{ij} may be complex due to the matrix elements.

The Brillouin zone integration is performed using a simple sampling method. It is also possible to use a more accurate tetrahedron method but the replacement of the δ function by a Gaussian is not possible anymore, resulting in a significantly more complicated programming.

Once we have obtain P, the response function R and the screened Coulomb interaction W can be calculated straightforwardly using equations (71) and (47) as matrix equations.

For simple metals and semiconductors (sp systems), the screened interaction is dominated by a plasmon excitation. For these systems, W is often approximated by a plasmon-pole approximation (Lundqvist 1967, Overhauser 1971, Hedin and Lundqvist 1969) which replaces the plasmon excitation by a delta function. The generalization of the plasmon-pole approximation was made by Hybertsen and Louie (1986) where each matrix component of the inverse dielectric function in plane-wave basis is written (for positive frequency)

$$\mathrm{Im}\epsilon^{-1}_{\mathbf{GG}'}(\mathbf{q}, \omega) = A_{\mathbf{GG}'}(\mathbf{q})\,\delta\left(\omega - \omega_{\mathbf{GG}'}(\mathbf{q})\right) \tag{144}$$

The weight $A_{\mathbf{GG}'}$ and the plasmon energy $\omega_{\mathbf{GG}'}$ are determined from the static- and f-sum rule. A modification of this approximation avoiding the possibility of complex plasmon energies was proposed by von der Linden and Horsch (1988) using the concept of dielectric bandstructure (Baldereschi and Tosatti 1979). The plasmon-pole approximation is not expected to work well for 3d systems since in these systems the single-particle excitations merge with the plasmon excitations already for small wave vectors as may be seen by comparing Figures 1 and 2.

An interesting mixed-space approach for calculating the polarization function was recently proposed by Blase *et al.* (1995). The polarization function is written as

$$P(\mathbf{r}, \mathbf{r}', \omega) = \sum_{\mathbf{q}} \exp[i\mathbf{q}\cdot(\mathbf{r} - \mathbf{r}')]P_{\mathbf{q}}(\mathbf{r}, \mathbf{r}', \omega) \tag{145}$$

where

$$P_{\mathbf{q}}(\mathbf{r}, \mathbf{r}', \omega) = \sum_{\text{spin}} \sum_{\mathbf{k}n}^{\text{occ}} \sum_{n'}^{\text{unocc}} u^*_{\mathbf{k}n}(\mathbf{r}) u_{\mathbf{k}+qn'}(\mathbf{r}) u^*_{\mathbf{k}+\mathbf{q}n'}(\mathbf{r}') u_{\mathbf{k}n}(\mathbf{r}') \times \left\{ \frac{1}{\omega - \varepsilon_{\mathbf{k}+\mathbf{q}n'} + \varepsilon_{\mathbf{k}n} + i\delta} - \frac{1}{\omega + \varepsilon_{\mathbf{k}+\mathbf{q}n'} - \varepsilon_{\mathbf{k}n} - i\delta} \right\} \tag{146}$$

The function $P_{\mathbf{q}}(\mathbf{r}, \mathbf{r}', \omega)$ is periodic in $\mathbf{r}$ and $\mathbf{r}'$ separately and it need be calculated within a unit cell only which distinguishes itself from the direct real-space approach where one of the position variables is not restricted to the central cell. This approach scales as N^3, similar to localized basis methods. It was found that the crossover

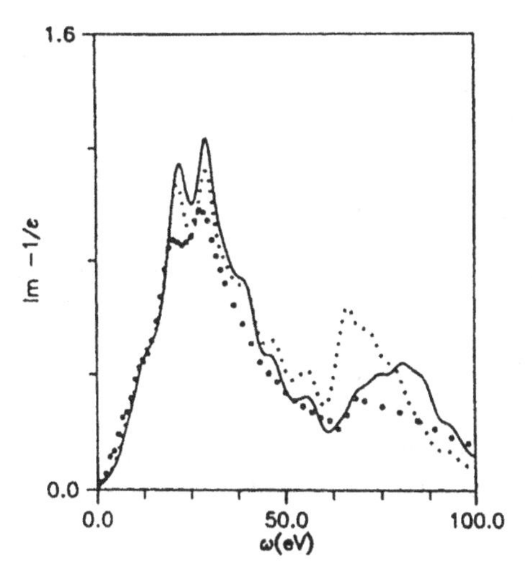

Figure 1: *The loss spectra of Ni with (solid line) and without (dotted line) local field compared with the experimental spectrum (full circles). Both theoretical spectra are calculated with 4s, 4p, 3d, 4f, and 5g LMTO orbitals, including an empty sphere at (0.5 0.5 0.5)a and core excitations (Aryasetiawan and Gunnarsson 1994b).*

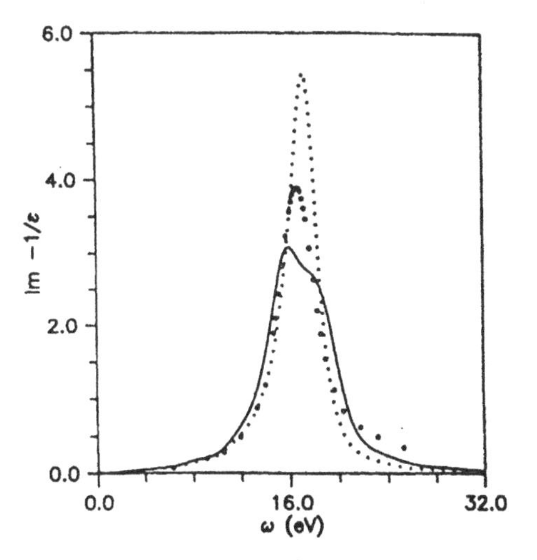

Figure 2: *The loss spectra of Si with (solid line) and without (dotted line) local field compared with the experimental spectrum (full circles). Both theoretical spectra are calculated with 3s, 3p, 3d, and 4f LMTO orbitals including core excitations (Aryasetiawan and Gunnarsson 1994b).*

between the mixed-space and reciprocal-space methods occurs for unit cells as small as that of Si. The real-space or the mixed-space approach is suitable for systems with large unit cell and a large variation in the electron density, or systems with a large empty space.

3.5 Evaluation of the self-energy Σ

Taking the matrix element of the bare exchange in a Bloch state $\psi_{\mathbf{q}n}$ we get from equation (85)

$$\Sigma^x_{\mathbf{q}n} = -\sum_{\mathbf{k}}\sum_{n'\leq\mu}\sum_{ij}\langle\psi_{\mathbf{q}n}\psi_{\mathbf{k}-\mathbf{q}n'}|B_{\mathbf{k}i}\rangle\, v_{ij}(\mathbf{k})\,\langle B_{\mathbf{k}j}|\psi_{\mathbf{k}-\mathbf{q}n'}\psi_{\mathbf{q}n}\rangle \tag{147}$$

obtained by expanding the Coulomb potential like in equation (107). From equation (86) the correlation part of Im Σ^c is given by

$$\Gamma_{\mathbf{q}n}(\omega) = \begin{cases} \sum_{\mathbf{k}}\sum_{n'\leq\mu}\sum_{ij}\langle\psi_{\mathbf{q}n}\psi_{\mathbf{k}-\mathbf{q},n'}|B_{\mathbf{k}i}\rangle\, D_{ij}(\mathbf{k},\varepsilon_{\mathbf{k}-\mathbf{q},n'}-\omega) & \\ \qquad\times\langle B_{\mathbf{k}j}|\psi_{\mathbf{k}-\mathbf{q},n'}\psi_{\mathbf{q}n}\rangle\,\theta(\varepsilon_{\mathbf{k}-\mathbf{q},n'}-\omega) & \text{for } \omega\leq\mu \\ \sum_{\mathbf{k}}\sum_{n'>\mu}\sum_{ij}\langle\psi_{\mathbf{q}n}\psi_{\mathbf{k}-\mathbf{q},n'}|B_{\mathbf{k}i}\rangle\, D_{ij}(\mathbf{k},\omega-\varepsilon_{\mathbf{k}-\mathbf{q},n'}) & \\ \qquad\times\langle B_{\mathbf{k}j}|\psi_{\mathbf{k}-\mathbf{q},n'}\psi_{\mathbf{q}n}\rangle\,\theta(\omega-\varepsilon_{\mathbf{k}-\mathbf{q},n'}) & \text{for } \omega>\mu \end{cases} \tag{148}$$

The real part of Σ^c is obtained from the Hilbert transform (principal value integral) in equation (76). As in the case of the polarization, care must be taken when $\omega' = \omega$ by expanding Γ in Taylor series around ω.

The quasiparticle energy can now be calculated as follows:

$$\begin{aligned} E_{\mathbf{q}n} &= \varepsilon_{\mathbf{q}n} + \Delta\Sigma_{\mathbf{q}n}(E_{\mathbf{q}n}) \\ &= \varepsilon_{\mathbf{q}n} + \Delta\Sigma_{\mathbf{q}n}(\varepsilon_{\mathbf{q}n}) + (E_{\mathbf{q}n}-\varepsilon_{\mathbf{q}n})\frac{\partial\Delta\Sigma_{\mathbf{q}n}(\varepsilon_{\mathbf{q}n})}{\partial\omega} \end{aligned} \tag{149}$$

where

$$\Delta\Sigma_{\mathbf{q}n}(\omega) = \langle\psi_{\mathbf{q}n}|\text{Re }\Sigma(\omega) - V^{xc}|\psi_{\mathbf{q}n}\rangle \tag{150}$$

The self-energy correction $\Delta\Sigma$ is obtained from first order perturbation theory from equation (34) and the Kohn-Sham equation:

$$(H_0+\Sigma)\Psi = E\Psi \tag{151}$$

$$(H_0+V^{xc})\psi = \varepsilon\psi \tag{152}$$

where H_0 is the kinetic energy plus the Hartree potential. The self-energy correction to $\varepsilon_{\mathbf{q}n}$ is given by

$$\begin{aligned} \Delta\varepsilon_{\mathbf{q}n} &= E_{\mathbf{q}n} - \varepsilon_{\mathbf{q}n} \\ &= Z_{\mathbf{q}n}\Delta\Sigma_{\mathbf{q}n}(\varepsilon_{\mathbf{q}n}) \end{aligned} \tag{153}$$

where

$$Z_{\mathbf{q}n} = \left[1 - \frac{\partial \Delta\Sigma_{\mathbf{q}n}(\varepsilon_{\mathbf{q}n})}{\partial\omega}\right]^{-1} < 1 \tag{154}$$

is the quasiparticle weight.

The frequency integration of the self-energy may also be performed along the imaginary axis (Godby, Schlüter, and Sham 1988) plus contributions from the poles of the Green function. From equation (96) we have

$$\Sigma^c_{\mathbf{q}n}(\omega) = \sum_{\mathbf{k}}\sum_{n'}^{\text{occ}}\sum_{ij}\langle\psi_{\mathbf{q}n}\psi_{\mathbf{k}-\mathbf{q},n'}|B_{\mathbf{k}i}\rangle\ \langle B_{\mathbf{k}j}|\psi_{\mathbf{k}-\mathbf{q},n'}\psi_{\mathbf{q}n}\rangle \times \frac{i}{2\pi}\int_{-\infty}^{\infty} d\omega'\ \frac{W^c_{ij}(\mathbf{k},\omega')}{\omega+\omega'-\varepsilon_{\mathbf{k}-\mathbf{q},n'}-i\delta} \tag{155}$$

$$+\sum_{\mathbf{k}}\sum_{n'}^{\text{unocc}}\sum_{ij}\langle\psi_{\mathbf{q}n}\psi_{\mathbf{k}-\mathbf{q},n'}|B_{\mathbf{k}i}\rangle\ \langle B_{\mathbf{k}j}|\psi_{\mathbf{k}-\mathbf{q},n'}\psi_{\mathbf{q}n}\rangle \times \frac{i}{2\pi}\int_{-\infty}^{\infty} d\omega'\ \frac{W^c_{ij}(\mathbf{k},\omega')}{\omega+\omega'-\varepsilon_{\mathbf{k}-\mathbf{q},n'}+i\delta} \tag{156}$$

We consider the integration along the imaginary axis with $\omega' \to i\omega''$, ω'' real, and along the path C (Figure 3):

$$\begin{aligned}
&\frac{i}{2\pi}\int_{-\infty}^{\infty} d\omega'\ \frac{W^c_{ij}(\mathbf{k},\omega')}{\omega+\omega'-\varepsilon_{\mathbf{k}-\mathbf{q},n'}\pm i\delta} \\
&= -\int_0^{\infty}\frac{d\omega''}{2\pi}W^c_{ij}(\mathbf{k},i\omega'')\left\{\frac{1}{\omega+i\omega''-\varepsilon_{\mathbf{k}-\mathbf{q},n'}}+\frac{1}{\omega-i\omega''-\varepsilon_{\mathbf{k}-\mathbf{q},n'}}\right\} \\
&\quad + \frac{i}{2\pi}\int_C d\omega'\ \frac{W^c_{ij}(\mathbf{k},\omega')}{\omega+\omega'-\varepsilon_{\mathbf{k}-\mathbf{q},n'}\pm i\delta} \\
&= -\int_0^{\infty} d\omega''\ W^c_{ij}(\mathbf{k},i\omega'')\frac{1}{\pi}\frac{\omega-\varepsilon_{\mathbf{k}-\mathbf{q},n'}}{(\omega-\varepsilon_{\mathbf{k}-\mathbf{q},n'})^2+\omega''^2} \\
&\quad \pm\ W^c_{ij}[\mathbf{k},\pm(\omega-\varepsilon_{\mathbf{k}-\mathbf{q},n'})]\ \theta[\pm(\omega-\varepsilon_{\mathbf{k}-\mathbf{q},n'})]\ \theta[\pm(\omega-\mu)] \\
&\quad \times\ \theta[\pm(\varepsilon_{\mathbf{k}-\mathbf{q},n'}-\mu)]
\end{aligned} \tag{157}$$

The first term is the contribution along the imaginary axis and the second from the poles of G. The integrand in the first term is very peaked around $\omega'' = 0$ when $\omega - \varepsilon_{\mathbf{k}-\mathbf{q},n'}$ is small. To handle this problem, we add and subtract the following term (Gunnarsson 1992)

$$\begin{aligned}
&\int_0^{\infty} d\omega'\ W^c_{ij}(\mathbf{k},0)\, e^{-\alpha^2\omega'^2}\frac{1}{\pi}\frac{\omega-\varepsilon_{\mathbf{k}-q,n'}}{(\omega-\varepsilon_{\mathbf{k}-q,n'})^2+\omega'^2} \\
&\quad = W^c_{ij}(\mathbf{k},0)\frac{\pi}{2}e^{\alpha^2(\omega-\varepsilon_{\mathbf{k}-\mathbf{q},n'})^2}\operatorname{erfc}[\alpha(\omega-\varepsilon_{\mathbf{k}-\mathbf{q},n'})]
\end{aligned} \tag{158}$$

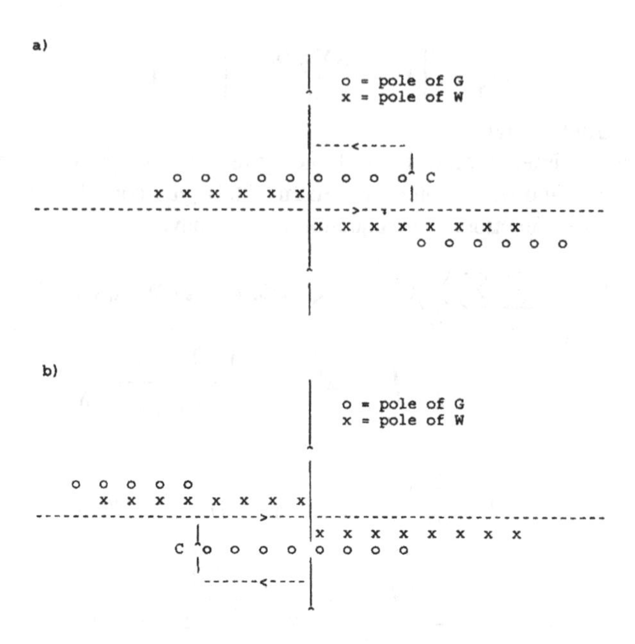

Figure 3: *The analytic structure of* $\Sigma^c = iGW^c$ *for* $\omega > E_F$ *(a) and* $\omega \leq E_F$ *(b). Frequency integration of the self-energy along the real axis from* $-\infty$ *to* ∞ *is equivalent to the integration along the imaginary axis including the path C.*

When this term is subtracted from the integrand in the first term, the resulting integrand is smooth and a Gaussian quadrature may be used.

A recent attempt to calculate the self-energy directly in real space (space-time method) was made by Rojas, Godby, and Needs (1995) with application to Si. Applications of this method to systems containing 3d or 4f orbitals may not be advantageous due to a large number of mesh points that would be required for these systems.

4 Applications

The GWA has been applied to a wide range of systems:

- Atoms (Shirley and Martin 1993).
- Simple metals (Northrup, Hybertsen, and Louie 1987, 1989, Shung, Sernelius, and Mahan 1987, Shung and Mahan 1988, Surh, Northrup, and Louie 1988).
- Semiconductors and wide-gap insulators (see references in Table 2).

- 3d systems (Aryasetiawan 1992a, Aryasetiawan and von Barth 1992b, Aryasetiawan and Gunnarsson 1995, Massidda *et al.* 1995a,b, 1997).
- Surfaces (Hybertsen and Louie 1988b, Zhu *et al.* 1989, Bechstedt and Del Sole 1990, Eguiluz *et al.* 1992, Deisz, Eguiluz, and Hanke 1993, Northrup 1993, Kress, Fiedler, and Bechstedt 1994, Rohlfing, Krüger, and Pollmann 1995b, 1996, Deisz and Eguiluz 1997).
- Interfaces (Charlesworth *et al.*. 1993, Godby and Sham 1994). Valence band off-set of AlAs-GaAs(001) (Zhang *et al.* 1988).
- Metal-insulator transitions (Godby and Needs 1989) and carrier-induced band-gap narrowing in Si (Oschlies A, Godby R W, and Needs R J 1992, 1995).
- Na_4 clusters (Onida *et al.* 1995).
- F-center defect in LiCl (Surh, Chacham, and Louie 1995).
- Two-dimensional crystals (Engel, Kwon, and Martin 1995).
- Fullerenes (Shirley and Louie 1993, Gunnarsson 1997a). For a review of fullerenes we refer to Erwin and Pickett (1991) and Gunnarsson (1997b).

In practically all of these applications, the GWA improves the LDA or the Hartree-Fock results. In this contribution we concentrate on applications to more strongly correlated systems of 3d materials. For a review of applications to other systems, we refer to a recent article (Aryasetiawan and Gunnarsson, 1998). Here we give the summary of the main results for simple metals and semiconductors.

4.1 Simple metals and semiconductors

Applications of the GWA to simple metals were naturally among the first to real systems since simple metals closely resemble the electron gas. As an example, we consider Na. *GW* calculation on the electron gas with Na average density reduces the band width by 0.27 eV from the free-electron value of 3.2 eV (Hedin 1965) but it is still in significant discrepancy compared with the experimental value of 2.50–2.65 eV (Kowalczyk *et al.* 1973, Jensen and Plummer 1985, Lyo and Plummer 1988). A similar discrepancy is observed in potassium (Itchkawitz *et al.* 1990). *GW* calculation on real Na by Northrup, Hybertsen, and Louie (1987, 1989) reduces the band width by 0.31 eV which is close to the electron gas result. An improvement is obtained when the dielectric function includes the LDA vertex correction $K^{xc} = \delta V^{xc}/\delta\rho$:

$$\epsilon^{-1} = 1 + v[1 - P(v + K^{xc})]^{-1}, \tag{159}$$

As pointed out by Hybertsen and Louie (1987a), this corresponds to the effective interaction between electrons rather than between a test charge and electrons. Using the above dielectric function narrows the band width by 0.57 eV and when the LDA eigenvalues are replaced by the quasiparticle energies the narrowing is 0.71 eV.

Table 2: *Minimum band gaps of semiconductors and insulators which have been calculated within the GWA. The energy is in eV.*

	LDA	GWA	Expt.
AlAs	1.37	2.18[b]	2.32[α]
$Al_{0.5}Ga_{0.5}As$	1.12	2.06[e]	2.09[α]
AlN (wurtzite)	3.9	5.8[d]	6.2[α]
AlN (zinc-blende)	3.2	4.9[d]	
AlP	1.52	2.59[e]	2.50[α]
AlSb	0.99	1.64[e]	1.68[α]
CdS (zinc-blende)	1.37[i], 0.83[g]	2.83[i], 2.45[g]	2.55[η,α]
CdS (wurtzite)	1.36	2.79[i]	2.59[α]
CdSe (zinc-blende)	0.76	2.01[i]	1.90[α]
CdSe (wurtzite)	0.75	1.91[i]	1.97[α]
CdTe (zinc-blende)	0.80	1.76[i]	1.92[α]
CdTe (wurtzite)	0.85	1.80[i]	1.60[α]
Diamond	3.90	5.6[a], 5.33[b], 5.67[c]	5.48[α]
GaAs	0.67	1.58[b], 1.32[c], 1.22[e]	1.52[α], 1.63[γ]
GaN (wurtzite)	2.3	3.5[d]	3.5[α]
GaN (zinc-blende)	2.1	3.1[d]	3.2[δ], 3.3[ε]
GaP	1.82	2.55[e]	2.39[α]
GaSb	-0.10	0.62[e]	0.80[α]
Ge	<0	0.75[a], 0.65[c]	0.744[α]
InAs	-0.39	0.40[e]	0.41[α]
$In_{0.53}Ga_{0.47}As$	0.02	0.80[e]	0.81[α]
InP	0.57	1.44[e]	1.42[α]
InSb	-0.51	0.18[e]	0.23[α]
LiCl	6.0	9.1[a]	9.4[α]
Li_2O	5.3	7.4[h]	6.6[θ]
MgO	5.0	7.7[f]	7.83[ζ]
Si	0.52	1.29[a], 1.24[b], 1.25[c]	1.17[β]
SiC(β)	1.31	2.34[c]	2.39[α]
ZnS (zinc-blende)	2.37	3.98[i]	3.80[α]
ZnS (wurtzite)	2.45	4.03[i]	3.92[α]
ZnSe (zinc-blende)	1.45	2.84[i]	2.96[α]
ZnSe (wurtzite)	1.43	2.75[i]	2.87[α]
ZnTe (zinc-blende)	1.33	2.57[i]	2.71[α]
ZnTe (wurtzite)	1.48	2.67[i]	

[α] *Numerical Data and Functional Relationships in Science and Technology* 1982
[β] Baldini and Bosacchi 1970
[γ] Aspnes 1976
[δ] Lei *et al.* 1992a,b, Eddy *et al.* 1993

[ε]Paisley *et al.* 1989, Sitar *et al.* 1992
[ζ]Whited, Flaten and Walker 1973
[η]Cardona, Weinstein, and Wolff 1965
[θ]Rauch 1940

[a]Hybertsen and Louie 1986
[b]Godby, Schlüter, and Sham 1988
[c]Rohlfing, Krüger, and Pollman 1993
[d]Rubio *et al.* 1993
[e]Zhu and Louie 1991 (The number in the bracket corresponds to a calculation using a model dielectric matrix)
[f]Schönberger and Aryasetiawan 1995
[g]Rohlfing, Krüger, and Pollmann 1995a
[h]Albrecht, Onida, and Reining 1997
[i]Zakharov *et al.* 1994

It was shown later on by Mahan and Sernelius (1989) that if vertex corrections were also included in the self-energy, which is a consistent procedure (Rice 1965, Ting, Lee, and Quinn 1975), the band width was similar to the GW result without vertex corrections. In other words, vertex corrections in the response function and in the self-energy cancel each other. A similar effect was found in calculations of the optical spectra of Si (Bechstedt *et al.* 1997, Del Sole and Girlanda 1996).

It was also observed experimentally the presence of almost dispersionless states near the Fermi level (Jensen and Plummer 1985, Itchkawitz *et al.* 1990) which were speculated to arise from charge density waves (Overhauser 1985). Calculations by Shung and Mahan (1986, 1988) and by Shung, Sernelius, and Mahan (1987) show that these dispersionless states can be explained by taking into account surface and quasiparticle life-time effects.

Applications to semiconductors were first made by Hybertsen and Louie (1986) and Godby, Schlüter, and Sham (1988). Later on similar results by Hamada, Hwang, and Freeman (1990) and Rohlfing, Krüger, and Pollman (1993). The main problem with the LDA is a systematic underestimation of the bandgaps from 30% to as much as 100% as may be seen in Table 2. The band-gap problem seems to originate from the error in the LDA. Exchange-correlation potentials V^{xc} calculated from GW self-energies using a procedure due to Sham and Schlüter (1985) turn out to be similar to the LDA V^{xc} (Godby, Schlüter, and Sham 1988) which indicates that even the exact V^{xc} probably does not give the correct gap. But we would like to emphasize that this is still an open question. The V^{xc} may be a non-analytic function of the particle number (Almbladh and von Barth 1985b, Perdew and Levy 1983, Sham and Schlüter 1983).

GW band gaps are in most cases within 0.1–0.2 eV of the experimental values. The quasiparticle energies are also accurate to within the same error. These works showed the feasibility of performing *ab initio* many-body calculations on real systems and they have initiated subsequent applications to more complex systems. Unlike in simple metals, the charge density in semiconductors is rather inhomogeneous due directional valence bonding. The response function in semiconductors is therefore quite different from that of simple metals. In a plane-wave basis, this means that the dielectric matrix has significant off-diagonal components. The non-locality (state-dependence) as well as the energy dependence of the self-energy are important in producing the correct band gap and the quasiparticle band structure (Hybertsen and Louie 1986, Godby, Schlüter, and Sham 1988). Non-locality is expected to be important when the extent of the wavefunction is comparable to or smaller than the range of non-locality, i.e. the range $|\mathbf{r} - \mathbf{r}'|$ within which $\Sigma(\mathbf{r}, \mathbf{r}', \omega)$ is non zero. The mechanism for the increase in the band gap by the self-energy may be understood from the different character of the valence and conduction states. The top of the valence band is of bonding p whereas the bottom of the conduction band is of antibonding p. Non-locality has a larger effect on the conduction state than on the valence state since the former has a shorter wavelength due to the extra node. The matrix element $\langle\psi|\Sigma|\psi^M\rangle$ relative to $\langle\psi|V^{xc}|\psi\rangle$ is smaller in magnitude and therefore the conduction band is pushed upwards. The valence state is less affected because its wavelength is longer than that of the conduction state and the net effect is a widening of the gap (Godby, Schlüter, and Sham 1988).

One difficulty in the pseudopotential approach commonly used in semiconductors is the inclusion of core-polarization effects. Shirley, Zhu, and Louie (1992) incorporated these effects by employing the core polarization potential (CPP) method used in quantum chemistry (Müller, Flesch, and Meyer 1984). Applications to Si, Ge, AlAs, and GaAs improve the bandstructures and bandgaps significantly.

4.2 Transition metals

Simple metals and semiconductors are sp systems where the effective onsite Coulomb energy is relatively small compared with the band width. For these systems, long-range correlations in the form of screening determine the quasiparticle properties. One can therefore understand qualitatively the success of the GWA in these s-p systems since the GWA takes into account long-range correlations pretty well through the RPA screening. One justifiably suspects that the GWA would fail for systems with localized states such as those of the 3d and 4f systems where onsite interaction is strong. It is therefore of importance to apply the GWA to the 3d and 4f systems in order to investigate the capability and limitations of the theory. As it turned out, the GWA works a lot better than expected.

The major practical difference between applications to s-p systems and applications to 3d systems lies in the numerics. s-p systems are in most cases suitable for pseudopotential treatment and consequently it is appropriate to use plane waves as basis functions for both the band structure calculations and the calculations of the response functions and the self-energy. As discussed in the previous section, plane-wave basis makes a lot of things simpler. Unfortunately, it is not advantageous to use plane-wave basis for 3d and 4f systems because it would amount to a prohibitively large number of plane waves. A new scheme based on the LMTO described in the previous section is now available to treat these interesting and important systems. With the discovery of the high temperature superconductors, a lot of interests have emerged for understanding the electronic structure of the transition metal oxides (Mott-Hubbard insulators). A great deal of work has been carried out for these systems using model Hamiltonians but very little first-principles work has been performed so far. While model calculations give valuable insight into the physical mechanism, quantitative predictions invariably require first-principles calculations.

GW calculations for transition metals have not been extensively performed. We concentrate therefore on two materials Ni and NiO for which full *GW* calculations have been done in some details and on MnO for which a model *GW* calculation has been performed.

4.2.1 Nickel

The electronic structure of Ni has received a great deal of attention for many years. There are several reasons for this. Ni is a relatively simple system with one atom per unit cell and yet it shows several manifestations of magnetism and many-body effects. The electronic structure of Ni atom shows the problem of applying a single-particle theory in this system because the two lowest configurations $3d^9 4s$ and $3d^8 4s^2$ are almost degenerate, differing in energy by only 0.025 eV (Moore 1958). This problem is presumably carried over to the solid case. Indeed, the standard LDA calculation produces a band structure which markedly deviates from experiment. Some indications of the importance of many-body effects beyond the LDA are

- The occupied 3d band width in the LDA (4.5 eV) is about 30 % too large compared with experiment (3.3 eV) (Hüfner *et al.* 1972, Himpsel, Knapp, and Eastman 1979).
- The LDA exchange splitting (0.6 eV) is a factor of two too large compared with experiment (0.3 eV) (Himpsel, Knapp, and Eastman 1979, Eberhardt and Plummer 1980).
- The details of the LDA band structure is also in significant discrepancy compared with the photoemission data (Mårtensson and Nilsson 1984).
- The presence of the well-known 6 eV satellite which is understandably missing in single-particle theories (Hüfner *et al.* 1972, Hüfner and Wertheim 1973, Kemeny and Shevchik 1975).
- The quasiparticle width is unusually large, up to 2 eV at the bottom of the band (Eberhardt and Plummer 1980).

The discrepancies described above are mostly related to excited state properties whereas the LDA is a ground state theory. Ground state properties such as equilibrium lattice constant, bulk modulus, and magnetic moment are well reproduced by the LDA with an exception of the cohesive energy which is underestimated by $\approx$1 eV (Moruzzi, Janak, and Williams 1978).

The difficulties of a single-particle theory in describing the electronic structure of Ni can be qualitatively understood by making a comparison with Cu. In Ni, the ground state has one 3d hole whereas in Cu the 3d shell is fully occupied. Thus after a photoemission process, there are two 3d holes in Ni giving rise to a strong onsite correlation which cannot be satisfactorily treated within a single-particle theory. In contrast, there is only one 3d hole in Cu after photoemission and indeed the LDA band structure is good apart from a somewhat too high position (0.4–0.5 eV) of the 3d band relative to the 4s band (see e.g. Jones and Gunnarsson 1989).

The partially filled 3d shell and the localized nature of the 3d states suggest strong onsite correlations which may need treatment beyond an RPA-type approach like the GWA. On the other hand, the band width in Ni is not too small and the ratio between the so called Hubbard U and the band width is of order one. Moreover, Fermi surface is observed experimentally (Tsui 1967, Tsui and Stark 1968). The itinerant character of the 3d electrons should therefore be taken into account which is crucial when we consider screening of a photoemission hole since 3d electrons from neighbouring cells can participate in the screening whereas such possibility is absent in the atomic case.

GW calculations for Ni and Cu have been performed by Aryasetiawan (1992a) and Aryasetiawan and Gunnarsson (1997). In Ni the LDA bandstructure is much improved, in particular the 3d-band width is narrowed by almost 1 eV. The quasiparticle life-times are also in reasonable agreement with experiment but the exchange splittings remain essentially unchanged from their LDA values and the 6 eV satellite is not reproduced. For Cu, the good LDA band structure is essentially unaffected, as it should be, and the GWA lowers the position of the 3d band relative to the 4s band, improving the LDA result. The calculation for Ni has been performed using two completely different numerical methods. The first calculation was based on the LAPW method (Aryasetiawan 1992a) and the later calculation on the LMTO method. That the two methods give the same results to the accuracy of the calculations is rather assuring.

Let us examine the self-energy as a function of frequency for the state Γ'_{25} corresponding to the lower 3d state at the Γ-point. We immediately see that the imaginary part of the self-energy, $\mathrm{Im}\Sigma^c$, is qualitatively different from those of simple metals or semiconductors. A typical $\mathrm{Im}\Sigma^c$ for a simple metal or a semiconductor is characterized by a large plasmon peak. The position of the peak is typically 1.5 plasmon energy below the main quasiparticle peak which is in disagreement with experiment. (See section 5). Apart from this strong peak, $\mathrm{Im}\Sigma^c$ is structureless. In fact, the structure of $\mathrm{Im}\Sigma^c$ as a function of frequency is determined by the imaginary part of the screened interaction which is in turn proportional to the loss spectra as may be seen in equation (86). In simple metals and semiconductors, the loss spectra consist of one plasmon peak but otherwise they are rather structureless. This is reflected in $\mathrm{Im}\,\Sigma^c$ and this is the reason why the plasmon-pole approximation works rather well for simple metals and semiconductors. In contrast, there is no well-defined plasmon excitation in transition metals. Rather, the plasmon merges with the single-particle excitations forming a broad spectrum which is illustrated in Figure 1 for the loss spectra. It is therefore questionable if the plasmon-pole approximation would work in transition metals. The loss spectrum shows a two-peak structure at about 20 and 30 eV which has been found to be characteristic of 3d transition metals starting from V. The double structure is probably due to band structure effects and it is reflected also in $\mathrm{Im}\,\Sigma^c$. The energy of the second peak coincides rather closely with an estimate based on the electron gas formula which gives a plasmon energy of 30.8 eV when the 3d electrons are included in the density.

Around the Fermi level $\mathrm{Im}\,\Sigma^c$ shows a quadratic behaviour (Fermi liquid) which becomes linear rather quickly. Unlike in simple metals or semiconductors, $\mathrm{Im}\,\Sigma^c$ in the quasiparticle energy range is rather large which indicates strong correlations between the quasiparticles and the surrounding resulting in larger decay probability. For the bottom of the band, a calculated value of $\approx$2 eV is obtained for the quasiparticle width which agrees reasonably well with the experimental finding.

Noteworthy is the behaviour of $\mathrm{Im}\ \Sigma^c$ at large frequencies. The hole (negative frequency) and particle (positive frequency) parts show a similar asymptotic behaviour and they therefore tend to cancel each other when one performs a Hilbert transform to obtain the real part of the self-energy. This justifies a posteriori the use of an energy cut-off in the calculation of the response function. It also agrees with our physical intuition that the main contribution to the self-energy should come from energies up to the plasmon energy.

The real part of the self-energy, which is the Hilbert transformed of the imaginary part, shows a large derivative at around the Fermi level which implies a large renormalization of the quasiparticle weight (equation (67)). Typical values for the renormalization factor is 0.5 for the 3d states. This is somewhat smaller than in the electron gas (0.7) (Hedin and Lundqvist 1969) or semiconductors (0.8) (Hybertsen and Louie 1986) which reflects a larger loss of single-particle character of the quasiparticles. The 4s states on the other hand have a renormalization factor ≈ 0.7, comparable to those in the alkalis and semiconductors. It is to be expected since the 3d states are more correlated than the 4s-4p states.

The self-energy correction is strongly state dependent as evident from Figure 4. The self-energy correction can be positive or negative and its magnitude varies throughout the Brillouin zone. For example, at the X-point the bottom of the 3d band experiences a self-energy correction of 0.8 eV while the top of the band is almost unchanged which results in band narrowing. Another illustration of the state dependence of the self-energy correction is provided by the lowest band. At the

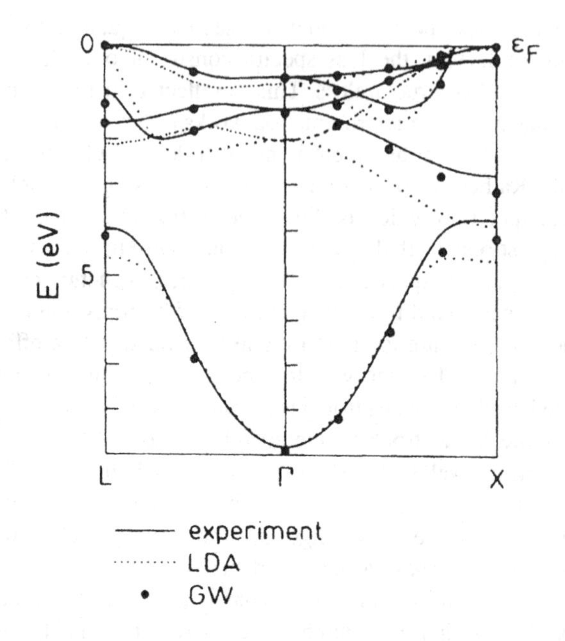

Figure 4: *The bandstructure of Ni along* ΓX *and* ΓL *averaged over the majority and minority channels. The solid curves are the experiment and the dotted curves are the LDA (Mårtensson and Nilsson 1984). The filled circles are the quasiparticle energies in the GWA (Aryasetiawan 1992).*

bottom the band is of 4s character and the self-energy correction is approximately zero whereas at the band edges the states are mixtures of 4s and 3d orbitals and the self-energy correction is positive. It also illustrates the well-known fact that free-electron-like states are well described by the LDA but localized d states are less satisfactorily described. Thus, the self-energy correction in Ni is in strong contrast to those in semiconductors which are approximately a scissor operator which increases the band gap by shifting the conduction band upwards and leaving the valence band unchanged (Hybertsen and Louie 1986, Godby, Schlüter, and Sham 1988).

The mechanism for band narrowing may be understood qualitatively as follows. For states lying a few eV below the Fermi level, the hole part of Im Σ^c has larger weight than the particle part. This simply reflects the fact that the hole (occupied) states have larger overlap and correlation with other occupied states resulting in larger correlation part of the self-energy than for the particle part. As we go towards the Fermi level, the hole and particle parts become of almost equal weight. This is illustrated in Figures 5 and 6. The bottom of the 3d band experiences therefore a larger positive self-energy correction than the top one. This is compensated

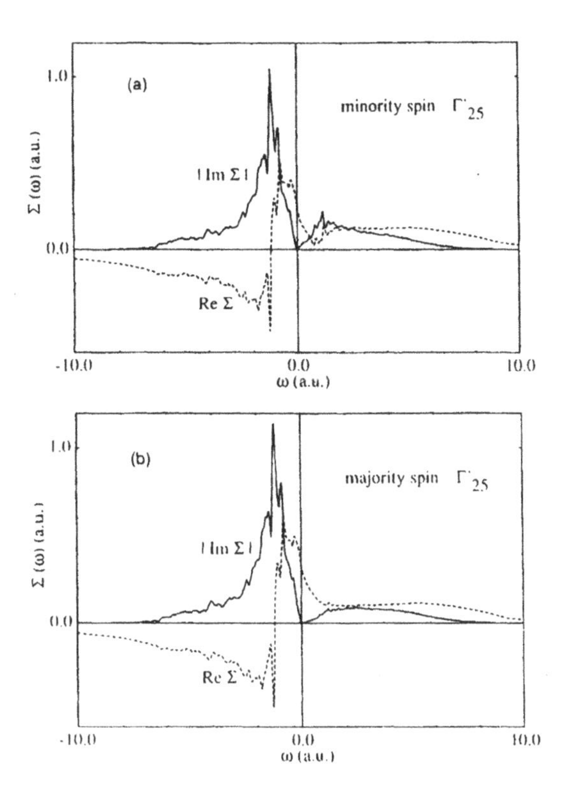

Figure 5: *Ni self-energy in the GWA. (a) The real and imaginary parts of the correlation part of the self-energy for the minority spin state* Γ'_{25}. *(b) The real and imaginary parts of the correlation part of the self-energy for the majority spin state* Γ'_{25} *(Aryasetiawan 1992).*

to some extent by the exchange part of the self-energy but the net effect leads to a band narrowing.

In Table 3 the quasiparticle energies at some high symmetry points are shown. The LDA eigenvalues are brought closer to the experimental photoemission data. In particular, the bottom of the 3d band is raised by almost 1 eV and a discrepancy of only 0.3–0.4 eV remains in the band width. The exchange splittings remain essentially unchanged from their LDA values but the discrepancy between experiment and the LDA is only 0.3 eV which is of the order of the numerical accuracy (estimated to be 0.1–0.2 eV from convergence test). But since the exchange splitting is the difference between eigenvalues, one would expect a cancellation of errors. The results seem then to indicate inadequacy in the GWA

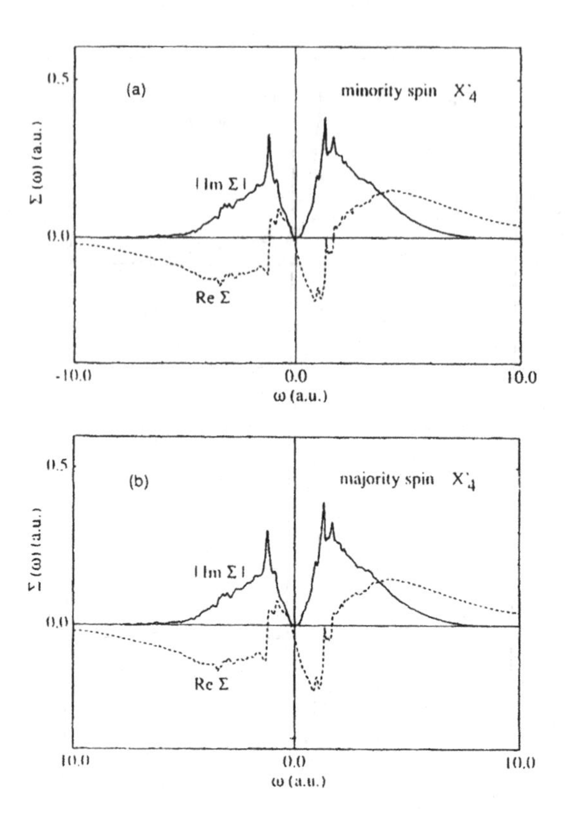

Figure 6: *Ni self-energy in the GWA. (a) The real and imaginary parts of the correlation part of the self-energy for the minority spin state X'_4. (b) The real and imaginary parts of the correlation part of the self-energy for the majority spin state X'_4 (Aryasetiawan 1992).*

itself. *GW* calculations on transition metal atoms also show that strong correlations among 3d electrons of opposite spin are not well accounted for by the GWA (Shirley and Martin 1993).

In Table 3 the quasiparticle widths or the inverse life-times are also listed. The agreement with available experimental data is reasonably good. The largest discrepancy is for the state X_1 and K_1 where the *GW* results are up to a factor of two larger. The large width at the bottom of the 3d band implies a strong interaction between the quasiparticles and the rest of the system. The quasiparticle wave functions have large overlap with each other and with other excitations so that there is a large probability of decaying to other states, resulting in short life-time.

Table 3: *Ni quasiparticle energies and widths in eV within the GWA at high symmetry points for majority and minority spin (alternately) along with the LDA eigenvalues and experiment (Aryasetiawan 1992).* $< E_{\mathbf{k}n} >$ *is the average quasiparticle energy. The experimental data are taken from Eberhardt and Plummer (1980) (shifted down by 0.3 eV for the purpose of comparison) and those in the brackets from Mårtensson and Nilsson (1984).*

$\mathbf{k}:n$	$\varepsilon_{\mathbf{k}n}^{LDA}$	$E_{\mathbf{k}n}$	$< E_{\mathbf{k}n} >$	exp	FWHM	
					GW	exp
Γ_1	−9.26	−9.0				2.1
	−9.24	−9.0	−9.0	−9.1±0.2	2.1	1.8
Γ'_{25}	−2.35	−1.7				0.8
	−1.76	−1.1	−1.4	−1.4±0.2		0.4
Γ_{12}	−1.23	−0.9				0.3
	−0.59	−0.3	−0.6	−0.7±0.1		0.2
X_1	−4.93	−4.3				2.9
	−4.54	−3.9	−4.1	−3.6±0.2 (−3.8)	2.5 ;	1.25
X_3	−4.28	−3.5				2.2
	−3.80	−2.9	−3.2	−3.1±0.2 (−2.8)	1.5 ;	1.4
X_2	−0.60	−0.5				0.1
	+0.07	+0.2	−0.3	−1.15±0.1 (−0.2)		0.1
L_1	−4.95	−4.3				2.5
	−4.60	−4.0	−4.1	−3.9±0.2		2.0
L_3	−2.44	−1.9				1.0
	−1.86	−1.3	−1.6	−1.6±0.1	0.6	0.9
L'_2	−0.74	−1.3				0.02
	−0.73	−1.3	−1.3	−1.3± (−1.0)0.1		0.02
W'_2	−4.00	−3.4				2.2
	−3.58	−3.1	−3.3	−2.9±0.2		1.9
W_3	−3.09	−2.6				1.4
	−2.61	−2.0	−2.3	−2.0±0.2	0.8	1.3
W_1	−1.40	−1.1				0.8
	−0.79	−0.5	−0.8	−0.95±0.1	0.2	0.8
K_1	−4.18	−3.6				2.4
	−3.76	−3.2	−3.4	−3.4±0.2	2.0	1.3
K_1	−3.80	−3.2				1.9
	−3.36	−2.6	−2.9	−2.85±0.1	1.2	1.0
K_3	−2.13	−1.8				0.4
	−1.65	−1.2	−1.5	−1.2±0.2	0.8	0.8
K_3	−1.14	−1.0				0.3
	−0.51	−0.4	−0.7	−0.75±0.1		0.2

The spectral function is shown in Figure 17. As mentioned before, the 6 eV satellite is not reproduced in the GWA. Satellite-like structure at around 6 eV actually arises from quasiparticle peaks as pointed out by Kanski, Nilsson, and Larsson (1980). The spiky structure is due to a small number of **k**-points used in the calculations and it would be smoothened out when more **k**-points are used. The double-peak structure at the plasmon frequency can also be seen. It should be emphasized that the satellite is due to many-body effects as shown by 3p-resonance photoemission measurements. At 67 eV photon energy corresponding to the binding energy of the 3p core, the photoemission spectra exhibit an asymmetric (Fano) resonant enhancement and the main 3d line shows a strong antiresonance (Guillot *et al.* 1997). This is explained as an Auger process where a 3p core electron is excited to fill the empty 3d states followed by a super Coster-Kronig decay

$$3p^6 3d^9 4s + \hbar\omega \rightarrow 3p^5 3d^{10} 4s \rightarrow 3p^6 3d^8 4s + e$$

That the 6 eV satellite is missing in the GWA can be seen directly in the imaginary part of the self-energy. For the existence of a satellite, there should be a strong peak at around 5–6 eV reflecting the presence of a stable excitation. But as can be seen in Figures 5 and 6, such peak is missing.

The standard explanation for the presence of the satellite is that a photoemission d hole causes a strong perturbation due to its localized nature, exciting another d electron to an empty state just above the Fermi level (shake-up effects). In atomic picture, the state with two d holes correspond to the configuration $3d^7 4s^2$ which is separated from the main line configuration $3d^8 4s$ by more than 6 eV. In the solid, however, this value is considerably reduced by metallic screening. The two holes scatter each other repeatedly forming a "bound state" at 6 eV (Liebsch 1979, 1981, Penn 1979). In a simple picture, the photon energy is used to emit a d electron and to excite another into an empty d state so that the emitted electron appears to have a lower binding energy (satellite). The excited electrons mainly come from the bottom of the d band since they have the largest mixing with the s-p states and therefore, according to the dipole selection rule, a large transition to the empty d states. This then has been argued as the source of band narrowing. The *GW* calculations, however, suggest that the largest contribution to band narrowing comes mainly from long-range screening rather than short-range two-hole interactions.

According to the above picture, the reduction in the exchange splittings is related to the satellite. For simplicity we consider a two-level ferromagnetic model with fully occupied majority channel and one occupied minority channel. A photoemission hole in the majority channel can induce another hole in the minority channel by exciting an electron to the unoccupied state. But a hole in the minority channel cannot induce another hole in the majority channel because there is no empty state available in the majority channel to which electrons can be excited. Thus, the effects of two-hole interactions are larger for the majority than for the minority channel resulting in a reduction in the exchange splitting. This picture is confirmed by calculations based on the Hubbard model within the T-matrix approach (Liebsch 1979, 1981). According to this model, the reduction in the band width originates from the two-hole interactions which is not in agreement with the *GW* results. These two seemingly conflicting results may be reconciled as follows. In the T-matrix calculations, it was found that there was no value of U that gave the correct satellite position and the band width (Liebsch 1979, 1981). To get the correct band width, the value of U was such that the satellite energy became too large. If we assume that the T-matrix theory gives the correct physical description for the satellite, the appropriate value of U would not give a large reduction in the band width, but this is taken care of by the GWA. Thus one concludes that self-energy calculations which include the diagrams of the GWA and T-matrix theory may give both the correct band width and the satellite structure. Diagrammatic comparison between the GWA and the T-matrix theory reveals that hole-hole interactions described by the T-matrix are not included in the GWA except to second order only. Direct comparison between *GW* calculations on real systems and Hubbard model calculations is, however, difficult if not impossible. This is due the Hubbard U which cannot be easily related to the screened interaction in the *GW* calculations.

4.2.2 Nickel Oxide

There is a renewed interest in the late transition-metal monoxides MnO, FeO, CoO, and NiO because the electronic structure of these monoxides is very similar to that of the parent compounds of the high-temperature superconductors. Both are antiferromagnetic insulators and of Mott-Hubbard type where the band gap arises from Coulomb correlation rather than the filling of single-particle bands like in ordinary semiconductors or insulators. A system with an onsite Coulomb energy larger than the single-particle bandwidth tends to become an insulator and that single-particle theory is likely to make qualitatively wrong predictions (Mott 1949). Indeed, the LDA predicts NiO to be a metal when the calculation is performed in a paramagnetic state (Mattheiss 1972a,b). A more convincing evidence is provided by CoO, where the number of electrons in the paramagnetic structure is odd, making it impossible for any single-particle theory to predict CoO as an insulator without doubling the unit cell. The experimental crystal structure of NiO and CoO, however, is antiferromagnetic of type II. Slater (1974) suggested that a gap could be opened up by an interplay between antiferromagnetism and crystal-field splittings. A detailed work along this direction can be found in a paper by Terakura *et al.* (1984). The LDA does give a gap for the antiferromagnetic structure but the gap is only 0.2 eV in large discrepancy compared with the experimental gap of 4.0 eV (Powell and Spicer 1970, Hüfner *et al.* 1984, Sawatzky and Allen 1984). The difference in the magnetic energy between the paramagnetic and antiferromagnetic states is only a fraction of an eV, which is much smaller than the band gap. Therefore, the band gap should not depend on whether the calculation is performed in a paramagnetic or an antiferromagnetic structure, as confirmed by the small gap in the LDA.

The free-electron like O p band is well described by the LDA, which is not surprising, but the magnetic moment is too small (1.0 μ_B) compared to experiment (1.7–1.9 μ_B). In principle, it should be possible to obtain the correct magnetic moment within the Kohn-Sham DFT since it is a groundstate property.

The basic physics of the Mott-Hubbard insulators was elucidated by Mott several decades ago (Mott 1949). In the tight-binding picture appropriate to describe localized 3d states, hopping matrix elements cause the formation of a band of width w centred around the atomic eigenvalue. The formation of a band favours electron hopping because the electrons can occupy states with lower energy but it costs a Coulomb energy U for an electron to hop from one site to the neighbouring site. If U is larger than w, the gain in kinetic energy is overwhelmed by the loss in Coulomb energy and the system prefers to be an insulator with a gap approximately given by U, splitting the so called lower and upper Hubbard bands. The original Mott picture is essentially correct, but there are a number of experimental data which cannot be explained in the Mott picture. The value of U, for instance, is estimated to be 8-10 eV which is much larger than the experimental gap.

Fujimori, Minami, and Sugano (1984) and Sawatzky and Allen (1984) clarified these discrepancies. Studies based on the cluster approach and Anderson impurity model show that the gap in NiO is a charge-transfer gap. If in a photoemission process an electron is removed from a Ni site, the resulting state has an extra hole and a high energy because of an increase in the Coulomb interaction among the holes. An electron from an O site may fill the hole on the Ni site which costs some energy transfer. But it is still favourable because it leads to a state with a lower energy. The final state therefore has the hole residing on the O site and consequently the states at the top of the valence band have a large O p character (d^8L^-). The lowest conduction state is of d character as in the Mott picture and the gap is formed between the valence O p and conduction Ni d states. Much of the Ni d weight goes into a satellite located below the O p, in contrast to the Mott and the Slater pictures where the Ni 3d weight is located mainly above the O p.

This model is able to explain experimental data which would otherwise be difficult to explain by the Mott and Slater pictures. The most convincing evidence supporting the charge-transfer picture is the 2p resonant photoemission experiment in CuO (Tjeng *et al.* 1991) which has a similar electronic structure as that of NiO. In this experiment, a 2p core electron is excited and the remaining hole is subsequently filled by a valence electron. Since dipole transition matrix element is largest between p and d states, resonance in the valence energy region can be identified as the position of d states which turns out to be below the O p band, rather than above as in the Mott and Slater pictures.

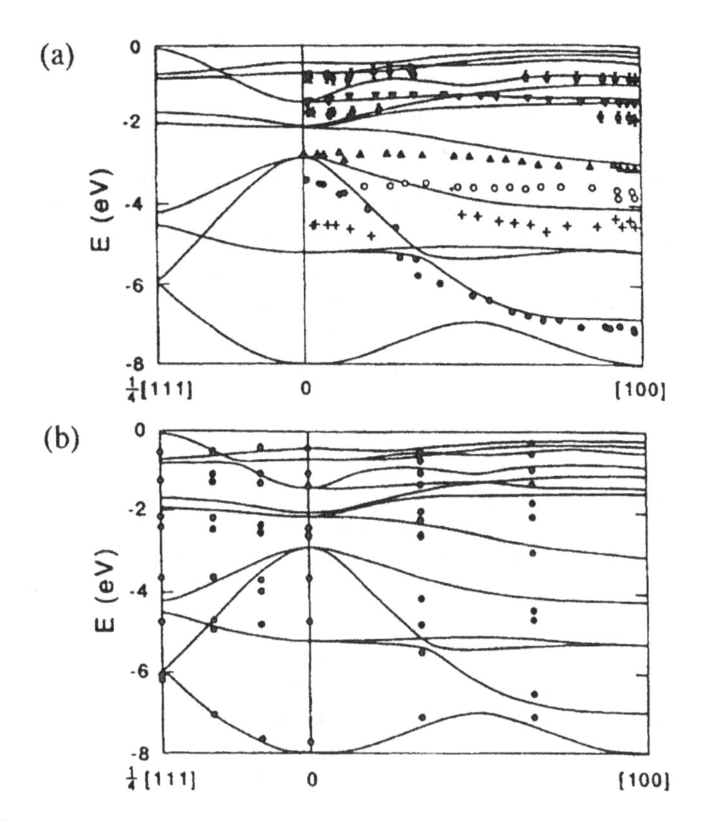

Figure 7: *NiO bandstructure. (a) Comparison between the LDA (solid line) and the experimental bandstructure (Shen et al. 1990, 1991a,b). (b) Comparison between the LDA (solid line) and the quasiparticle bandstructure in the GWA (Aryasetiawan and Gunnarsson 1995).*

The LDA antiferromagnetic bandstructure is shown in Figure 7. The highest valence state is formed by the majority e_g and minority t_{2g} Ni states and the lowest conduction band is formed by the minority e_g state. As can be seen, the LDA gap is only 0.2 eV. Since the true gap (4 eV) is much larger than the LDA gap, one suspects that the true quasiparticle wavefunctions for the gap states may be significantly different from the LDA wavefunctions. One anticipates the need to improve the starting single-particle Hamiltonian. Indeed, starting from the LDA antiferromagnetic bandstructure with a gap of 0.2 eV, a one-iteration GW calculation shifts the Ni e_g conduction upwards giving a gap of $\approx$1 eV, still far from the true value. An obvious thing to do would be to replace the LDA eigenvalues by the quasiparticle energies and repeat the self-energy calculations to self-consistency. This procedure, however, does not produce the desired effect of opening up the gap. A better and more physical procedure is to construct a new single-particle Hamiltonian $H = H_{\mathrm{LDA}} + (\Sigma - V^{xc})$ and use this as new zeroth order Hamiltonian in a subsequent GW calculation. The procedure should be repeated to self-consistency. We observe that the effect of the self-energy correction in the first iteration, i.e. with $H = H^{\mathrm{LDA}}$, is to raise the unoccupied e_g state. The new Hamiltonian may therefore be approximated by adding a non-local potential $\Delta|e_g\rangle\langle e_g|$ to the LDA Hamiltonian: $H = H_{\mathrm{LDA}} + \Delta|e_g\rangle\langle e_g|$. where Δ is chosen to reproduce the 1 eV gap obtained in the first iteration. Δ is changed subsequently until self-consistency is achieved, i.e. until the resulting gap remains unchanged, The final self-consistent gap was found to be $\approx$5.5 eV. This procedure modifies the eigenvalues as well as the wave functions used to construct the zeroth order Green function G^0.

The raising of the unoccupied majority e_g band by the self-energy correction reduces the hybridization with the O p band and has the effects of raising the bottom of the O p band and pushing down the top of the O p band at the Γ -point resulting in better agreement with photoemission data. In addition, the width of the unoccupied e_g band is reduced. The reduction in hybridization also reduces the magnitude of the exchange interaction of the e_g band with the occupied states which has the consequence of widening the gap. Thus, it is important that the wave functions are also modified in the self-consistent procedure. The final position of the unoccupied e_g band is just below the Ni 4s. As a check, the calculation has also been performed in the ferromagnetic state. A gap of $\approx$ 5.2 eV was obtained, close to the antiferromagnetic value. In contrast to the Slater model, the gap does not depend on the antiferromagnetic ordering and the results correctly predict that NiO remains an insulator above the Néel temperature. The GW calculation clearly improves the LDA gap markedly and it is in reasonable agreement with the experimental value of 4.0 eV. An estimate of the magnetic moment yields a value of 1.6 μ_B in good agreement with the experimental value of 1.7-1.9 μ_B (Alperin 1962, Fender *et al* 1968, Cheetham and Hope 1983).

To study the character of the states at the top of the valence band, the spectral function has been calculated. Projection of the spectrum into the Ni 3d and O p orbitals shows that there is an increase of the O p character at the top of the valence band and but the main character is primarily Ni 3d. The satellite at -8 eV (Shen *et al.* 1990, Shen *et al.* 1991a, Shen *et al.* 1991b) is not reproduced at the final self-consistent spectrum. It is often thought that the GWA cannot produce low energy satellites since the theory is based on the RPA screening which gives mainly a plasmon satellite of high energy. A recent more detailed study of the spectral function, however, reveals an interesting behaviour of the satellite structure. The satellite structure turns out to be rather sensitive to the starting Hamiltonian (Aryasetiawan and Karlsson 1996). Starting from the LDA Hamiltonian in fact gives a satellite at about -10 eV but this satellite diminishes in intensity as the gap opens up. The origin of this behaviour can be traced back

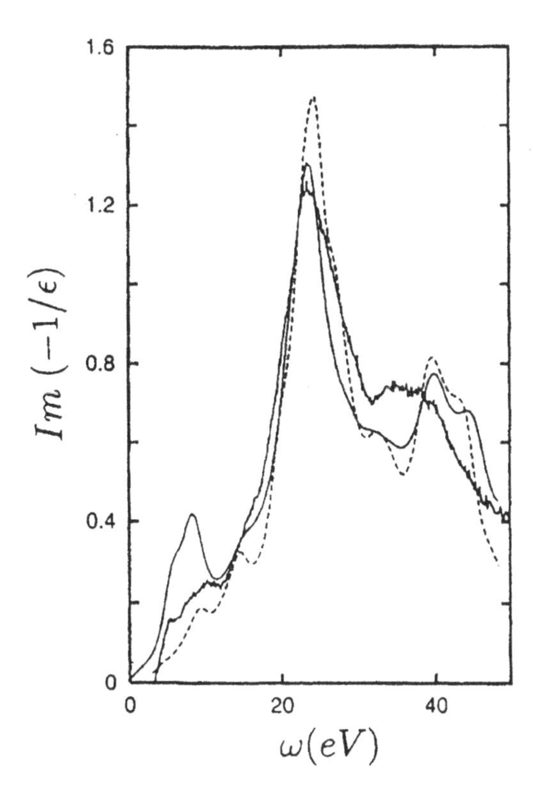

Figure 8: *The energy-loss spectra of NiO. The smooth solid curve corresponds to the calculated spectrum with virtually no gap in the LDA Hamiltonian and the dashed one with ≈5 eV gap. The other curve is the experiment (Aryasetiawan et al 1994c).*

to the presence of a plasmon-like peak at low energy which is related to the incorrect LDA bandstructure as shown in Figure 8. As the gap opens up, this peak structure becomes broadened and consequently the satellite structure diminishes as illustrated in Figure 9. One can still see the remaining structure when the spectra is magnified ten times. As in the case of Ni, it appears that the satellite structure is due to short-range correlations which are not properly taken into account by the RPA. T-matrix approach could be appropriate and might remove some Ni dweight from the top of the valence band to the satellite region but it is not clear how this could increase the charge transfer from the O p.

Very recent calculation based on a model GW (Massidda *et al.* 1997) is in reasonable agreement with the full calculation described above. An increase in the O p character at the top of the valence band was found when the calculation

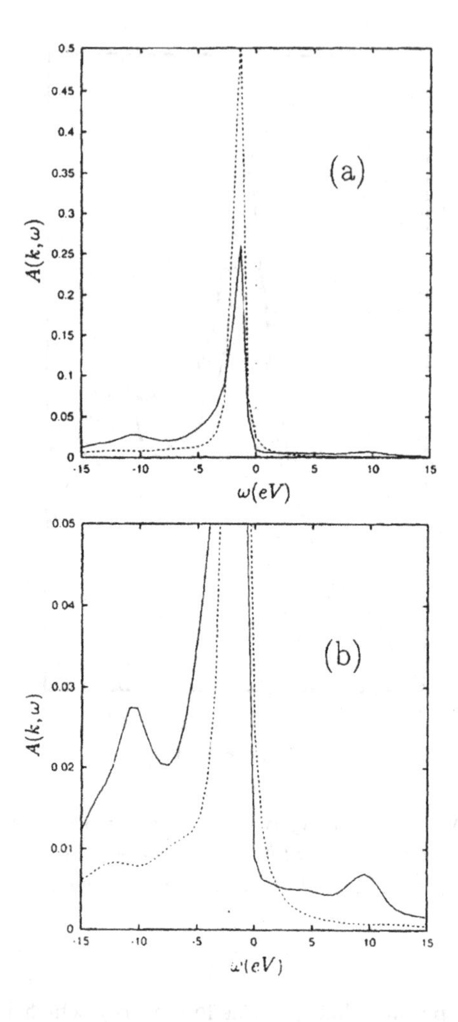

Figure 9: *The spectral function of NiO at the* Γ *point for a Ni 3d state. The solid line corresponds to the case where the starting LDA Hamiltonian has virtually no gap and the dashed line to the case with* $\approx$*5 eV gap. (b) The same as in (a) but magnified 10 times (Aryasetiawan and Karlsson 1996).*

was performed fully self-consistently, which is rather encouraging. However, the O p band, which is nicely reproduced by the LDA, is placed to high although the dispersion is still similar to that of the LDA. One reason for this could be due to the lack of energy dependence in the model. The result of this model GW calculation is in some respects similar to that of the LDA+U. This appears to support the recent attempt to relate the GWA to the LDA+U scheme, where it was suggested that the LDA+U may be regarded as a static approximation to the GWA (Anisimov, Aryasetiawan, and Liechtenstein 1997).

4.2.3 Manganese Oxide

A calculation on MnO based on a simplified *GW* scheme has been performed by Massidda *et al.* (1995a). The electronic structure of MnO is the simplest among the transition metal oxides and in some respects similar to NiO. As in the case of NiO, the LDA gives a too small band gap of 1.0 eV compared with the experimental value of 3.8–4.2 eV. The magnetic moment is also somewhat too small (LDA 4.3 μ_B, experiment 4.6–4.8 μ_B) although the relative discrepancy is not as large as in NiO. The larger LDA band gap is due to the fact that in MnO the majority spin is fully occupied and the minority spin is empty resulting in a large magnetic moment so that the exchange splitting is also large and dominates the ligand-field splitting and band broadening due to intersublattice coupling. In NiO the magnetic moment is smaller and the exchange splitting is comparable to the ligand-field splitting and band broadening.

The semiempirical model *GW* calculation gives a gap of 4.2 eV which compares well with the experimental value of 3.8–4.2 eV. The LDA magnetic moment is also improved to 4.52 μ_B. There is an increase of O p character and a decrease of Mn 3d character at the top of the valence band, which are percentagewise large but small in absolute term, so that the main character is still primarily Mn 3d. The results are qualitatively similar to the full *GW* calculations on NiO described above. Calculation on $CaCuO_2$ using the model GW scheme has also been done (Massidda *et al.* 1997).

4.2.4 3d semicore states

It is well-known that the LDA eigenvalues for localized states are usually too high compared to experiment. The discrepancies can be a few eV. For example, the Zn semicore 3d states in ZnSe are too high by 2.5 eV, the Ga semicore 3d states in GaAs by 4 eV and the Ge 3d semicore states by as large as 5 eV. In this respect it is worth noting that the self-energy error in the LDA *total energy* is rather small because the LDA exchange-correlation hole satisfies the charge sum-rule. On the other hand, the error in the eigenvalues can be large since the fact that the exchange-correlation hole satisfies the sum rule does not necessarily imply a good exchange-correlation potential (Gunnarsson, Lundqvist, and Wilkins 1974). In atoms, a large part of the discrepancy is attributed to the absence of self-interaction in the LDA. Indeed, self-interaction corrected calculations improve the eigenvalues substantially. In solids, additional complication arises because apart from self-interaction there is the effect of screening or polarization. The discrepancies in solids are known to be smaller than in atoms. For instance, the Zn atom the discrepancy is 6.5 eV. One would expect that the self-interaction correction should be approximately the same in solids as in atoms since the 3d states are very localized. The difference in the discrepancies must then be due to the cancellation of errors in self-interaction (exchange) and correlations.

It is not a priori clear that the GWA can describe semicore states. Unlike core states where the screening is mainly due to the valence states the semicore states can themselves participate in the screening. *GW* calculations for semicore states in ZnSe, GaAs, Ge (Aryasetiawan and Gunnarsson 1996) improve the LDA results significantly. A discrepancy of 0.5–1.0 eV (too high) still persists. Calculations using the ΔSCF method (Hedin and Johansson 1969) in the Slater transition state approach (Slater 1974) improve the LDA results substantially too although the eigenvalues become too low by about 1 eV. The ΔSCF method avoids the problems with the self-interaction, exchange-correlation potential, and the neglect of relaxation in the normal LDA calculations. Essential to the results is that there is a transfer of screening charge to the atom with the semicore hole (Lang and Williams 1977, Zunger and Lindefelt 1983). The difference between the *GW* and ΔSCF results could be due to the error in the LDA which tends to overestimate the exchange-correlation energy between a 3d electron and the 3s-3p core (Gunnarsson and Jones 1980) which is about 1 eV in the series K-Cu (Harris and Jones 1978). It could also be due to the difference in the RPA screening and the static LDA screening. Transition state calculations for a number of Zn compounds have also been performed by Zhang, Wei, and Zunger (1995). We observe, however, that the Slater transition state approach requires that the semicore states are sufficiently localized to form a bound state in the transition state calculations. The bound state has no dispersion, in contrast to the *GW* results which show the full band structure. Thus, if a bound state is not formed in the transition state calculation, the result would be identical to that of the LDA. The problem can be illustrated for Cu metal where the 3d band is 0.5 eV too high. The transition state would give a single number rather than a band assuming that a bound state is formed in the first place. A *GW* calculation on the other hand lowers the position of the band by 0.2–0.3 eV while maintaining the LDA band structure (Aryasetiawan and Gunnarsson 1997).

A model *GW* calculation for ZnO by Massidda *et al.* (1995b) also improved the LDA result from −5.4 eV to −6.4 eV but a significant error remains when compared to experimental results −8.6 eV and −7.5 eV.

4.3 f system: Gd

Application of the GWA in f systems has only been made to Gd (Aryasetiawan and Karlsson 1996). Gd does not exhibit the Kondo resonance and it is probably the simplest f-system for which a single-particle treatment still gives reasonable results. To describe the Kondo resonance, an RPA-type approach like the GWA is not expected to be successful. In the LDA, the majority f channel of Gd is fully occupied and the minority f channel is fully unoccupied. The majority f state lies at -4.5 eV and the minority at 0.5 eV with respect to the Fermi level. The separation between the occupied and unoccupied f states is then 5.0 eV which is in poor agreement with the experimental value of 12.5 eV.

The GW spectral function shows satellite structures at the positions of the peaks in the experimental spectra but some quasiparticle weight remains at the LDA eigenvalue. However, the weight of the quasiparticle is $Z = 0.3$, which is unusually small. In simple metals and semiconductors $Z = 0.6 - 0.7$ and in Ni $Z = 0.5$.

The structure in the spectral function is determined by the structure in the self-energy as can be seen in equation (86). Peaks in the imaginary part of the self-energy indicates the presence of "boundstates" or satellites. Thus the peak at -7 eV gives rise to the satellite structure at -9 eV. The sharp increase in Im Σ^c starting at around -4.3 eV is due to the very localized character of the f states. As a consequence, the matrix elements in equation (86) are large and responsible for the presence of the peak at -7 eV. For $\omega \leq \mu$, the occupied s-p states, which lie almost above the f-states, make almost no contribution to Im Σ^c. The relevant quantity for the f states is $\langle ff|W|ff\rangle$ (where f denotes an f-orbital) which describes the screened Coulomb interaction among the f-electrons. The large peak at around -21 eV is due to plasmon excitation and it gives rise to a large plasmon satellite in the spectral function. Note, however, that the height of the peak is very sensitive to the value of Im Σ^c while the weight is not. The plasmon peak is expected to be smoothened out if g orbitals are included in the calculations.

To interpret the *GW* result, we consider how the exact self-energy correction to the LDA should behave. We expect a large imaginary part of the self-energy, Im Σ^c, around the LDA f eigenvalue, ϵ_f, which would imply a loss of the quasiparticle weight since Im Σ^c is proportional to the life-time of the quasiparticle. In order to shift the weight to the experimental position, we expect that Im Σ^c should be rather peaked around ϵ_f so that Re Σ^c has a strong variation around this energy resulting in a formation of a boundstate (satellite). The *GW* self-energy indeed shows this behaviour as can be seen in Figure 10 but the variation is not strong enough in order to completely remove the quasiparticle weight to the satellite region as shown in Figure 11. It is clear, however, that the GWA reproduces the correct trend as reflected in the small quasiparticle weight ($Z = 0.3$) and the formation of the satellite at approximately the right energy.

5 Vertex corrections: Beyond *GW*

The GWA has proven to be very successful for describing quasiparticle energies for *sp* and even 3*d* systems. Its description for satellite structures, however, is less satisfactory. A number of cases which reveal the shortcomings of the GWA in describing the satellite structure are

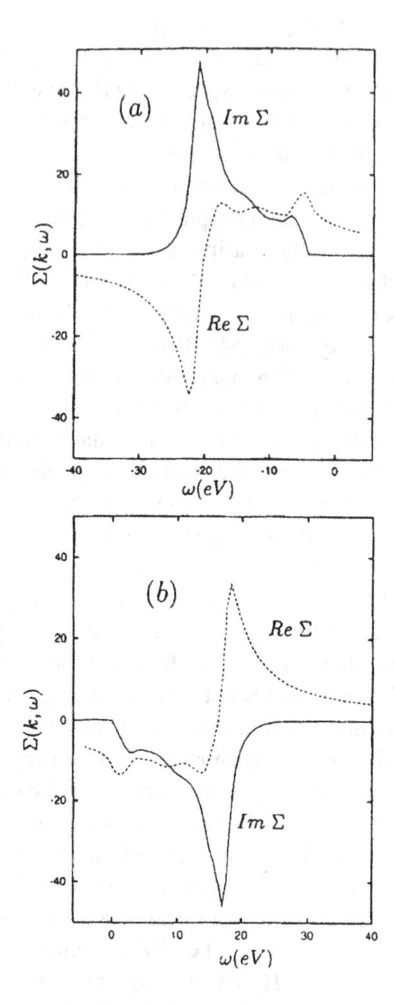

Figure 10: *The correlation part of the GW self-energy of Gd for an f state at the Γ point for $\omega < \mu$ (a) and $\omega > \mu$ (b). Note that the constant bare exchange contribution is not included in the figure (Aryasetiawan and Karlsson 1996).*

- The valence as well as the core photoemission spectra of the alkalis show a series of plasmon satellites which are located at multiple of plasmon energies below the main quasiparticle peak. The GWA gives only one plasmon satellite and its position is typically 1.5 plasmon energy below the quasiparticle peak (Hedin, Lundqvist, and Lundqvist 1970).
- The valence photoemission spectrum of Ni shows the presence of a satellite at 6 eV which is not obtained within the GWA.

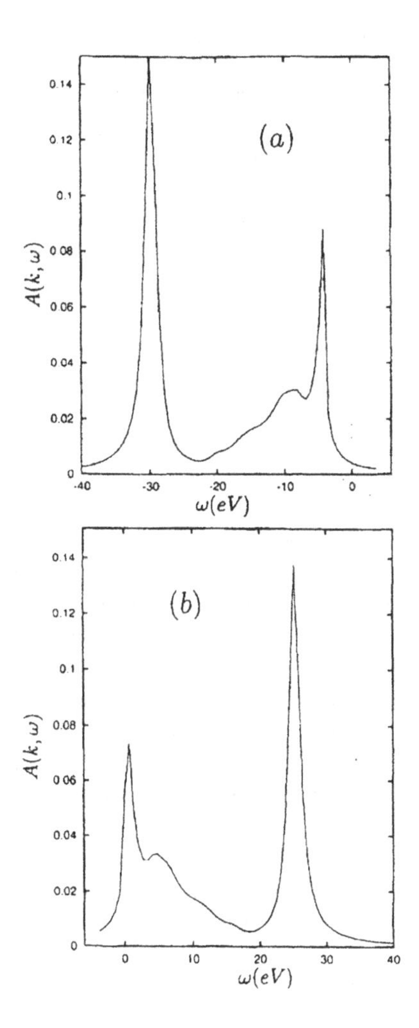

Figure 11: *The spectral function of Gd for an f state at the* Γ *point for* $\omega < \mu$ *(a) and* $\omega > \mu$ *(b) (Aryasetiawan and Karlsson 1996).*

- Similarly, the GWA appears to be insufficient for describing the satellite structure in transition metal oxides.

The presence of only one plasmon satellite in the GWA follows from the fact that the self-energy is of first order in W which contains one plasmon excitation through the RPA dielectric function. From the diagram, it is clear that a hole or an electron interacts with its surrounding by exchanging a plasmon. The too large plasmon energy can be understood qualitatively as an average of the first and the second plasmon energies since they carry most of the satellite weight. Because the peak in Im Σ^c is at one plasmon energy below the quasiparticle peak, the plasmon satellite in the spectral function ends up at $\approx$1.5 plasmon energy (Hedin, Lundqvist, and Lundqvist 1970). To get the correct satellite energy the peak in Im Σ^c should therefore lie somewhat closer to the quasiparticle energy.

The problem with the satellite structure in Ni or transition metal oxides is of a different nature since the satellite energy is much lower than the plasmon energy. In the atomic picture, the ground state of Ni is a mixture of the configurations $3d^9 4s$ and $3d^8 4s^2$. The final configurations after photoemission are $3d^8 4s$ and $3d^7 s^2$. The former corresponds to the main line (quasiparticle) and the latter to the satellite. The two configurations are separated in energy by 6 eV which is essentially the Coulomb energy of two *d* holes. This value is reduced by screening in the solid. The presence of satellite at a certain energy implies that the imaginary part of the self-energy should exhibit a peak at an energy slightly lower then the satellite energy. Such a two-hole excitation state is partly described by the GWA to second order in the bare interaction but the GWA mainly describes the coupling of the electrons to the plasmon excitation. A secondary plasmon-like excitation is possible within the RPA when the band structure gives rise to two well-defined peaks in Im ϵ^{-1}.

It is a major challenge to develop a theory beyond the GWA for real systems which can overcome the problems described above. Here we describe those approaches which are based on systematic diagrammatic expansions. One approach which has been applied with success to the alkalis is the cumulant expansion (Langreth 1970, Bergersen 1973, Hedin 1980). As described in the next section, this approach is suitable for dealing with systems which can be mapped into systems of electrons coupled to bosons (e.g. plasmons) where long-range correlations dominate. However, short-range correlations arising from multiple-hole interactions on the same site should be better treated within the T-matrix approach, described in a later section. In this contribution, we are concerned with vertex corrections related to the satellite description rather than quasiparticle description.

The effects of vertex corrections on the quasiparticles have been studied by a number of authors. These include a direct calculation of the second-order diagram (Daling and van Haeringen 1989, Daling *et al.* 1991, Bobbert and van Haeringen 1994) and vertex correction based on the LDA V^{xc} with application to Si (Del Sole *et al.* 1994). Vertex corrections in the electron gas (Mahan and Sernelius 1989, Frota and Mahan 1992) are discussed in a review article by Mahan (1994) where it is emphasized that it is important to include vertex corrections in both the response function and the self-energy in a consistent way (Ward 1950, Baym and Kadanoff 1961, Baym 1962, Rice 1965).

5.1 The cumulant expansion

One of the first applications of the cumulant expansion method was in the problems of X-ray spectra of core-electrons in metals (Nozières and de Dominicis 1969, Langreth, 1970). The problem is modelled by a Hamiltonian consisting of a core electron interacting with electron-hole excitations and a set of plasmons:

$$H = \varepsilon c^\dagger c + \sum_{\mathbf{k}\sigma} \varepsilon_{\mathbf{k}\sigma} c^\dagger_{\mathbf{k}\sigma} c_{\mathbf{k}\sigma} + \sum_{\mathbf{q}} \omega_{\mathbf{q}} b^\dagger_{\mathbf{q}} b_{\mathbf{q}} + \sum_{\mathbf{k}k'\sigma} V_{\mathbf{k}k'} c^\dagger_{\mathbf{k}\sigma} c_{\mathbf{k}\sigma'} c c^\dagger + \sum_{\mathbf{q}} c c^\dagger g_{\mathbf{q}} \left(b_{\mathbf{q}} + b^\dagger_{\mathbf{q}}\right) \tag{160}$$

where c is the annihilation operator for the core electron, $c^\dagger_{\mathbf{k}\sigma}$ is the creation operator for a conduction electron of wave vector $\mathbf{k}$, spin σ, and energy $\varepsilon_{\mathbf{k}\sigma}$ and $b^\dagger_{\mathbf{q}}$ is the creation operator for a plasmon of wave vector $\mathbf{q}$ and energy $\omega_{\mathbf{q}}$. The last two terms are the coupling of the core electron to the conduction electrons (electron-hole excitations) and to the plasmons respectively. The model Hamiltonian without the last term (Mahan, 1967a,b) was solved exactly by Nozières and de Dominicis (1969) and the full Hamiltonian was solved exactly by Langreth (1970). It can be shown (Langreth, 1970) that the cumulant expansion also gives the exact solution. We consider the case with coupling only to the plasmon field (Lundqvist, 1967) but not to the conduction electrons. The exact solution is given by (Langreth, 1970)

$$A_\pm(\omega) = \sum_{n=0}^{\infty} \frac{e^{-a} a^n}{n!} \delta\left(\omega - \varepsilon - \Delta\varepsilon \mp n\omega_p\right) \tag{161}$$

where $+$ refers to absorption spectrum and $-$ to emission spectrum. $a = \sum_{\mathbf{q}} g^2_{\mathbf{q}}/\omega^2_p$ and $\Delta\varepsilon = a\omega_p$ is the shift in core energy due to the interaction with the plasmon field. The spectrum consists therefore of the main quasiparticle peak at $\omega = \varepsilon + \Delta\varepsilon$ and a series of plasmon excitations at multiples of the plasmon energy below the quasiparticle peak. It has been assumed that the plasmon excitations have no dispersion although this assumption is not necessary.

The physics of the cumulant expansion when applied to valence electrons was discussed in detail by Hedin (1980). More recently, the cumulant expansion was calculated to higher order for a model Hamiltonian by Gunnarsson, Meden, and Schönhammer (1994).

5.1.1 Theory

In the cumulant expansion approach, the Green function for the hole ($t < 0$) is written as

$$\begin{aligned} G(k,t) &= i\theta(-t) \langle N|\hat{c}^\dagger_k(0)\hat{c}_k(t)|N\rangle \\ &= i\theta(-t)\, e^{-i\varepsilon_k t + C^h(k,t)} \end{aligned} \tag{162}$$

where k denotes all possible quantum labels. For $\omega \leq \mu$ only the first term in equation (59) contributes and the hole spectral function is

$$\begin{aligned} A(k, \omega \leq \mu) &= \frac{1}{\pi}\mathrm{Im}G(k,\omega) \\ &= \frac{1}{2\pi}\int_{-\infty}^{\infty} dt e^{i\omega t}\langle N|\hat{c}_k^{\dagger}(0)\hat{c}_k(t)|N\rangle \\ &= \frac{1}{\pi}\mathrm{Im}\, i\int_{-\infty}^{0} dt\, e^{i\omega t}e^{-i\varepsilon_k t + C^h(k,t)} \end{aligned} \tag{163}$$

$C^h(k,t)$ is called the cumulant. Expanding the exponential in powers of the cumulant we get

$$G(k,t) = G_0(k,t)\left[1 + C^h(k,t) + \frac{1}{2}[C^h(k,t)]^2 + \ldots\right] \tag{164}$$

where $G_0(k,t) = i\theta(-t)\exp(-i\varepsilon_k t)$. In terms of the self-energy, the Green function for the hole can be expanded as

$$G = G_0 + G_0\Sigma G_0 + G_0\Sigma G_0\Sigma G_0 + \ldots \tag{165}$$

We may group the cumulant into terms labelled by the order of the interaction n:

$$C^h = \sum_{n=1}^{\infty} C_n^h \tag{166}$$

The above equations for G may now be equated and the cumulant can be obtained by equating terms of the same order in the interaction. Thus to lowest order in the screened interaction W, the cumulant is obtained by equating

$$G_0 C^h = G_0\Sigma G_0 \equiv \Delta G^{(1)} \tag{167}$$

where $\Sigma = \Sigma_{GW} = iG_0W$. If G_0 corresponds to, e.g. the LDA G, then $\Sigma = \Sigma_{GW} - V^{xc}$. The explicit form of $\Delta G^{(1)}$ is

$$\Delta G^{(1)}(1,2) = \int d3d4 G_0(1,3)\,\Sigma(3,4)\,G_0(4,2) \tag{168}$$

Using

$$\begin{aligned} G_0(1,2) = & \; i\sum_n^{\text{occ}} \phi_n(\mathbf{r}_1)\phi_n^*(\mathbf{r}_2)e^{-i\varepsilon_n(t_1-t_2)}\theta(t_2-t_1) \\ & -i\sum_n^{\text{unocc}} \phi_n(\mathbf{r}_1)\phi_n^*(\mathbf{r}_2)e^{-i\varepsilon_n(t_1-t_2)}\theta(t_1-t_2) \end{aligned} \tag{169}$$

and taking matrix element with respect to an occupied state ϕ_k yields

$$\Delta G^{(1)}(k, t_1 - t_2) = -e^{-i\varepsilon_k(t_1-t_2)} \int_{t_1}^{\infty} dt_3 \int_{-\infty}^{t_2} dt_4 \, e^{i\varepsilon_k(t_3-t_4)} \Sigma(k, t_3 - t_4) \tag{170}$$

Without loss of generality we can set $t_2 = 0$ and $t_1 = t < 0$:

$$\Delta G^{(1)}(k, t) = -e^{-i\varepsilon_k t} \int_{t}^{\infty} dt_3 \int_{t_3}^{\infty} d\tau \, e^{i\varepsilon_k \tau} \Sigma(k, \tau) \tag{171}$$

The first-order cumulant is therefore

$$\begin{aligned} C^h(k, t) &= i \int_{t}^{\infty} dt' \int_{t'}^{\infty} d\tau \, e^{i\varepsilon_k \tau} \Sigma(k, \tau) \\ &= i \int_{t}^{0} dt' \int_{t'}^{\infty} d\tau \, e^{i\varepsilon_k \tau} \Sigma(k, \tau) + C^h(k, 0) \end{aligned} \tag{172}$$

$C^h(k, 0)$ is a constant which contributes to an asymmetric line shape of the quasiparticle.

The cumulant can be expressed as an integral over frequency by defining a Fourier transform

$$\begin{aligned} \Sigma(k, \tau) &= \int \frac{d\omega}{2\pi} e^{-i\omega\tau} \Sigma(k, \omega) \\ &= \int \frac{d\omega}{2\pi} e^{-i\omega\tau} \left\{ \int_{-\infty}^{\mu} d\omega' \frac{\Gamma(k, \omega')}{\omega - \omega' - i\delta} + \int_{\mu}^{\infty} d\omega' \frac{\Gamma(k, \omega')}{\omega - \omega' + i\delta} \right\} \\ &= i\theta(-\tau) e^{\delta\tau} \int_{-\infty}^{\mu} d\omega' e^{-i\omega'\tau} \Gamma(k, \omega') \\ &\quad - i\theta(\tau) e^{-\delta\tau} \int_{\mu}^{\infty} d\omega' e^{-i\omega'\tau} \Gamma(k, \omega') \end{aligned} \tag{173}$$

The second line uses the spectral representation of Σ in equation (77) and the last line is obtained from

$$\begin{aligned} \int \frac{d\omega}{2\pi} \frac{e^{-i\omega\tau}}{\omega - \omega' - i\delta} &= e^{-i\omega'\tau} \int \frac{d\omega''}{2\pi} \frac{e^{-i\omega''\tau}}{\omega'' - i\delta}, \quad \omega'' = \omega - \omega' \\ &= i\theta(-\tau) e^{-i\omega'\tau} e^{\delta\tau} \end{aligned} \tag{174}$$

Similarly for the other integral. The self-energy becomes

$$\Sigma(k, \tau) = i\theta(-\tau) e^{\delta\tau} \int_{-\infty}^{\mu} d\omega \, e^{-i\omega\tau} \Gamma(k, \omega) - i\theta(\tau) e^{-\delta\tau} \int_{\mu}^{\infty} d\omega \, e^{-i\omega\tau} \Gamma(k, \omega) \tag{175}$$

The cumulant can now be expressed as, keeping in mind that $t < 0$,

$$
\begin{aligned}
C_1^h(k,t) &= i\int_t^0 dt' \int_{t'}^{\infty} d\tau\, e^{i\varepsilon_k\tau}\Sigma(k,\tau) \\
&= i\int_t^0 dt' \int_{t'}^{\infty} d\tau\, e^{i\varepsilon_k\tau}\Big\{ i\theta(-\tau)\, e^{\delta\tau}\int_{-\infty}^{\mu} d\omega\, e^{-i\omega\tau}\Gamma(k,\omega) \\
&\qquad -i\theta(\tau)\, e^{-\delta\tau}\int_{\mu}^{\infty} d\omega\, e^{-i\omega\tau}\Gamma(k,\omega)\Big\} \\
&= -\int_t^0 dt' \int_{t'}^{0} d\tau \int_{-\infty}^{\mu} d\omega\, e^{i(\varepsilon_k-\omega-i\delta)\tau}\Gamma(k,\omega) \\
&\quad +\int_t^0 dt' \int_{0}^{\infty} d\tau \int_{\mu}^{\infty} d\omega\, e^{i(\varepsilon_k-\omega+i\delta)\tau}\Gamma(k,\omega) \\
&= i\int_t^0 dt' \int_{-\infty}^{\mu} d\omega \frac{1-e^{i(\varepsilon_k-\omega-i\delta)t'}}{\varepsilon_k-\omega-i\delta}\Gamma(k,\omega) \\
&\quad +i\int_t^0 dt' \int_{\mu}^{\infty} d\omega \frac{\Gamma(k,\omega)}{\varepsilon_k-\omega+i\delta} \\
&= i\int_t^0 dt' \left\{ \int_{-\infty}^{\mu} d\omega \frac{\Gamma(k,\omega)}{\varepsilon_k-\omega-i\delta} + \int_{\mu}^{\infty} d\omega \frac{\Gamma(k,\omega)}{\varepsilon_k-\omega+i\delta} \right\} \\
&\quad +\int_{-\infty}^{\mu} d\omega \frac{e^{i(\varepsilon_k-\omega-i\delta)t}-1}{(\varepsilon_k-\omega-i\delta)^2}\Gamma(k,\omega) \\
&= -i\Sigma(k,\varepsilon_k)t + \left.\frac{\partial\Sigma^h(k,\omega)}{\partial\omega}\right|_{\omega=\varepsilon_k} \\
&\quad +\int_{-\infty}^{\mu} d\omega \frac{e^{i(\varepsilon_k-\omega-i\delta)t}}{(\varepsilon_k-\omega-i\delta)^2}\Gamma(k,\omega)
\end{aligned}
\tag{176}
$$

The last line is obtained from

$$
\begin{aligned}
\int_{-\infty}^{\mu} d\omega' \frac{\Gamma(k,\omega')}{(\varepsilon_k-\omega'-i\delta)^2} &= -\frac{\partial}{\partial\omega}\int_{-\infty}^{\mu} d\omega' \left.\frac{\Gamma(k,\omega')}{(\omega-\omega'-i\delta)}\right|_{\omega=\varepsilon_k} \\
&= -\left.\frac{\partial\Sigma^h(k,\omega)}{\partial\omega}\right|_{\omega=\varepsilon_k}
\end{aligned}
\tag{177}
$$

where

$$
\Sigma^h(k,\omega) = \int_{-\infty}^{\mu} d\omega' \frac{\Gamma(k,\omega')}{\omega-\omega'-i\delta}
\tag{178}
$$

We can also evaluate $C^h(k, 0)$:

$$\begin{aligned} C^h(k,0) &= i\int_0^\infty dt' \int_{t'}^\infty d\tau\, e^{i\varepsilon_k\tau} \left\{ i\theta(-\tau)\, e^{\delta\tau} \int_{-\infty}^{\mu} d\omega\, e^{-i\omega\tau}\Gamma(k,\omega) \right. \\ &\qquad \left. -i\theta(\tau)\, e^{-\delta\tau} \int_{\mu}^{\infty} d\omega\, e^{-i\omega\tau}\Gamma(k,\omega) \right\} \\ &= \int_0^\infty dt' \int_{t'}^\infty d\tau \int_\mu^\infty d\omega\, e^{i(\varepsilon_k-\omega+i\delta)\tau}\Gamma(k,\omega) \\ &= i\int_0^\infty dt' \int_\mu^\infty d\omega \frac{e^{i(\varepsilon_k-\omega+i\delta)t'}}{\varepsilon_k-\omega+i\delta}\Gamma(k,\omega) \\ &= -\int_\mu^\infty d\omega \frac{\Gamma(k,\omega)}{(\varepsilon_k-\omega+i\delta)^2} \\ &= \left.\frac{\partial \Sigma^p(k,\omega)}{\partial\omega}\right|_{\omega=\varepsilon_k} \end{aligned} \tag{179}$$

where

$$\Sigma^p(k,\omega) = \int_\mu^\infty d\omega' \frac{\Gamma(k,\omega')}{\omega-\omega'+i\delta} \tag{180}$$

The total cumulant can be conveniently divided into a quasiparticle part which is linear in t and a satellite part:

$$C^h = C^h_{QP} + C^h_S \tag{181}$$

Thus collecting the linear terms in t from equations (176) and (179) we get the cumulant contribution to the quasiparticle:

$$C^h_{QP}(k,t) = (i\alpha_k + \gamma_k) + (-i\Delta\varepsilon_k + \eta_k)t \tag{182}$$

where

$$i\alpha_k + \gamma_k = \left.\frac{\partial\Sigma(k,\omega)}{\partial\omega}\right|_{\omega=\varepsilon_k}, \quad \Delta\varepsilon_k = \mathrm{Re}\Sigma(k,\varepsilon_k), \quad \eta_k = \mathrm{Im}\Sigma(k,\varepsilon_k) \tag{183}$$

The contribution to the satellite is

$$C^h_S(k,t) = \int_{-\infty}^{\mu} d\omega \frac{e^{i(\varepsilon_k-\omega-i\delta)t}}{(\varepsilon_k-\omega-i\delta)^2}\Gamma(k,\omega) \tag{184}$$

A similar derivation can be carried out for the particle Green function

$$G(k, t>0) = -i\theta(t)\, e^{-i\varepsilon_k t + C^p(k,t)} \tag{185}$$

where k labels an unoccupied state. The result is

$$C^p_{QP}(k,t) = (i\alpha_k + \gamma_k) + (-i\Delta\varepsilon_k - \eta_k)t \tag{186}$$

$$C^p_S(k,t) = \int_{\mu}^{\infty} d\omega \frac{e^{i(\varepsilon_k-\omega+i\delta)t}}{(\varepsilon_k - \omega + i\delta)^2} \Gamma(k,\omega) \tag{187}$$

It is physically appealing to extract the quasiparticle part from the Green function:

$$\begin{aligned} G^h_{QP}(k,t) &= i\theta(-t)\, e^{-i\varepsilon_k t + C^h_{QP}(k,t)} \\ &= i\theta(-t)\, e^{i\alpha_k+\gamma_k} e^{(-iE_k+\eta_k)t}, \quad E_k = \varepsilon_k + \Delta\varepsilon_k \end{aligned} \tag{188}$$

The spectral function for this quasiparticle can be calculated analytically:

$$A_{QP}(k,\omega<\mu) = \frac{e^{-\gamma_k}}{\pi} \frac{\eta_k \cos\alpha_k - (\omega - E_k)\sin\alpha_k}{(\omega - E_k)^2 + \eta_k^2} \tag{189}$$

Thus we can see that the quasiparticle peak is essentially determined by the GW value.

From equation (162) we have for $t < 0$

$$\langle N | \hat{c}^\dagger_k(0) \hat{c}_k(t) | N \rangle = e^{-i\varepsilon_k t + C^h(k,t)} \tag{190}$$

By analytical continuation to $t > 0$ and using equation (163) the spectral function can be rewritten as

$$A(k,w) = \frac{1}{2\pi} \int_{-\infty}^{\infty} dt\, e^{i\omega t} e^{-i\varepsilon_k t + C^h(k,t)} \tag{191}$$

where C^h for positive t is obtained from $C^{h*}(-t) = C^h(t)$ since $A(k,\omega)$ must be real.

The total spectra can be written as a sum of A_{QP} and a convolution between the quasiparticle and the satellite part:

$$\begin{aligned} A(k,\omega) &= A_{QP}(k,\omega) + \frac{1}{2\pi} \int_{-\infty}^{\infty} dt\, e^{i\omega t} e^{-i\varepsilon_k t + C^h_{QP}(k,t)} \left[e^{C^h_S(k,t)} - 1 \right] \\ &= A_{QP}(k,\omega) + A_{QP}(k,\omega) * A_S(k,\omega) \end{aligned} \tag{192}$$

where

$$\begin{aligned} A_S(k,\omega) &= \frac{1}{2\pi} \int dt\, e^{i\omega t} \left\{ e^{C^h_S(k,t)} - 1 \right\} \\ &= \frac{1}{2\pi} \int dt\, e^{i\omega t} \left\{ C^h_S(k,t) + \frac{1}{2!} \left[C^h_S(k,t) \right]^2 + \dots \right\} \end{aligned} \tag{193}$$

The second term $A_{QP} * A_S$ is responsible for the satellite structure. The Fourier transform of C_S^h can be done analytically

$$C_S^h(k,\omega<0) = \int_{-\infty}^{\infty} \frac{dt}{2\pi} e^{i\omega t} C_S^h(k,t) \tag{194}$$

$$\begin{aligned} &= \int_{-\infty}^{0} \frac{dt}{2\pi} \left\{ e^{i\omega t} C_S^h(k,t) + e^{-i\omega t} C_S^h(k,-t) \right\} \\ &= \int_{-\infty}^{0} \frac{dt}{2\pi} \left\{ e^{i\omega t} C_S^h(k,t) + \text{c.c.} \right\} \\ &= \int_{-\infty}^{0} \frac{dt}{2\pi} \int_{-\infty}^{\mu} d\omega' \frac{e^{i(\varepsilon_k+\omega-\omega'-i\delta)t}}{(\varepsilon_k-\omega'-i\delta)^2} \Gamma(k,\omega') + \text{c.c.} \end{aligned} \tag{195}$$

Integrating over t gives

$$\begin{aligned} C_S^h(k,\omega<0) &= \frac{1}{\pi} \text{Im} \int_{-\infty}^{\mu} d\omega' \frac{\Gamma(k,\omega')}{(\varepsilon_k-\omega'-i\delta)^2(\varepsilon_k+\omega-\omega'-i\delta)} \\ &= -\frac{1}{\pi}\frac{\partial}{\partial\omega''} \text{Im} \int_{-\infty}^{\mu} d\omega' \frac{\Gamma(k,\omega')}{(\omega''-\omega'-i\delta)(\varepsilon_k+\omega-\omega'-i\delta)} \Bigg|_{\omega''=\varepsilon_k} \\ &= -\frac{\partial}{\partial\omega''} \left\{ \frac{\Gamma(k,\varepsilon_k+\omega)}{\omega''-\omega-\varepsilon_k} + \frac{\Gamma(k,\omega'')}{\varepsilon_k+\omega-\omega''} \right\} \Bigg|_{\omega''=\varepsilon_k} \\ &= \frac{\Gamma(k,\varepsilon_k+\omega) - \Gamma(k,\varepsilon_k) - \omega\Gamma'(k,\varepsilon_k)}{\omega^2} \end{aligned} \tag{196}$$

As follows from equations 189 and 192, the quasiparticle energy in the cumulant expansion is essentially determined by E_k, which is the quasiparticle energy in the GWA.

Let us apply the cumulant method to the Hamiltonian

$$H = \varepsilon c^\dagger c + \omega_p b^\dagger b + g c c^\dagger \left(b^\dagger + b\right) \tag{197}$$

which is a simplified version of the model Hamiltonian discussed previously. First we must calculate the self-energy to first order in the plasmon propagator

$$D(\omega) = \frac{1}{\omega-\omega_p+i\delta} - \frac{1}{\omega+\omega_p-i\delta} \tag{198}$$

The effective interaction between the core electron and its surrounding is $g^2 D$ (e.g. Inkson, 1984). The self-energy of the core electron as a result of the coupling to the plasmon is then

$$\begin{aligned}\Sigma(\omega) &= \frac{i}{2\pi}\int d\omega' G(\omega+\omega') g^2 D(\omega') \\ &= \frac{ig^2}{2\pi}\int d\omega' \frac{1}{\omega+\omega'-\varepsilon-i\delta}\left\{\frac{1}{\omega'-\omega_p+i\delta}-\frac{1}{\omega'+\omega_p-i\delta}\right\} \\ &= \frac{g^2}{\omega+\omega_p-\varepsilon-i\delta} \end{aligned} \tag{199}$$

The spectral function for Σ is given by

$$\Gamma(\omega) = g^2\delta(\omega-\varepsilon+\omega_p) \tag{200}$$

and the self-energy correction to the core eigenvalue is $\mathrm{Re}\Sigma(\varepsilon)$ which according to equation (199) is given by

$$\Delta\varepsilon = \frac{g^2}{\omega_p} \tag{201}$$

The derivative of Σ at $\omega = \varepsilon$ is

$$\left.\frac{\partial\Sigma}{\partial\omega}\right|_{\omega=\varepsilon} = -\left(\frac{g}{\omega_p}\right)^2 \tag{202}$$

$C^h(k,0) = 0$ and $C^h_S(k,t)$ becomes

$$\int_{-\infty}^{\mu} d\omega \frac{e^{i(\varepsilon-\omega-i\delta)t}}{(\varepsilon-\omega-i\delta)^2}\Gamma(\omega) = \left(\frac{g}{\omega_p}\right)^2 e^{i\omega_p t} \tag{203}$$

The spectral function is thus given by

$$\begin{aligned}A(\omega) &= \frac{1}{\pi}\mathrm{Im}\, i\int_{-\infty}^{0} dt\, e^{i\omega t} e^{-i\varepsilon t + C^h(t)} \\ &= \frac{1}{\pi}\mathrm{Im}\, i e^{-(g/\omega_p)^2}\int_{-\infty}^{0} dt\, e^{i(\omega-\varepsilon-\Delta\varepsilon)t}\sum_{n=0}^{\infty}\frac{[C^h(t)]^n}{n!} \\ &= e^{-(g/\omega_p)^2}\sum_{n=0}^{\infty}\frac{1}{n!}\left(\frac{g}{\omega_p}\right)^{2n}\delta(\omega-\varepsilon-\Delta\varepsilon+n\omega_p)\end{aligned} \tag{204}$$

which is precisely the exact solution (Langreth 1970).

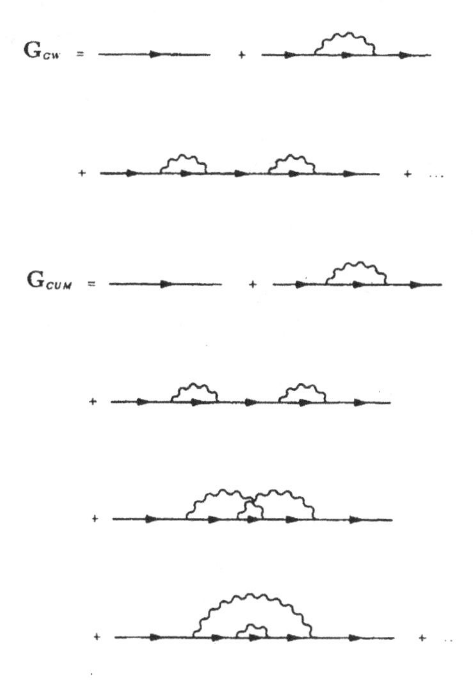

Figure 12: *Diagrammatic expansion for the Green function to second order in the GWA and the cumulant expansion respectively. The solid line represents the noninteracting Green function G_0 and the wiggly line represents the screened interaction W.*

5.1.2 Comparison between the cumulant expansion and the GWA

To identify vertex corrections contained in the cumulant expansion, we compare it with the GWA. Direct comparison in the self-energy diagrams is, however, difficult if not impossible. This is because the cumulant expansion is an expansion in the Green function, rather than in the self-energy. It is therefore more appropriate to compare the Green functions in the two approximations. In Figure 12 the Green function diagrams are shown to second order in the screened interaction, which should be sufficient for our purpose. The cumulant expansion diagrams are obtained by considering the three possible time-orderings of the integration time variables t' in $C^2(k,t)$ with $C(k,t)$ given by equation (172). As can be seen in the figure, the cumulant expansion contains second-order diagrams which are not included in the GWA. It is these additional diagrams that give rise to the second plasmon satellite and they are quite distinct from the second-order diagram common to both approximations. The interpretation of the latter diagram is that a hole emits a plasmon which is reabsorbed at a later time and the hole returns to its original state before plasmon emission. This process is repeated once at a later time. Thus there is only one plasmon coupled to the hole at one time. In contrast, the other two diagrams, not contained in the GWA, describe an additional plasmon emission before the first one is reabsorbed, giving two plasmons coupled to the hole simultaneously. Similar consideration can be extended to the higher-order diagrams.

The cumulant expansion contains only boson-type diagrams describing emission and reabsorbtion of plasmons but it does not contain diagrams corresponding to interaction between a hole and particle-hole pairs. This type of interaction is described by the ladder diagrams. For this reason, the cumulant expansion primarily corrects the satellite description whereas the quasiparticle energies are to a large extent determined by the GWA as mentioned before.

5.1.3 Applications

The cumulant expansion was applied recently to calculate the photoemission spectra in Na and Al (Aryasetiawan, Hedin, and Karlsson 1996). The experimental spectra consist of a quasiparticle peak with a set of plasmon satellites separated from the quasiparticle by multiples of the plasmon energy (Figure 13a). The spectra in the GWA shows only one plasmon satellite located at a too high energy, approximately 1.5 ω_p below the quasiparticle (13b)) which is similar to the core electron case. The cumulant expansion method remedies this problem and yields spectra in good agreement with experiment regarding the position of the satellites as can be seen in Figures 13b, 14 and 15. Interesting to observe is the presence of a double-peak structure in Figure 15. This structure evidently arises from bandstructure effects. The relative intensities of the satellites with respect to that of the quasiparticle are still in discrepancy. This is likely due to extrinsic effects corresponding to the interaction of the photoemitted electron with the bulk and the surface on its way out of the solid resulting in energy loss. These are not taken into account in the sudden approximation.

When applied to valence electrons with band dispersions the cumulant expansion does not yield the exact result anymore as in the core electron case. Surprisingly, the numerical results show that the cumulant expansion works well even in Al with a band width of $\approx$11 eV. Considering its simplicity, it is a promising approach for describing plasmon satellites.

5.2 T-matrix approximation

In many strongly correlated systems, such as transition metal oxides, the photoemission spectra often show a satellite structure a few eV below the main peak. The origin of this satellite is different from that of the plasmon-related

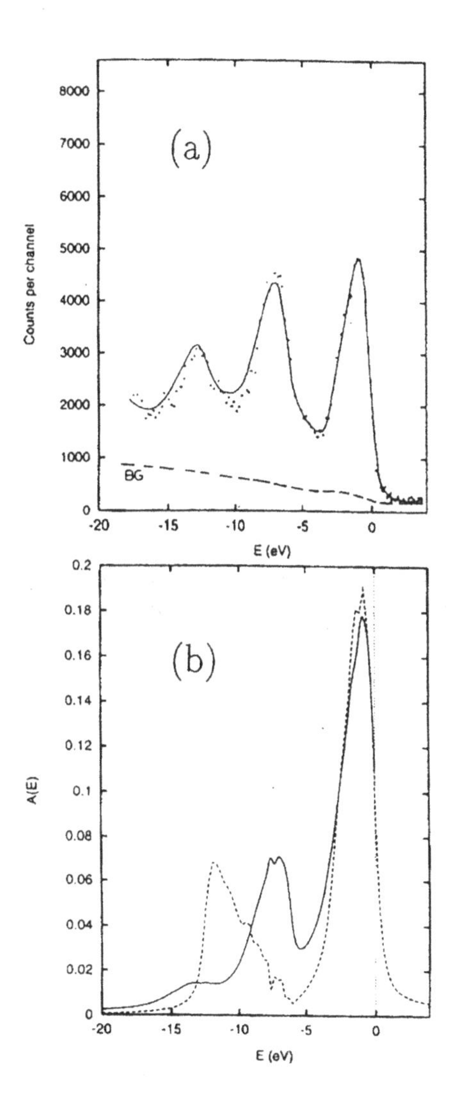

Figure 13: *(a) The experimental spectral function for Na (dots). The solid line is a synthetic spectrum obtained by convoluting the density of states from a bandstructure calculation and the the experimental core level spectrum. BG is the estimated background contribution. The data are taken from Steiner, Höchst, and Hüfner (1979). (b) The total spectral function of Na for the occupied states. The solid and dashed line correspond to the cumulant expansion and GWA respectively (Aryasetiawan, Hedin, and Karlsson 1996).*

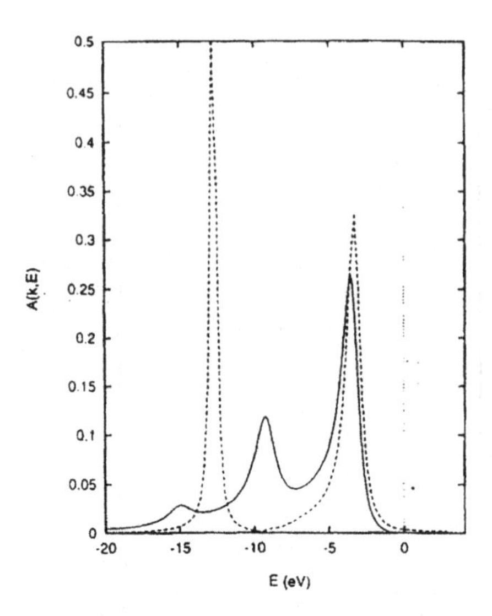

Figure 14: *The calculated spectral functions of Na at the Γ point. The solid and dashed lines correspond to the cumulant and GWA, respectively (Aryasetiawan, Hedin, and Karlsson 1996).*

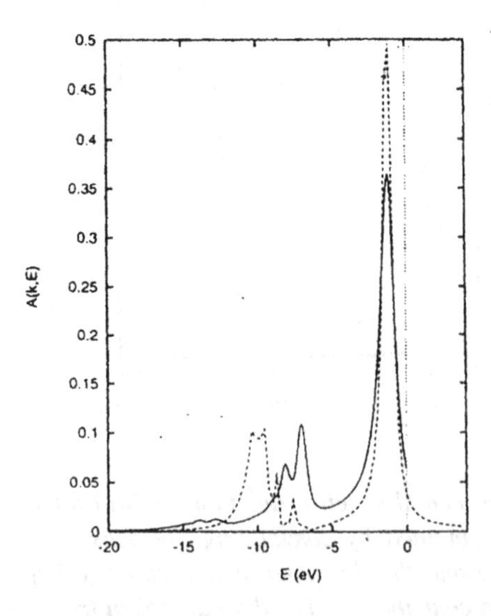

Figure 15: *The calculated spectral functions of Na for k=[1.0 1.0 0.5] $2\pi/a$. The solid and dashed lines correspond to the cumulant and GWA, respectively (Aryasetiawan, Hedin, and Karlsson 1996).*

satellite which is also found in sp-systems and which usually has a much higher energy. The additional satellite found in strongly correlated systems is due to the presence of two or more holes in a narrow band after a photoelectron is emitted, i.e. it is due to short-range rather than long-range correlations. An illustrative example is provided by Cu and Ni. In Cu, the $3d$-band is fully occupied so that after photoemission there is only one d-hole in the system corresponding to the $3d^9$ configuration resulting in no hole-hole correlation. Consequently there is no satellite either and a single-particle theory is sufficient to describe the electronic structure of Cu. On the other hand, the ground state of Ni already contains a configuration with one d-hole so that after photoemission, the final state contains a configuration with two d holes. Since the two holes are localized on the same atomic site, there will be a strong $d - d$ interaction resulting in the well-known 6 eV satellite.

So far there is no good *ab initio* scheme for dealing with short-range correlations. Most works have been based on model Hamiltonians which have given important insights into the underlying physics but which contain adjustable parameters, preventing direct quantitative comparison with experiment. The GWA takes into account long-range correlations through the RPA screening which determines to a large extent the quasiparticle energies. It is known that the GWA works well for quasiparticle energies but from a number of calculations (Aryasetiawan 1992a, Aryasetiawan and Gunnarsson 1995, Aryasetiawan and Karlsson 1996) it is clear that the GWA has shortcomings in describing satellite structures. Even the plasmon-related satellites are not well described as discussed in the previous section. A natural extension of the GWA is to include short-range hole-hole correlations. This type of interactions seems to be suitably described by the T-matrix approximation (Kanamori 1963). Previous calculations based on the Hubbard model showed qualitatively that the T-matrix theory was capable of yielding a satellite structure in Ni (Liebsch 1979, 1981, Penn 1979). Most other works in Ni have also been based on model Hamiltonians (e.g. Treglia, Ducastelle, and Spanjaard 1980, Steiner, Albers, and Sham 1992, Calandra and Manghi 1992, Igarashi *et al.* 1994). T-matrix calculation on a two-dimensional Hubbard model is also found to improve the satellite description of the GWA (Verdozzi, Godby, and Holloway 1995). In this section we develop a T-matrix theory for performing *ab initio* calculations on real systems (Springer, Aryasetiawan, and Karlsson 1998, Springer 1997). The method is applied to calculate the spectral function of Ni for which 6 eV satellite is not obtained in the GWA (Aryasetiawan 1992a).

An extension of the T-matrix theory including Fadeev's three-body interaction (Fadeev 1963) was made by Igarashi (1983, 1985) and by Calandra and Manghi (1992, 1994). The theory has been applied to Ni (Igarashi *et al.* 1994) and NiO (Manghi and Calandra 1994) within the Hubbard model.

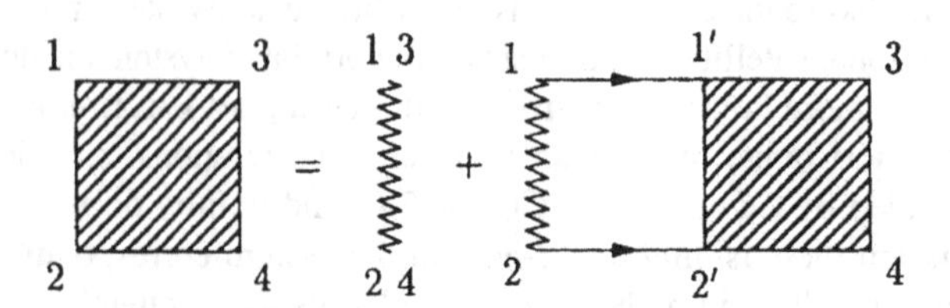

Figure 16: *Feynman diagrams for the T-matrix (square): The wiggly and the solid line with arrow represent the screened interaction U and the Green function G respectively.*

5.2.1 T-matrix

The T-matrix is defined by the following Bethe-Salpeter equation:

$$\begin{aligned} T &= U + UKT \\ &= [1 - UK]^{-1}U \end{aligned} \tag{205}$$

where U is a screened Coulomb interaction and K is a two-particle propagator. Diagrammatically, the multiple scattering processes are shown in Figure 16. It is evident from the figure that the T-matrix describes multiple scattering between two holes or electrons.

The full expression for equation (205) is

$$\begin{aligned} T_{\sigma\sigma'}(1,2|3,4) &= U(1-2)\delta(1-3)\delta(2-4) \\ &+ U(1-2)\int d1'd2'\; K_{\sigma\sigma'}(1,2|1',2')T_{\sigma\sigma'}(1',2'|3,4) \end{aligned} \tag{206}$$

where we have used a short-hand notation $1 \equiv (\mathbf{r}_1, t_1)$ and σ labels the spin. The kernel K or the two-particle propagator is given by

$$K_{\sigma\sigma'}(1,2|1',2') = iG_\sigma(1',1)G_{\sigma'}(2',2) \tag{207}$$

where G_σ is a time-ordered single-particle Green function

$$iG_\sigma(1',1) = \langle N|T[\hat{\phi}_\sigma(1')\hat{\phi}^\dagger_\sigma(1)]|N\rangle \tag{208}$$

We have assumed that U is an instantaneous interaction which means that $t_1 = t_2$, $t_3 = t_4$, and $t_{1'} = t_{2'}$. Without loss of generality, we may set $t_1 = t_2 = 0$ and $t_3 = t_4 = t$. Equation (206) then becomes

$$\begin{aligned} &T_{\sigma\sigma'}(\mathbf{r}_1,\mathbf{r}_2|\mathbf{r}_3,\mathbf{r}_4;t) = \\ &\quad U(\mathbf{r}_1-\mathbf{r}_2)\delta(\mathbf{r}_1-\mathbf{r}_3)\delta(\mathbf{r}_2-\mathbf{r}_4)\delta(t) \\ &\quad + U(\mathbf{r}_1-\mathbf{r}_2)\int d^3r'_1d^3r'_2\int dt' K_{\sigma\sigma'}(\mathbf{r}_1,\mathbf{r}_2|\mathbf{r}'_1,\mathbf{r}'_2;t') \\ &\qquad\qquad \times\; T_{\sigma\sigma'}(\mathbf{r}'_1,\mathbf{r}'_2|\mathbf{r}_3,\mathbf{r}_4;t-t') \end{aligned} \tag{209}$$

Fourier transformation of the above equation yields

$$\begin{aligned} T_{\sigma\sigma'}(\mathbf{r}_1, \mathbf{r}_2|\mathbf{r}_3, \mathbf{r}_4; \omega) = \\ & U(\mathbf{r}_1 - \mathbf{r}_2)\delta(\mathbf{r}_1 - \mathbf{r}_3)\delta(\mathbf{r}_2 - \mathbf{r}_4) \\ & + U(\mathbf{r}_1 - \mathbf{r}_2) \int d^3r_1' d^3r_2' \int K_{\sigma\sigma'}(\mathbf{r}_1, \mathbf{r}_2|\mathbf{r}'_1, \mathbf{r}'_2; \omega) \\ & \qquad \times T_{\sigma\sigma'}(\mathbf{r}'_1, \mathbf{r}'_2|\mathbf{r}_3, \mathbf{r}_4; \omega) \end{aligned} \tag{210}$$

where

$$K_{\sigma\sigma'}(\mathbf{r}_1, \mathbf{r}_2|\mathbf{r}'_1, \mathbf{r}'_2; \omega) = i \int \frac{d\omega'}{2\pi} G_\sigma(\mathbf{r}'_1, \mathbf{r}_1; \omega - \omega') G_{\sigma'}(\mathbf{r}'_2, \mathbf{r}_2; \omega') \tag{211}$$

The Fourier transforms are defined by

$$F(\omega) = \int_{-\infty}^{\infty} dt\, e^{-i\omega t} f(t) \tag{212}$$

$$f(t) = \int_{-\infty}^{\infty} \frac{d\omega}{2\pi}\, e^{i\omega t} F(\omega) \tag{213}$$

Using the spectral representation of G_σ

$$G_\sigma(\omega) = \int_{-\infty}^{\mu} d\omega' \frac{A_\sigma(\omega')}{\omega - \omega' - i\delta} + \int_{\mu}^{\infty} d\omega' \frac{A_\sigma(\omega')}{\omega - \omega' + i\delta} \tag{214}$$

with a non-interacting A_σ

$$A_\sigma(\mathbf{r}, \mathbf{r}'; \omega) = \sum_i \psi_{i\sigma}(\mathbf{r}) \psi^*_{i\sigma}(\mathbf{r}') \delta(\omega - \varepsilon_{i\sigma}) \tag{215}$$

an explicit form for the kernel K is given by

$$\begin{aligned} K_{\sigma\sigma'}(\mathbf{r}_1, \mathbf{r}_2|\mathbf{r}'_1, \mathbf{r}'_2; \omega) = \\ & -\sum_{ij}^{\text{occ}} \frac{\psi_{i\sigma}(\mathbf{r}'_1)\psi^*_{i\sigma}(\mathbf{r}_1)\psi_{j\sigma'}(\mathbf{r}'_2)\psi^*_{j\sigma'}(\mathbf{r}_2)}{\omega - \varepsilon_{i\sigma} - \varepsilon_{j\sigma'} - i\delta} \\ & + \sum_{ij}^{\text{unocc}} \frac{\psi_{i\sigma}(\mathbf{r}'_1)\psi^*_{i\sigma}(\mathbf{r}_1)\psi_{j\sigma'}(\mathbf{r}'_2)\psi^*_{j\sigma'}(\mathbf{r}_2)}{\omega - \varepsilon_{i\sigma} - \varepsilon_{j\sigma'} + i\delta} \end{aligned} \tag{216}$$

The first term on the right hand side is due to hole-hole scattering and the second to particle-particle scattering. This expression is similar to the RPA polarization propagator but the states are either both occupied or unoccupied.

5.2.2 T-matrix self-energy

The self-energy can be obtained from the T-matrix which consists of the direct term

$$\Sigma^d(4,2) = -i\int d1d3G(1,3)T(1,2|3,4) \tag{217}$$

and the exchange term

$$\Sigma^x(3,2) = i\int d1d4G(1,4)T(1,2|3,4) \tag{218}$$

Fourier transformation gives

$$\Sigma^d_\sigma(\mathbf{r}_4,\mathbf{r}_2;\omega) = -i\int d^3r_1d^3r_3$$
$$\int \frac{d\omega'}{2\pi}\sum_{\sigma'} G_{\sigma'}(\mathbf{r}_1,\mathbf{r}_3;\omega'-\omega)T_{\sigma'\sigma}(\mathbf{r}_1,\mathbf{r}_2|\mathbf{r}_3,\mathbf{r}_4;\omega') \tag{219}$$

$$\Sigma^x_\sigma(\mathbf{r}_3,\mathbf{r}_2;\omega) = i\int d^3r_1d^3r_4$$
$$\int \frac{d\omega'}{2\pi} G_{\sigma}(\mathbf{r}_1,\mathbf{r}_4;\omega'-\omega)T_{\sigma\sigma}(\mathbf{r}_1,\mathbf{r}_2|\mathbf{r}_3,\mathbf{r}_4;\omega') \tag{220}$$

We note that for the exchange term there is no summation over the spin since exchange between particles of opposite spins is zero.

The spectral representation of $T_{\sigma\sigma'}$ is given by

$$T_{\sigma\sigma'}(\omega) = \int_{-\infty}^{2\mu} d\omega' \frac{Q_{\sigma\sigma'}(\omega')}{\omega-\omega'-i\delta} + \int_{2\mu}^{\infty} d\omega' \frac{Q_{\sigma\sigma'}(\omega')}{\omega-\omega'+i\delta} \tag{221}$$

where

$$Q_{\sigma\sigma'}(\omega) = -\frac{1}{\pi}\text{Im } T_{\sigma\sigma'}(\omega)\ \text{sgn}(\omega-2\mu) \tag{222}$$

The analytic structure of T is determined by K in equation (216) which can be seen by considering equation (205). The first order term in the T-matrix is the Hartree-Fock term which is independent of frequency and therefore it does not influence the analytic structure. The T-matrix without the first order term is then given by

$$\begin{aligned} T &= UKU + UKT \\ &= [1-UK]^{-1}UKU \end{aligned} \tag{223}$$

Thus, the analytic structure is determined by K because it can be shown that $[1 - UK]^{-1}$ has no poles in the first and third quadrants (the poles of K lie in the second and fourth quadrants).

Using the spectral representations of G and T the self-energy can be written explicitly as

$$\begin{aligned}
&\text{Im } \Sigma^d_\sigma(\mathbf{r}_4, \mathbf{r}_2; \omega > \mu) = \\
&\quad \int d^3r_1 d^3r_3 \sum_{\mathbf{k}'n'\sigma'}^{\text{occ}} \psi_{\mathbf{k}'n'\sigma'}(\mathbf{r}_1)\psi^*_{\mathbf{k}'n'\sigma'}(\mathbf{r}_3) \\
&\quad \times \text{Im } T_{\sigma'\sigma}(\mathbf{r}_1, \mathbf{r}_2|\mathbf{r}_3, \mathbf{r}_4; \omega + \varepsilon_{\mathbf{k}'n'\sigma'})\, \theta(\omega + \varepsilon_{\mathbf{k}'n'\sigma'} - 2\mu)
\end{aligned} \tag{224}$$

$$\begin{aligned}
&\text{Im } \Sigma^d_\sigma(\mathbf{r}_4, \mathbf{r}_2; \omega \leq \mu) = \\
&\quad -\int d^3r_1 d^3r_3 \sum_{\mathbf{k}'n'\sigma'}^{\text{unocc}} \psi_{\mathbf{k}'n'\sigma'}(\mathbf{r}_1)\psi^*_{\mathbf{k}'n'\sigma'}(\mathbf{r}_3) \\
&\quad \times \text{Im } T_{\sigma'\sigma}(\mathbf{r}_1, \mathbf{r}_2|\mathbf{r}_3, \mathbf{r}_4; \omega + \varepsilon_{\mathbf{k}'n'\sigma'})\, \theta(-\omega - \varepsilon_{\mathbf{k}'n'\sigma'} + 2\mu)
\end{aligned} \tag{225}$$

A similar expression for the exchange part can be easily derived by interchanging $\mathbf{r}_3$ and $\mathbf{r}_4$ in the T-matrix. The real part of Σ is obtained from the spectral representation

$$\Sigma_\sigma(\omega) = \int_{-\infty}^{\mu} d\omega' \frac{\Gamma_\sigma(\omega')}{\omega - \omega' - i\delta} + \int_{\mu}^{\infty} d\omega' \frac{\Gamma_\sigma(\omega')}{\omega - \omega' + i\delta} \tag{226}$$

where

$$\Gamma_\sigma(\omega) = -\frac{1}{\pi} \text{Im } \Sigma_\sigma(\omega)\, \text{sgn}(\omega - \mu) \tag{227}$$

The screened potential U is in general frequency dependent. For a narrow band of width Δ, the time scale for a hole to hop from one site to a neighbouring site is determined by $1/\Delta$ which is large and in frequency space it means that the largest contribution to the T-matrix comes from $\omega \approx 0$ and therefore it is justified to use a static screened interaction. This static approximation will be used in the present work.

The one- electron spectral functions are obtained from

$$A_i(\omega) = \frac{1}{\pi} \frac{|\text{Re } \Sigma_i(\omega)|}{|\omega - \varepsilon_i - \text{Im } \Sigma_i(\omega)|^2 + |\text{Re } \Sigma_i(\omega)|^2} \tag{228}$$

with the shorthand notation $i \equiv \mathbf{k}n\sigma$. We have assumed in the above expression that the self-energy is diagonal in the LDA wavefunctions which are used to construct G.

5.2.3 Double counting and the total self-energy

To calculate the total self-energy, we add the T-matrix self-energy to the GW self-energy. However, this leads to double counting since the second order term in the T-matrix is already included in the GW self-energy. A straightforward subtraction of the second order term leads however to wrong analytic properties of the self-energy and consequently gives some negative spectral weight. To solve this double-counting problem, we proceed as follows. We divide the bare Coulomb interaction into the short-range screened potential U and a long-range part v_L ((von Barth 1995):

$$v = U + v_L \tag{229}$$

The correlation part of the GW self-energy is schematically given by

$$\Sigma^c_{GW} = GvRv \tag{230}$$

where R is the total RPA response function

$$R = (1 - Pv)^{-1}P \tag{231}$$

and P is the RPA polarization function. Using the above division of the Coulomb potential we obtain

$$\Sigma^c_{GW} = GURv_L + Gv_LRU + Gv_LRv_L + GURU \tag{232}$$

The last term is then subtracted out to avoid double counting. It has the same analytic structure as the GW self-energy, but since U is smaller than the bare v, this term is always smaller than the GW self-energy for all frequencies. This guarantees that the resulting self-energy has the correct analytic properties as in the GWA. Numerically this term turns out to be very small.

Thus, according to the above scheme the total correlated self-energy becomes:

$$\Sigma^c_{GWT} = \Sigma^c_{GW} + \Sigma^c_T - GURU \tag{233}$$

where Σ^c_T is given by

$$\Sigma^c_T = \Sigma^d + \Sigma^x \tag{234}$$

with Σ^d and Σ^x given by equations (219) and (220) respectively.

5.2.4 Numerical procedure

Since the T-matrix describes scattering between localized holes or particles, it is suitable to work with basis functions which are also localized such as the linear muffin-tin orbitals (LMTO). In the atomic sphere approximation, the LMTO basis functions are of the form (central cell)

$$\chi^{\sigma}_{RL}(\mathbf{r},\mathbf{k}) = \varphi^{\sigma}_{RL}(\mathbf{r}) + \sum_{R'L'} \dot{\varphi}^{\sigma}_{R'L'}(\mathbf{r})\, h^{\sigma}_{R'L',RL}(\mathbf{k}) \tag{235}$$

ϕ^{σ}_{RL} is a solution to the Schrödinger equation in atomic sphere R at an energy in the center of the L-band region and $\dot{\phi}^{\sigma}_{RL}$ is the corresponding energy derivative. $h^{\sigma}_{R'L',RL}$ is a constant matrix depending on the crystal structure as well as the potential. The Bloch states are expanded in the LMTO basis.

$$\psi_{\mathbf{k}n\sigma}(\mathbf{r}) = \sum_{RL} \chi^{\sigma}_{RL}(\mathbf{r},\mathbf{k})\, b_{RL}(\mathbf{k}n\sigma) \tag{236}$$

Inserting $\psi_{\mathbf{k}n\sigma}$ in equation (216), we obtain

$$\begin{aligned} K_{\sigma\sigma'}(\mathbf{r}_1,\mathbf{r}_2|\mathbf{r}'_1,\mathbf{r}'_2;\omega) = &- \sum_{\mathbf{k}n,\mathbf{k}'n'}^{\mathrm{occ}} \sum_{R_1L_1}\sum_{R_2L_2}\sum_{R_3L_3}\sum_{R_4L_4} \\ &\chi^{\sigma}_{R_1L_1}(\mathbf{r}'_1,\mathbf{k})\chi^{\sigma*}_{R_2L_2}(\mathbf{r}_1,\mathbf{k})\chi^{\sigma'}_{R_3L_3}(\mathbf{r}'_2,\mathbf{k}')\chi^{\sigma'*}_{R_4L_4}(\mathbf{r}_2,\mathbf{k}') \\ &\times \frac{b_{R_1L_1}(\mathbf{k}n\sigma)b^*_{R_2L_2}(\mathbf{k}n\sigma)b_{R_3L_3}(\mathbf{k}'n'\sigma')b^*_{R_4L_4}(\mathbf{k}'n'\sigma')}{\omega-\varepsilon_{\mathbf{k}n\sigma}-\varepsilon_{\mathbf{k}'n'\sigma'}-i\delta} \\ &+ \text{unoccupied term} \end{aligned} \tag{237}$$

So far the above expression is exact. The largest contribution to K arises from the onsite terms, i.e. within the same unit cell. Furthermore, since U is short-range, the largest contribution to the T-matrix also arises from the onsite terms. For practical purpose, we then neglect the $\mathbf{k}$ dependence of χ so that

$$\begin{aligned} K_{\sigma\sigma'}(\mathbf{r}_1,\mathbf{r}_2|\mathbf{r}'_1,\mathbf{r}'_2;\omega) = \\ \sum_{\alpha\beta\gamma\delta} \phi^{\sigma}_{\alpha}(\mathbf{r}'_1)\phi^{\sigma'}_{\beta}(\mathbf{r}_2)\, K_{\sigma\sigma'}(\alpha\beta|\gamma\delta;\ \omega)\, \phi^{\sigma}_{\gamma}(\mathbf{r}'_1)\phi^{\sigma'}_{\delta}(\mathbf{r}'_2) \end{aligned} \tag{238}$$

where

$$\begin{aligned} K_{\sigma\sigma'}(\alpha\beta|\gamma\delta;\ \omega) = &- \sum_{\mathbf{k}n,\mathbf{k}'n'}^{\mathrm{occ}} \frac{b^*_{\alpha}(\mathbf{k}n\sigma)b^*_{\beta}(\mathbf{k}'n'\sigma')b_{\gamma}(kn\sigma)b_{\delta}(k'n'\sigma')}{\omega-\varepsilon_{\mathbf{k}n\sigma}-\varepsilon_{\mathbf{k}'n'\sigma'}-i\delta} \\ &+ \text{unoccupied term} \end{aligned} \tag{239}$$

We have used a short-hand notation $\alpha \equiv RL$ and Q^{σ}_{RL} is real.

Using this expression for K in equation (210) and taking matrix element with respect to $\varphi^{\sigma}_{\alpha}(\mathbf{r}_1)\,\varphi^{\sigma}_{\beta}(\mathbf{r}_2)$ and $\varphi^{\sigma'}_{\gamma}(\mathbf{r}_3)\,\varphi^{\sigma'}_{\delta}(\mathbf{r}_4)$ we obtain a matrix equation for T

$$T_{\sigma\sigma'}(\alpha\beta|\gamma\delta;\ \omega) = U(\alpha\beta|\gamma\delta) + \sum_{\eta\nu,\lambda\rho} U(\alpha\beta|\eta\nu)\ K_{\sigma\sigma'}(\eta\nu|\lambda\rho;\ \omega)\ T_{\sigma\sigma'}(\lambda\rho|\gamma\delta;\ \omega) \tag{240}$$

which can be easily solved by inversion.

$$U(\alpha\beta|\eta\nu) = \int d1d2\varphi_{\alpha}(1)\varphi_{\beta}(2)v(1-2)\varphi_{\eta}(1)\varphi_{\nu}(2) \tag{241}$$

We note that although U depends on the spin through the basis, it has no physical spin dependence.

Having obtained T, it is straightforward to calculate the self-energy. From equations (224) and equations (225), the matrix element of the self-energy in a Bloch state $\psi_{\mathbf{k}n\sigma}$ is given by

$$\begin{aligned}\text{Im}\ \Sigma^{d}_{\sigma}(\mathbf{k}n;\omega > \mu) = \\ \sum_{\mathbf{k}'n'\sigma'}^{\text{occ}} \sum_{\alpha\beta,\gamma\delta} b_{\alpha}(\mathbf{k}'n'\sigma')b_{\beta}(\mathbf{k}n\sigma)b^{*}_{\gamma}(\mathbf{k}'n'\sigma')b^{*}_{\delta}(\mathbf{k}n\sigma) \\ \times \text{Im}\ T_{\sigma'\sigma}(\alpha\beta|\gamma\delta;\ \omega+\varepsilon_{\mathbf{k}'n'\sigma'})\ \theta(\omega+\varepsilon_{\mathbf{k}'n'\sigma'}-2\mu)\end{aligned} \tag{242}$$

$$\begin{aligned}\text{Im}\ \Sigma^{d}_{\sigma}(\mathbf{k}n;\omega \le \mu) = \\ -\sum_{\mathbf{k}'n'\sigma'}^{\text{unocc}} \sum_{\alpha\beta,\gamma\delta} b_{\alpha}(\mathbf{k}'n'\sigma')b_{\beta}(\mathbf{k}n\sigma)b^{*}_{\gamma}(\mathbf{k}'n'\sigma')b^{*}_{\delta}(\mathbf{k}n\sigma) \\ \times \text{Im}\ T_{\sigma'\sigma}(\alpha\beta|\gamma\delta;\ \omega+\varepsilon_{\mathbf{k}'n'\sigma'})\ \theta(-\omega-\varepsilon_{\mathbf{k}'n'\sigma'}+2\mu)\end{aligned} \tag{243}$$

The real part of Σ is obtained from equation (226). It is straightforward to derive the corresponding expressions for the exchange part.

5.2.5 Application to Ni with a model screened interaction

As a first application, we use the T-matrix theory to calculate the self-energy of ferromagnetic Ni. In the case of Ni, the T-matrix should describe repeated scattering of two 3d holes on the same site. The mechanism for the origin of the 6 eV satellite is discussed in the previous section on Ni.

A model potential of the form

$$U(|\mathbf{r}-\mathbf{r}'|) = \frac{\mathrm{erfc}(\alpha|\mathbf{r}-\mathbf{r}'|)}{|\mathbf{r}-\mathbf{r}'|} \tag{244}$$

is used to calculate the T-matrix. For the T-matrix self-energy, the position of the peaks in Im Σ is rather insensitive to the screening parameter. However, the intensity varies with α and consequently the satellite position, which is determined by Re Σ, is modified accordingly. Thus, if we were allowed to adjust the screening parameter α, the position of the satellite could be shifted arbitrarily. It is therefore crucial to determine the screened interaction U from a parameter-free scheme. According to constrained LDA calculations (Gunnarsson *et al.* 1989) $U_{dd} \approx$ 5.5 eV (Steiner, Albers, and Sham 1992). Consequently in our calculations we choose $\alpha = 1.2$ which is equivalent to $U_{dd} \approx 5.5$ eV.

The calculated spectral functions are displayed in Figure 17 and compared with the GW spectra. The main peak ≈ 3 eV below the Fermi level is the quasiparticle peak. Two satellite structures originating from the T-matrix self-energy, absent in the GWA, can be observed below the main peak and just above the Fermi level. The position of the satellite below the Fermi level is, however, somewhat too low. This is likely due to the difficulties in determining the correct screened interaction in the T-matrix and the neglect of particle-hole interaction in the present scheme.

A new interesting feature is the presence of a peak structure just above the Fermi energy which arises from particle-particle scattering. At first sight, these scattering processes are expected to be insignificant, since the number of unoccupied states is small which leads to a small T-matrix for positive energies. However, our results point to the importance of matrix element and bandstructure effects, usually neglected in the Hubbard model. As may be seen by an examination of equation (224) there is a sum over occupied states which amplifies the small contribution from the T-matrix. It may be possible to measure the satellite structure above the Fermi level in an angular resolved inverse photoemission experiment by choosing certain **k**-vectors, where the quasiparticle peak is well separated from the satellite.

There is a significant difference between the majority and minority self-energy as reflected in the spectral function in Figure 17. The probability of creating a hole in the majority channel is larger than in the minority channel, since the former is fully occupied. The virtual hole can mainly be created in the minority channel due to the presence of $3d$ unoccupied states just above the Fermi level. This implies that the majority $3d$ self-energy will be larger than the minority one which reduces the exchange splitting and the bandwidth, improving the GW result. It is found that the T-matrix self-energy reduces the exchange splitting by ≈ 0.3 eV (Springer, Aryasetiawan, and Karlsson 1997, Springer 1998) for states at the bottom of the 3d band and thus also improving the GW bandwidth. But contrary to model calculations (Liebsch 1979, 1981), the T-matrix has a significantly smaller effect on the bandwidth.

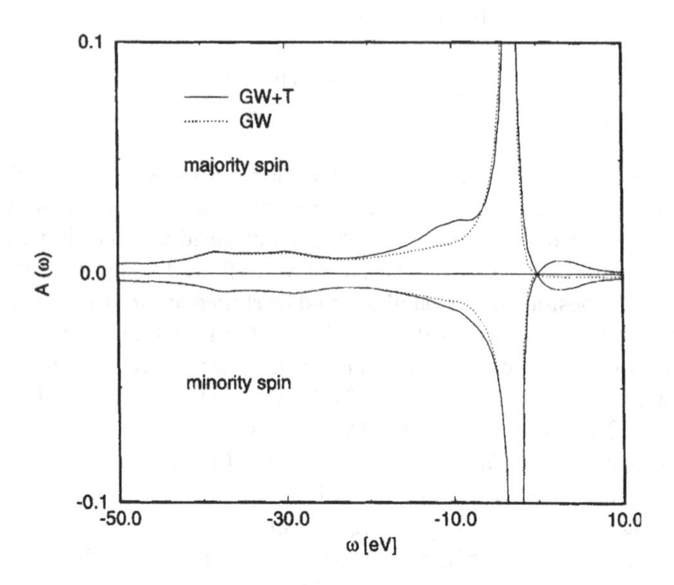

Figure 17: *Ni spectral functions at the X point for the second band (Springer, Aryasetiawan, and Karlsson 1998).*

6 Self-consistency and total energies

6.1 Self-consistency

The current practice of performing GW calculations is to choose a non-interacting Hamiltonian (usually the LDA) and to construct the Green function G_0 from the eigenstates of the Hamiltonian and to calculate the response function. These are then used to calculate the self-energy which is taken as the final result. This procedure is non-self-consistent since the interacting Green function G obtained from the Dyson equation $G = G_0 + G_0\Sigma_0 G$ is not necessarily the same as G_0. To achieve self-consistency, the interacting Green function should be used to form a new polarization function $P = -iGG$, a new screened interaction W, and a new self-energy Σ as in Hedin's equations (48)–(52). This process is repeated until G obtained from the Dyson equation is the same as G used to calculate the self-energy. Self-consistency guarantees that the final result is independent of the starting Green function G_0. Moreover, a self-consistent GW scheme ensures conservation of particle number and energy. This is a consequence of a general theorem due to Baym and Kadanoff (1961) and Baym (1962) who proved that approximations for Σ are conserving if Σ is obtained as a functional derivative of an energy functional Φ with respect to G :

$$\Sigma = \frac{\delta \Phi [G]}{\delta G} \tag{245}$$

Conservation of particle number means that the continuity equation

$$-\partial_t n(\mathbf{r}, t) = \nabla \cdot \mathbf{j}(\mathbf{r}, t) \tag{246}$$

is satisfied when n and $\mathbf{j}$ is obtained from the self-consistent Green function. Furthermore

$$N = \frac{1}{\pi} tr \int_{-\infty}^{\mu} d\omega \, \mathrm{Im}\, G(\omega) \tag{247}$$

gives the correct total number of particles. Conservation of energy means that the energy change when an external potential is applied to the system is equal to the work done by the system against the external potential when calculated using the self-consistent G.

The first self-consistent calculation GW calculation was probably by de Groot, Bobbert, and van Haeringen (1995) for a model quasi-one dimensional semiconducting wire. The relevance of this model to real solids is, however, unclear. Self-consistent calculations for the electron gas were performed recently by von Barth and Holm (1996) and by Shirley (1996). The calculations were done for two cases: in the first case the screened interaction W is fixed at the RPA level, $W = W_0$, (calculated using the non-interacting G_0) and only the Green function is allowed to vary to self-consistency and in the other case both G and W are allowed to vary (full self-consistency case) (Holm and von Barth 1998). The results of these studies are

- The band width is increased from its non-self-consistent value, worsening the agreement with experiment (Figure 20).
- The weight of the quasiparticles is increased, reducing the weight in the plasmon satellite.
- The quasiparticles are slightly narrowed, increasing their life-time.
- The plasmon satellite is broadened and shifted towards the Fermi level (Figure 18). In the full self-consistent case, the plasmon satellite almost disappears (Figure 19).

The main effects of self-consistency are mainly due to allowing the quasiparticle weight Z to vary. The increase in band width is disturbing and can be understood as follows. We consider the first case with fixed $W = W_0$ for simplicity. First we note that the GW result for the band width after one iteration is close to the free electron one. This means that there is almost a complete cancellation between exchange and correlation. After one iteration the quasiparticle weight is reduced

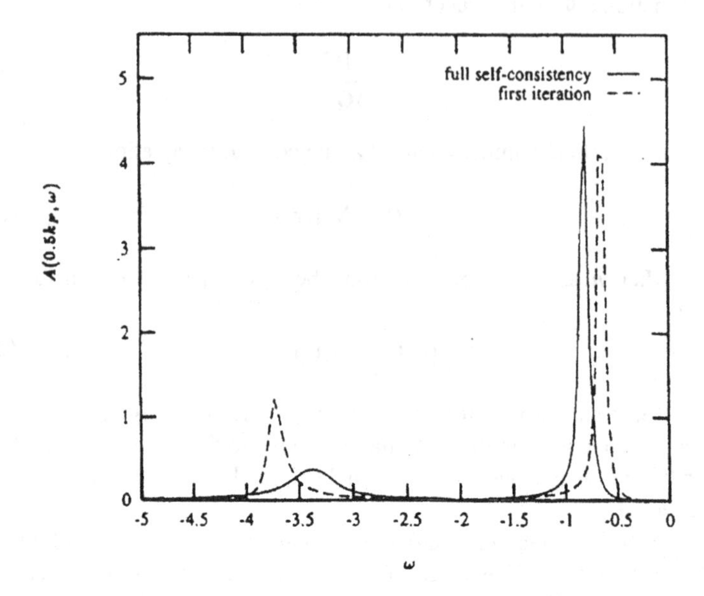

Figure 18: *The partially self-consistent spectral function $A(k = 0.5k_F, \omega)$ compared to that of the first iteration for $r_s = 4$ (von Barth and Holm 1996).*

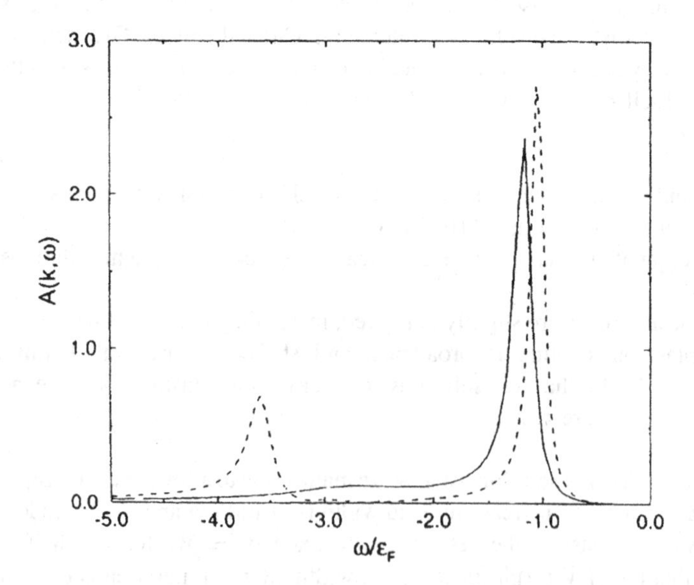

Figure 19: *The fully self-consistent spectral function $A(k = 0.5k_F, \omega)$ compared to that of the first iteration for $r_s = 4$ (Holm and von Barth 1998).*

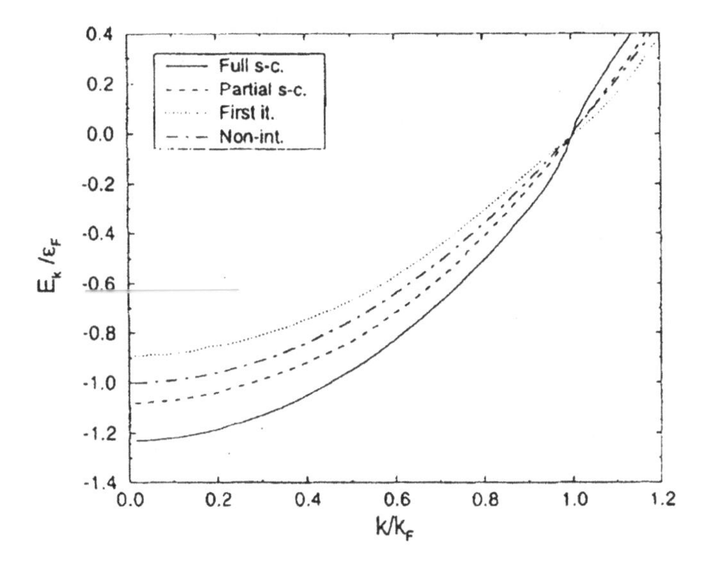

Figure 20: *The quasiparticle dispersions for $r_s = 4$ corresponding to full self-consistency (solid line), partial self-consistency (dashed line), first iteration (dotted line) and the free electron (dashed-dotted line) (Holm and von Barth 1996).*

to typically 0.7 and the rest of the weight goes to the plasmon satellite. The new Im Σ, calculated from G obtained from the first iteration has its weight reduced at low energy and increased at high energy compared to the non-self-consistent result (Figure 21). This is due to the sum-rule (von Barth and Holm 1996)

$$\int_{-\infty}^{\infty} d\omega |\mathrm{Im}\Sigma^c(\mathbf{k},\omega)| = \sum_{\mathbf{q}} \int_0^{\infty} d\omega |\mathrm{Im}W(\mathbf{q},\omega)| \tag{248}$$

which shows that the left hand side is a constant depending only on the prechosen W_0 but independent of $\mathbf{k}$ and self-consistency. For a state at the Fermi level, self-consistency has little effects since Im Σ has almost equal weights for the hole ($\omega \leq \mu$) and the particle part ($\omega > \mu$) which cancel each other when calculating Re Σ_c, as can be seen in equation (76) and as illustrated in Figure 22. But for the state at the bottom of the valence band, Im Σ has most of its weight in the hole part (see e.g. Figure 5) so that the shifting of the weight in Im Σ to higher energy causes Re Σ_c to be less positive than its non-self-consistent value. A similar effect is found for the exchange part which becomes less negative but because the bare Coulomb interaction has no frequency dependence, the renormalization factor has a smaller effect on Σ^x so that the reduction in Σ^x is less than the reduction in Σ^c.

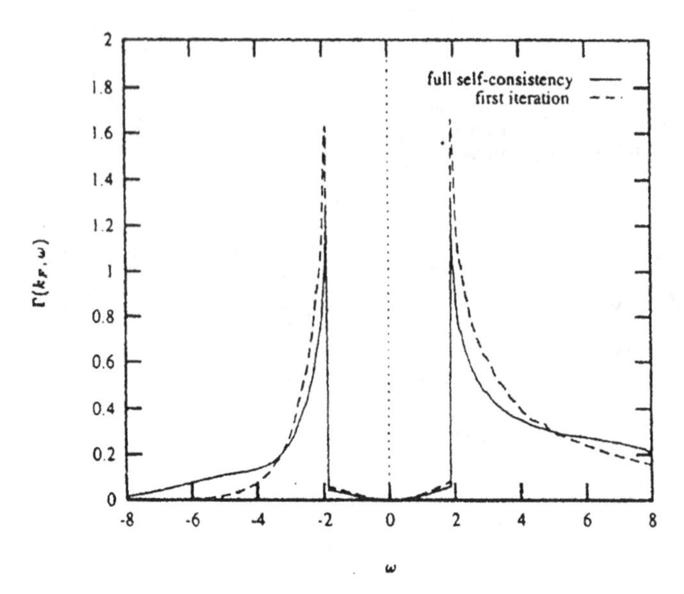

Figure 21: *The spectral function of the partially self-consistent self-energy,* $\Gamma = |Im\Sigma|/\pi$, *for* $k = k_F$ *compared to that of the first iteration (Holm and von Barth 1996).*

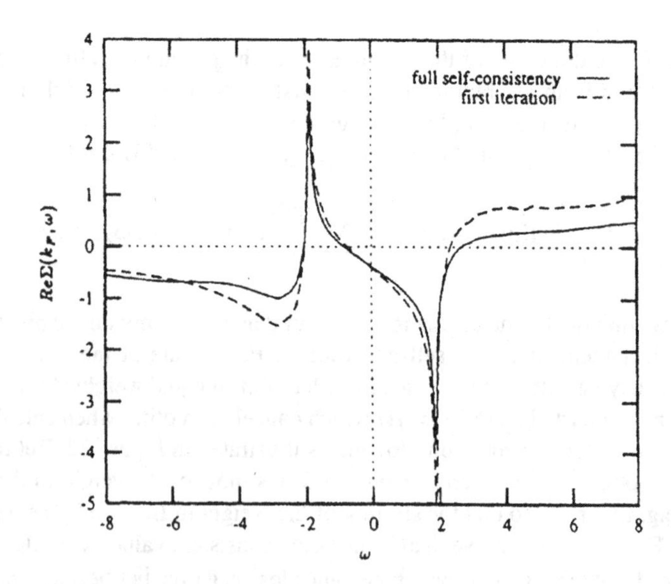

Figure 22: *The real part of the partially self-consistent self-energy for* $k = k_F$ *compared to that of the first iteration (Holm and von Barth 1996).*

The net effect is then an increase in the band width. The shifting of the weight in ImΣ to higher energy has immediate consequences of increasing the life-time and the renormalization weight Z (through a decrease in $\partial \text{Re}\Sigma_c/\partial\omega|$) of of the quasiparticles and of broadening the plasmon satellite, compared to the results of one iteration.

When W is allowed to vary (full self-consistency) the band width becomes even more widened and the plasmon satellite becomes broad and featureless, in contradiction to experiment. The quasiparticle weight is increased further (Holm and von Barth 1998). These can be traced back to the disappearance of a well-defined plasmon excitation in W. The reason for this is that the quantity $P = -iGG$ no longer has a physical meaning of a response function, rather it is an auxiliary quantity needed to construct W. Indeed, it does not satisfy the usual $f-$ sum rule. The equation $\epsilon(\mathbf{q}, \omega_p) = 0$ determining the plasmon energy is not satisfied any more since the Green function now always has weight around $\omega = \omega_p$. This has the effect of transferring the weight in Im Σ even further to higher energy with the consequences discussed in the previous paragraph.

A self-consistent calculation for the electron gas within the cumulant expansion has been performed by Holm and Aryasetiawan (1997). The result for the quasiparticle energy is similar to the self-consistent GWA since the quasiparticle energy in the cumulant expansion is essentially determined by that of the GWA as discussed in the cumulant expansion section. The satellite part is little affected by self-consistency.

6.2 Total energies

So far, the GWA has been applied mainly to calculate single-particle excitation spectra, but it is also possible to calculate the total energy (Holm and von Barth 1998, Farid, Godby, and Needs 1990). Holm and von Barth (1998) calculated the total energy using the self-consistent Green function and self-energy in the Galitskii-Migdal formulation (1958):

$$E = -\frac{i}{2}\int d^3r \lim_{t'\to t^+}\lim_{\mathbf{r}'\to\mathbf{r}}\left[i\frac{\partial}{\partial t} + h_0(\mathbf{r})\right] G(\mathbf{r}t, \mathbf{r}'t') \tag{249}$$

where h_0 is the kinetic energy operator plus a local external potential. The total energy turns out to be very accurate in comparison with the quantum Monte Carlo (QMC) results of Ceperley and Alder (1980). For $r_s = 2$ and 4 QMC gives 0.004 and -0.155 Rydberg respectively while self-consistent GW gives 0.005 and -0.156 Rydberg. This is rather surprising since the GWA represents only the first term in the self-energy expansion. The reason for the accurate results is not fully understood. Applications to other more realistic systems are necessary to show if the results are of general nature. It is probably related to the fact that the self-consistent GW

scheme is energy conserving and it is partly explained by consideration of the so called Luttinger-Ward energy functional (1960)

$$\Omega[G] = \mathrm{Tr}\left\{\ln(\Sigma[G] - G_0^{-1}) + \Sigma[G]G\right\} - \Phi[G] \tag{250}$$

which is variational with respect to G and it is stationary when G is equal to the self-consistent G which obeys the Dyson equation:

$$\frac{\delta\Omega}{\delta G} = 0 \text{ when } G = G_0 + G_0\Sigma G \tag{251}$$

$\Omega = E - \mu N$ is the grand canonical potential whose stationary value corresponds to the energy (minus μN) obtained from the Galitskii-Migdal formula and Φ is an energy functional consisting of irreducible energy diagrams. It is not clear, however, why the first order energy diagram (giving the GW self-energy upon taking functional derivative with respect to G) appears to represent a very good energy functional. Furthermore, the chemical potential calculated from $\mu = \partial E/\partial N$ is in agreement with the value obtained from $\mu = k_F^2/2 + \Sigma(k_F, 0)$. The particle density $n = 2\sum_{\mathbf{k}} \int_{-\infty}^{\mu} d\omega A(\mathbf{k}, \omega)$ yields $n = k_F^3/(3\pi^2)$, i.e. particle number is conserved, as proven by Baym. It can also be shown that with a fixed $W = W_0$ particle number is also conserved (Holm and von Barth 1998). The conclusion is that fully self-consistent GW calculations for quasiparticle energies should be avoided. It is more fruitful to construct vertex corrections (beyond GW) or to perform partially self-consistent GW calculations where the choice of G and W is physically motivated. For instance, one can fix W at the RPA level and modify G by using quasiparticle energies but keeping the renormalization factor equal to one. Or one could choose a single-particle Hamiltonian such that the GW quasiparticle energies are consistent with the single-particle eigenvalues.

Recently, Almbladh and Hindgren (Hindgren 1997) discovered that the GW total energies of the electron gas calculated using the Luttinger-Ward formalism (1960) are in very good agreement with the QMC results of Ceperley and Alder (1980) and the self-consistent results of Holm and von Barth (1998). Due to the variational nature of the Luttinger-Ward functional, the Green function need not be the self-consistent one. A generalization of the Luttinger-Ward formalism by treating both G and W as independent parameters has also been developed (Almbladh, von Barth, and van Leeuwen 1997). Using the non-interacting $G = G_0$ and the plasmon-pole approximation for W gives almost as good results (Hindgren 1997) as the self-consistent results of Holm and von Barth (1998) or the QMC results of Ceperley and Alder (1980). These results are very encouraging and applications to real inhomogeneous systems are now in progress to test the validity of the scheme.

References

[1] Albrecht, S., Onida, G. and Reining, L. (1997) *Phys. Rev. B* **55**, 10278–81.
[2] Almbladh, C.-O. and Hedin, L. (1983) *Handbook on Synchroton Radiation* **1**, 686, ed. E. E. Koch (North-Holland).
[3] Almbladh, C-O. and von Barth, U. (1985a) *Phys. Rev. B* **31**, 3231.
[4] Almbladh, C.-O. and von Barth, U. (1985b) *Density Functional Methods in Physics*, Vol. 123 of *NATO Advanced Study Institute, Series B*, eds. Dreizler R M. and da Providencia J (Plenum, New York).
[5] Almbladh, C.-O., von Barth, U. and van Leeuwen, R. (1997) private communication.
[6] Alperin, H.A. (1962) *J. Phys. Soc. Jpn. Suppl. B* **17**, 12.
[7] Andersen, O.K. (1975) *Phys. Rev. B* **12**, 3060.
[8] Anisimov, V., I., Zaanen, J. and Andersen, O.K. (1991) *Phys. Rev. B* **44**, 943.
[9] Anisimov, V.I., Solovyev, I.V., Korotin, M.A., Czyzyk, M.T. and Sawatzky, G.A. (1993) *Phys. Rev. B* **48**, 16929.
[10] Anisimov, V.I., Aryasetiawan, F. and Lichtenstein, A.I., *J. Phys.: Condens. Matter* **9**, 767–808 (1997).
[11] Arai, M. and Fujiwara, T. (1995) *Phys. Rev. B* **51**, 1477–89.
[12] Aryasetiawan, F., von Barth, U., Blaha, P. and Schwarz, K. (1990) (unpublished).
[13] Aryasetiawan, F. (1992a) *Phys. Rev. B* **46**, 13051–64.
[14] Aryasetiawan, F. and von Barth, U. (1992b) *Physica Scripta* **T45**, 270-1.
[15] Aryasetiawan, F. and Gunnarsson, O. (1994a) *Phys. Rev. B* **49**, 16214–22.
[16] Aryasetiawan, F. and Gunnarsson, O. (1994b) *Phys. Rev.B* **49**, 7219.
[17] Aryasetiawan, F., Gunnarsson, O., Knupfer. and Fink, J. (1994)c *Phys. Rev.B* **50**, 7311–21.
[18] Aryasetiawan, F. and Gunnarsson, O. (1995) *Phys. Rev. Lett.* **74**, 3221–24.
[19] Aryasetiawan, F. and Karlsson, K. (1996) *Phys. Rev. B* **54**, 5353–7.
[20] Aryasetiawan, F. and Gunnarsson, O. (1996) *Phys. Rev. B* **54**, 17564–8.
[21] Aryasetiawan, F., Hedin, L. and Karlsson, K. (1996) *Phys. Rev. Lett.* **77**, 2268–71.
[22] Aryasetiawan, F. and Gunnarsson, O. (1997) (unpublished).
[22a]Aryasetiawan, F. and Gunnarsson, O. (1998) *Rep. Prog. Phys.* **61**, 237–312.
[23] Ashcroft, N.W. and Mermin, N.D. (1976) *Solid State Physics* (Saunders College).
[24] Aspnes, D.E. (1976) *Phys. Rev. B* **14**, 5331.
[25] Baldereschi, A. and Tosatti, E. (1979) *Solid State Commun.* **20**, 131.
[26] Baldini, G. and Bosacchi, B. (1970) *Phys. Status Solidi* **38**, 325.
[27] Bates, D.R. (1947) *Proc. R. Soc. London Ser. A* **188**, 350.
[28] Baym, G. and Kadanoff, L.P. (1961) *Phys. Rev.* **124**, 287.
[29] Baym, G. (1962) *Phys. Rev.* **127**, 1391.
[30] Bechstedt, F. and Del Sole, R. (1988) *Phys. Rev. B* **38**, 7710-6.

[31] Bechstedt, F. and Del Sole, R. (1990) *Solid State Commun.* **74**, 41–4.

[32] Bechstedt, F., Del Sole, R., Cappellini, G. and Reining, L. (1992) *Solid State Commun.* **84**, 765–70.

[33] Bechstedt, F., Tenelsen, K., Adolph, B. and Del Sole, R. (1997) *Phys. Rev. Lett.* **78**, 1528–31.

[34] Becke, A.D. (1988) *Phys. Rev. A* **38**, 3098.

[35] Becke, A.D. (1992) *J. Chem. Phys.* **96**, 2155.

[36] Becke, A.D. (1996) *J. Chem. Phys.* **104**, 1040-6.

[37] Bergersen, B. (1973) *Can. J. Phys.* **51**, 102.

[38] Biermann, L. (1943) *Z. Astrophys.* **22**, 157.

[39] Biermann, L. and Lübeck, K. (1948) *Z. Astrophys.* **25**, 325.

[40] Blase, X., Rubio, A., Louie, S.G. and Cohen, M.L. (1995) *Phys. Rev. B* **52** R2225–8.

[41] Bobbert, P.A. and van Haeringen, W. (1994) *Phys. Rev. B* **49**, 10326–31.

[42] Bohm, D. and Pines, D. (1953) *Phys. Rev.* **92**, 609.

[43] Bonacic-Koutecky, V., Fantucci, P. and Koutecky, J. (1990a) *J. Chem. Phys.* **96**, 3802.

[44] Bonacic-Koutecky, V., Fantucci, P. and Koutecky, J. (1990b) *Chem. Phys. Lett.* **166**, 32.

[45] Bonacic-Koutecky, V., Pittner, J., Scheuch, C., Guest, M.F. and Koutecky, J. (1992) *J. Chem. Phys.* **96**, 7938.

[46] Boys, S.F. (1950) *Proc. R. Soc. London A* **200**, 542.

[47] Bylander, D.M. and Kleinman, L. (1995a) *Phys. Rev. Lett.* **74**, 3660.

[48] Bylander, D.M. and Kleinman, L. (1995b) *Phys. Rev. B* **52**, 14566.

[49] Calandra, C. and Manghi, F. (1992) *Phys. Rev. B* **45**, 5819.

[50] Cardona, M., Weinstein, M. and Wolff, G.A. (1965) *Phys. Rev.* **140** A633.

[51] Causa, M. and Zupan, A. (1994) *Chem. Phys. Lett.* **220**, 145–53.

[52] Ceperley, D.M. and Alder, B.J. (1980) *Phys. Rev. Lett.* **45**, 566.

[53] Charlesworth, J.P.A., Godby, R.W. and Needs, R.J. (1993) *Phys. Rev. Lett.* **70**, 1685–8.

[54] Cheetham, A.K. and Hope, D.A.O. (1983) *Phys. Rev. B* **27**, 6964.

[55] Cowan, R.D. (1967) *Phys. Rev.* **163**, 54.

[56] Dal Corso, A., Pasquarello, A., Baldereschi, A. and Car, R. (1996) *Phys. Rev. B* **53**, 1180-5.

[57] Daling, R. and van Haeringen, W. (1989) *Phys. Rev. B* **40**, 11659–65.

[58] Daling, R., Unger, P., Fulde, P. and van Haeringen, W. (1991) *Phys. Rev. B* **43**, 1851–4.

[59] de Groot, H.J., Bobbert, P.A. and van Haeringen, W. (1995) *Phys. Rev. B* **52**, 11000.

[60] Deisz, J.J., Eguiluz, A. and Hanke, W. (1993) *Phys. Rev. Lett.* **71**, 2793–96.

[61] Deisz, J.J. and Eguiluz, A. (1997) *Phys. Rev. B* **55**, 9195–9.

[62] Del Sole, R., Reining, L. and Godby, R.W. (1994) *Phys. Rev. B* **49**, 8024–8.

[63] Del Sole, R. and Girlanda, R. (1996) *Phys. Rev. B* **54**, 14376–80.

[64] Dreizler, R.M. and Gross, E.K.U. (1990) *Density Functional Theory* (Springer-Verlag, New York).
[65] DuBois, D.F. (1959a) *Ann. Phys.* **7**, 174.
[66] DuBois, D.F. (1959b) *Ann. Phys.* **8**, 24.
[67] Dufek, P., Blaha, P., Sliwko, V. and Schwarz, K. (1994) *Phys. Rev. B* **49**, 10170-5.
[68] Dykstra, C.E. (1988) *Ab Initio Calculation of the Structure. and Properties of Molecules* (Elsevier, Amsterdam).
[69] Eberhardt, W. and Plummer, E.W. (1980) *Phys. Rev. B* **21**, 3245.
[70] Eddy, C.R., Moustakas, T.D. and Scanlon, J. (1993) *J. Appl. Phys.* **73**, 448.
[71] Eguiluz, A.G., Heinrichsmeier, M., Fleszar, A. and Hanke, W. (1992) *Phys. Rev. Lett.* **68**, 1359–62.
[72] Engel, G.E., Kwon, Y. and Martin R.M. (1995) *Phys. Rev. B* **51**, 13538–46.
[73] Erwin, S.C. and Pickett, W.E. (1991) *Science* **254**, 842.
[74] Farid, B., Godby, R.W. and Needs, R.J. (1990) *20th International Conference on the Physics of Semiconductors* editors Anastassakis, E.M. and Joannopoulos, J.D., Vol. 3, 1759–62 (World Scientific, Singapore).
[75] Fender, B.E.F., Jacobson, A.J. and Wedgwood, F.A. (1968) *J. Chem. Phys.* **48**, 990.
[76] Fetter, A.L. and Walecka, J.D. (1971) *Quantum Theory of Many-Particle Systems* (McGraw-Hill).
[77] Fock, V. (1930) *Z. Phys.* **61**, 126.
[78] Frota, H.O. and Mahan, G.D. (1992) *Phys. Rev. B* **45**, 6243.
[79] Fujimori, A., Minami, F. and Sugano, S. (1984) *Phys. Rev.* B **29**, 5225.
[80] Galitskii, V.M. and Migdal, A.B. (1958) *Sov. Phys. JETP* **7**, 96.
[81] Galitskii, V.M. (1958) *Sov. Phys. JETP* **7**, 104.
[82] Gell-Mann, M. and Brueckner, K. (1957) *Phys. Rev.* **106**, 364.
[83] Godby, R.W., Schlüter, M. and Sham, L.J. (1986) *Phys. Rev. Lett.* **56**, 2415–8.
[84] Godby, R.W., Schlüter, M. and Sham, L.J. (1987a) *Phys. Rev. B* **36**, 6497–500.
[85] Godby, R.W., Schlüter, M. and Sham, L.J. (1987b) *Phys. Rev. B* **35**, 4170-1.
[86] Godby, R.W., Schlüter, M. and Sham, L.J. (1988) *Phys. Rev. B* **37**, 10159–75.
[87] Godby, R.W. and Needs, R.J. (1989) *Phys. Rev. Lett.* **62**, 1169–72 .
[88] Godby, R.W. and Sham, L.J. (1994) *Phys. Rev. B* **49**, 1849–57.
[89] Gross, E.K.U., Dobson, J.F. and Petersilka, M. (1996) *Density Functional Theory*, edited by Nalewajski, R.F. (Springer).
[90] Gu, Z.-q. and Ching, W.Y. (1994) *Phys. Rev. B* **49**, 10958–67.
[91] Guillot, C., Ballu, Y., Paigné, J., Lecante, J., Jain, K.P., Thiry, P., Pinchaux, R., Pétroff, Y. and Falicov, L.M. (1977) *Phys. Rev. Lett.* **39**, 1632.
[92] Gunnarsson, O., Lundqvist, B.I. and Wilkins, J.W. (1974) *Phys. Rev. B* **10**, 1319.
[93] Gunnarsson, O. and Jones, R.O. (1980) *J. Chem. Phys.* **72**, 5357.
[94] Gunnarsson, O., Andersen, O.K., Jepsen, O. and Zaanen, J. (1989) *Phys. Rev. B* **39**, 1708 .

[95] Gunnarsson, O., Meden, V. and Schönhammer, K. (1994) *Phys. Rev. B* **50**, 10462.
[96] Gunnarsson, O. (1997a) *J. Phys.: Cond. Matt.* **9**, 5635.
[97] Gunnarsson, O. (1997b) *Rev. Mod. Phys.* **69**, 575–606.
[98] Gygi, F. and Baldereschi, A. (1989) *Phys. Rev. Lett.* **62**, 2160.
[99] Hamada, N., Hwang, M. and Freeman, A.J. (1990) *Phys. Rev. B* **41**, 3620-6.
[100] Hanke, W. and Sham, L.J. (1988) *Phys. Rev. B* **38**, 13361–70.
[101] Hanke, W. and Sham, L.J. (1975) *Phys. Rev. B* **12**, 4501–11.
[102] Harris, J. and Jones, R.O. (1978) *J. Chem. Phys.* **68**, 3316.
[103] Hartree, D.R. (1928) *Proc. Camb. Phil. Soc.* **24**, 89, 111, 426.
[104] Hedin, L. (1965) *Phys. Rev.* **139** A796.
[105] Hedin, L. and Johansson, A. (1969) *J. Phys. B* **2**, 1336.
[106] Hedin, L. and Lundqvist, S. (1969) *Solid State Physics vol. 23*, eds. H. Ehrenreich, F. Seitz. and D. Turnbull (Academic, New York).
[107] Hedin, L., Lundqvist, B.I. and Lundqvist, S. (1970) *J. Res. Natl. Bur. Stand. Sect. A* **74A**, 417.
[108] Hedin, L. and Lundqvist, B.I. (1971) *J. Phys. C* **4**, 2064–83.
[109] Hedin, L. (1980) *Physica Scripta* **21**, 477–80.
[110] Hedin, L. (1995) *Int. J. Quantum Chem.* **56**, 445–52.
[111] Himpsel, F.J., Knapp, J.A. and Eastman, D.E. (1979) *Phys. Rev. B* **19**, 2919.
[112] Hindgren, M. (1997) (PhD thesis, University of Lund).
[113] Hohenberg, P. and Kohn, W. (1964) *Phys. Rev.* **136** B864.
[114] Holm, B. and von Barth, U. (1998) *Phys. Rev. B* **57**, 2108.
[115] Holm, B. and Aryasetiawan, F. (1997) *Phys. Rev. B* **56**, 12825.
[116] Hüfner, S., Wertheim, G.K., Smith, N.V. and Traum, M.M. (1972) *Solid State Commun.* **11**, 323.
[117] Hüfner, S. and Wertheim, G.K. (1973) *Phys. Rev. B* **8**, 4857.
[118] Hüfner, S., Osterwalder, J., Riester, T. and Hullinger, F. (1984) *Solid State Commun.* **52**, 793.
[119] Hybertsen, M.S. and Louie, S.G. (1985a) *Phys. Rev. Lett.* **55**, 1418–21.
[120] Hybertsen, M.S. and Louie, S.G. (1985b) *Phys. Rev. B* **32**, 7005–8.
[121] Hybertsen, M.S. and Louie, S.G. (1986) *Phys. Rev. B* **34**, 5390-413.
[122] Hybertsen, M.S. and Louie, S.G. (1987a) *Phys. Rev. B* **35**, 5585–601, 5602–10.
[123] Hybertsen, M.S. and Louie, S.G. (1987b) *Comments Condens. Matter Phys.* **13**, 223.
[124] Hybertsen, M.S. and Louie, S.G. (1988a) *Phys. Rev. B* **37**, 2733–6.
[125] Hybertsen, M.S. and Louie, S.G. (1988b) *Phys. Rev. B* **38**, 4033–44.
[126] Igarashi, J. (1983) *J. Phys. Soc. Jpn.* **52**, 2827.
[127] Igarashi, J. (1985) *J. Phys. Soc. Jpn.* **54**, 260.
[128] Igarashi, J., Unger, P., Hirai, K. and Fulde, P. (1994) *Phys. Rev. B* **49**, 16181.
[129] Inkson, J.C. (1984) *Many-body Theory of Solids* (Plenum Press, New York,. (1984)) chap. 11.2.

[130] Ishii, Y., Ohnishi, S. and Sugano, S. (1986) *Phys. Rev. B* **33**, 5271.
[131] Itchkawitz, B.S., I.-W. Lyo and E.W. Plummer (1985) *Phys. Rev. B* **41**, 8075.
[132] Jenkins, S.J., Srivastava, G.P. and Inkson, J.C. (1993) *Phys. Rev. B* **48**, 4388–97.
[133] Jensen, E. and E.W. Plummer (1985) *Phys. Rev. Lett.* **55**, 1912.
[134] Jones, R.O. and Gunnarsson, O. (1989) *Rev. Mod. Phys.* **61**, 689–746.
[135] Kanamori, J. (1963) *Prog. Theor. Phys.* **30**, 275.
[136] Kanski, J., Nilsson, P.O. and Larsson, C.G. (1980) *Solid State Commun.* **35**, 397.
[137] Kemeny, P.C. and Shevchik (1975) *Solid State Commun.* **17**, 255.
[138] Kohn, W. and Sham, L.J. (1965) *Phys. Rev.* **140** A1133.
[139] Kotani, T. (1995) *Phys. Rev. Lett.* **74**, 2989.
[140] Kotani, T. and Akai, H. (1996) *Phys. Rev. B* **54**, 16502.
[141] Kowalczyk, S.P., Ley, L., McFeely, F.R., Pollak, R.A. and Shirley, D.A. (1973) *Phys. Rev. B* **8**, 3583.
[142] Kress, C., Fiedler, M. and Bechstedt, F. (1994) in *Proceedings of the Fourth International Conference on the Formation of Semiconductor Interfaces* edited by B Lengeler, H Lüth, W Mönch and J Pollmann (World Scientific, Singapore,. (1994)), p. 19.
[143] Lang, N.D. and Kohn, W. (1973) *Phys. Rev. B* **7**, 3541–50.
[144] Lang, N.D. and Williams, A.R. (1977) *Phys. Rev. B* **16**, 2408.
[145] Langreth, D.C. (1970) *Phys. Rev. B***1**, 471.
[146] Langreth, D.C. and Mehl, M.J. (1983) *Phys. Rev. B* **28**, 1809.
[147] Lebedev, V.I. (1975) *Zh. Vychisl. Mat. Mat. Fiz.* **15**, 48 (Note erratum in Lebedev. (1976)).
[148] Lebedev, V.I. (1976) *Zh. Vychisl. Mat. Mat. Fiz.* **16**, 293.
[149] Lebedev, V.I. (1977) *Sibirsk. Mat. Zh.* **18**, 132.
[150] Lebedev, V.I. (1978) *Proc. Conf. Novosibirsk*, edited by Sobolev S L (Nauka Sibirsk. Otdel., Novosibirsk,. (1980)).
[151] Lei, T., Moustakas, T.D., Graham, R.J., He, Y. and Berkowitz, S.J. (1992a) *J. Appl. Phys.* **71**, 4933.
[152] Lei, T., Fanciulli, M., Molnar, R.J., Moustakas, T.D., Graham, R.J. and Scanlon, J. (1992b) *Appl. Phys. Lett* **59**, 944 .
[153] Levine, Z.H. and Louie, S.G., *Phys. Rev. B* **25**, 6310.
[154] Lichtenstein, A.I., Zaanen, J. and Anisimov, V.I. (1995) *Phys. Rev. B* **52** R5467.
[155] Liebsch, A. (1979) *Phys. Rev. Lett.* **43**, 1431–4.
[156] Liebsch, A. (1981) *Phys. Rev. B* **23**, 5203–12.
[157] Lindgren, I. (1971) *Int. J. Quantum Chem.* **5**, 411.
[158] Lindhard, J. (1954) *Kgl. Danske Videnskap. Selskap, Mat.-fys. Medd.* **28**, 8.
[159] Lof, R.W., van Veenendaal, M.A., Koopmans, B., Jonkman, H.T. and Sawatzky, G.A. (1992) *Phys. Rev. Lett.* **68**, 3924.
[160] Lundqvist, B.I. (1967a) *Phys. Kondens. Mater.* **6**, 193.

[161] Lundqvist, B.I.. (1967b) *Phys. Kondens. Mater.* **6**, 206.
[162] Lundqvist, B.I. (1968) *Phys. Kondens. Mater.* **7**, 117.
[163] Lundqvist, B.I. and Samathiyakanit, V. (1969) *Phys. Kondens. Mater.* **9**, 231.
[164] Lundqvist, B.I. (1969) *Phys. Stat. Sol.* **32**, 273.
[165] Luttinger, J.M. and Ward, J.C. (1960) *Phys. Rev* **118**, 1417–27.
[166] Lyo, I.-W. and Plummer, E.W. (1988) *Phys. Rev. Lett.* **60**, 1558.
[167] Mahan, G.D. (1967a) *Phys. Rev.* **153**, 882.
[168] Mahan G D. (1967)b *Phys. Rev.* **163**, 612.
[169] Mahan, G.D. and Sernelius, B.E. (1989) *Phys. Rev. Lett.* **62**, 2718.
[170] Mahan, G.D. (1990) *Many-particle Physics* (Plenum Press, New York).
[171] Mahan, G.D. (1994) *Comments Cond. Mat. Phys.* **16**, 333.
[172] Manghi, F. and Calandra, C. (1994) *Phys. Rev. Lett.* **73**, 3129–32.
[173] Mårtensson, H. and Nilsson, P.O. (1984) *Phys. Rev. B* **30**, 3047.
[174] Massidda, S., Continenza, A., Posternak, M. and Baldereschi, A. (1995a) *Phys. Rev. Lett.* **74**, 2323–6.
[175] Massidda, S., Resta, R., Posternak, M. and Baldereschi, A. (1995b) *Phys. Rev. B* **52** R16977.
[176] Massidda, S., Continenza, A., Posternak, M. and Baldereschi, A. (1997) *Phys. Rev. B* **55**, 13494–502.
[177] Mattheiss, L.F. (1972)a *Phys. Rev. B* **5**, 290.
[178] Mattheiss, L.F. (1972)b *Phys. Rev. B* **5**, 306.
[179] Minnhagen, P. (1974) *J. Phys. C* **7**, 3013.
[180] Moore, C.E., *Atomic Energy Levels*, Natl. Bur. Stand. Ref. Data Ser., Natl. Bur. Stand. (U.S) Circ. No. 467 (U.S. GPO, Washington, D., C., 1949, 1952, 1958).
[181] Moruzzi, V.L., Janak, J.F. and Williams, A.R. (1978) *Calculated Electronic Properties of Metals* (Pergamnon, New York).
[182] Mott, N.F. (1949) *Phys. Soc. London Sect. A* **62**, 416.
[183] Müller, W., Flesch, J. and Meyer, W. (1984) *J. Chem. Phys.* **80**, 3297–310.
[184] Northurp, J.E., Hybertsen, M.S. and Louie, S.G. (1987) *Phys. Rev. Lett.* **59**, 819.
[185] Northrup, J.E., Hybertsen, M.S. and Louie, S.G. (1989) *Phys. Rev. B* **39**, 8198.
[186] Northrup, J.E. (1993) *Phys. Rev. B* **47**, 10032–5.
[187] Nozières, P. (1964) *Theory of Interacting Fermi Systems*, (Benjamin, New York).
[188] Nozières, P. and de Dominicis, C.J. (1969) *Phys. Rev.* **178**, 1097.
[189] *Numerical Data. and Functional Relationships in Science. and Technology* 1982, edited by Hellwege K-H. and Madelung, O., Landolt-Börnstein, New Series, Group III, Vols. 17a. and 22a (Springer, Berlin) .
[190] Onida, G., Reining, L., Godby, R.W., Del Sole, R. and Andreoni, W. (1995) *Phys. Rev. Lett.* **75**, 818–21.

[191] Ortuno, M. and Inkson, J.C. (1979) *J. Phys. C* **12**, 1065–71.
[192] Oschlies, A., Godby, R.W. and Needs, R.J. (1992) *Phys. Rev. B* **45**, 13741–4.
[193] Oschlies, A., Godby, R.W. and Needs, R.J. (1995) *Phys. Rev. B* **51**, 1527–35.
[194] Overhauser, A.W. (1971) *Phys. Rev. B* **3**, 1888.
[195] Overhauser, A.W. (1985) *Phys. Rev. Lett.* **55**, 1916.
[196] Paisley, M.J., Sitar, Z., Posthill, J.B. and Davis, R.F. (1989) *J. Vac. Sci. Technol. A* **7**, 701.
[197] Palummo, M., Del Sole, R., Reining, L., Bechstedt, F. and Cappellini, G. (1995) *Solid State Commun.* **95**, 393–8.
[198] Penn, D.R. (1962) *Phys. Rev.* **128**, 2093.
[199] Penn, D.R. (1979) *Phys. Rev. Lett.* **42**, 921–4.
[200] Perdew, J.P. and Zunger, A. (1981) *Phys. Rev. B* **23**, 5048.
[201] Perdew, J.P. and Levy, M. (1983) *Phys. Rev. Lett.* **51**, 1884.
[202] Perdew, J.P., Burke, K. and Ernzerhof, M. (1997) *Phys. Rev. Lett.* **77**, 3865–8.
[203] Philipsen, P.H.T. and Baerends, E.J. (1996) *Phys. Rev. B* **54**, 5326–33.
[204] Pickett, W.E. and Wang, C.S. (1984) *Phys. Rev. B* **30**, 4719–33.
[205] Pickett, W.E. (1989) *Rev. Mod. Phys.* **62**, 433.
[206] Pines, D. and Bohm, D. (1952) *Phys. Rev.* **85**, 338.
[207] Pines, D. (1961) *Elementary Excitations in Solids* (W A Benjamin, New York).
[208] Powell, R.J. and Spicer, W.E. (1970) *Phys. Rev. B* **2**, 2182.
[209] Quinn, J.J. and Ferrel, R.A. (1958) *Phys. Rev.* **112**, 812.
[210] Rauch, W. (1940) *Z. Phys.* **116**, 652.
[211] Rice, T.M. (1965) *Ann. Phys.* **31**, 100.
[212] Rohlfing, M., Krüger, P. and Pollmann, J. (1993) *Phys. Rev. B* **48**, 17791–805.
[213] Rohlfing, M., Krüger, P. and Pollmann, J. (1995a) *Phys. Rev. Lett.* **19**, 3489–92.
[214] Rohlfing, M., Krüger, P. and Pollmann, J. (1995b) *Phys. Rev. B* **52**, 1905–17.
[215] Rohlfing, M., Krüger, P. and Pollmann, J. (1996) *Phys. Rev. B* **54**, 13759–66.
[216] Rojas, H.N., Godby, R.W. and Needs, R.J. (1995) *Phys. Rev. Lett.* **74**, 1827–1830.
[217] Rubio, A., Corkill, J.L., Cohen, M.L., Shirley, E.L. and Louie, S.G. (1993) *Phys. Rev. B* **48**, 11810-6.
[218] Runge, E. and Gross, E.K.U. (1984) *Phys. Rev. Lett.* **52**, 997.
[219] Saito, S. and Cohen, M.L. (1988) *Phys. Rev. B* **38**, 1123.
[220] Saito, S., Zhang, S.B., Louie, S.G. and Cohen, M.L. (1989) *Phys. Rev. B* **40**, 3643–6.
[221] Satpathy, S., Antropov, V.P., Andersen, O.K., Jepsen, O., Gunnarsson, O. and Liechtenstein, A.I. (1992) *Phys. Rev. B* **46**, 1773.
[222] Sawatzky, G.A. and Allen, J.W. (1984) *Phys. Rev. Lett.* **53**, 2339.
[222a] Schindlmayr, A. and Godby, R.W. (1998) *Phys. Rev. Lett.* **80**, 1702.

[223] Schönberger, U. and Aryasetiawan, F. (1995) *Phys. Rev. B* **52**, 8788–93.

[224] Seidl, A., Görling, A., Vogl, P., Majewski, J.A. and Levy, M. (1996) *Phys. Rev. B* **53**, 3764–74.

[225] Sham, L.J. and Kohn, W. (1966) *Phys. Rev.* **145**, 561–7.

[226] Sham, L.J. and Schlüter, M. (1983) *Phys. Rev. Lett.* **51**, 1888.

[227] Sham, L.J. and Schlüter, M. (1985) *Phys. Rev. B* **32**, 3883.

[228] Shen, Z.X., Shih, C.K., Jepsen, O., Spicer, W.E., Lindau, I. and Allen, J.W. (1990) *Phys. Rev. Lett.* **64**, 2442.

[229] Shen, Z.X., List, R.S., Dessau, D.S., Wells, B.O., Jepsen O, Arko, A.J., Bartlet, R., Shih, C.K., Parmigiani, F., Huang, J.C. and Lindberg, P.A.P. (1991a) *Phys. Rev. B* **44**, 3604.

[230] Shen, Z.X. *et al.*. (1991b) *Solid State Commun.* **79**, 623.

[231] Shirley, E.L., Zhu, X. and Louie. S.G. (1992) *Phys. Rev. Lett.* **69**, 2955–8.

[232] Shirley, E.L. and Martin, R.M. (1993) *Phys. Rev. B* **47**, 15404–27.

[233] Shirley, E.L. and Louie, S.G. (1993) *Phys. Rev. Lett.* **71**, 133.

[234] Shirley, E.L. (1996) *Phys. Rev. B* **54**, 7758–64.

[235] Shung, K.W.K., Sernelius, B.E. and Mahan, G.D. (1987) *Phys. Rev. B* **36**, 4499.

[236] Shung, K.W.K. and Mahan, G.D. (1988) *Phys. Rev. B* **38**, 3856.

[237] Sitar, Z. *et al.*. (1992) *J. Matter. Sci. Lett.* **11**, 261.

[238] Slater, J.C. (1951a) *Phys. Rev.* **81**, 385.

[239] Slater, J.C, (1951b) *Phys. Rev.* **82**, 538.

[240] Slater, J.C. (1953) *Phys. Rev.* **91**, 528.

[241] Slater, J.C. (1974) *Quantum Theory of Molecules. and Solids, Vol. IV, The Self-consistent Field for Molecules. and Solids* (McGraw-Hill, New York).

[242] Springer, M., Svendsen, P.S. and von Barth, U. (1996) *Phys. Rev. B* **54**, 17392–401.

[243] Springer, M., Aryasetiawan, F. and Karlsson, K. (1998) *Phys. Rev. Lett.* **80**, 2389.

[244] Springer, M., Thesis. (1997) (Lund University).

[245] Steiner, P., Höchst, H. and Hüfner S. (1979) in *Photoemission in Solids II*, edited by L. Ley. and M. Cardona, Topics in Applied Physics Vol. 27 (Springer-Verlag, Heidelberg).

[246] Steiner, M.M., Albers, R.C. and Sham, L.J. (1992) *Phys. Rev. B* **45**, 13272.

[247] Sterne, P.A. and Inkson, J.C. (1984) *J. Phys. C* **17**, 1497–510.

[248] Strinati, G., Mattausch, H.J. and Hanke, W. (1982) *Phys. Rev. B* **25**, 2867–88.

[249] Surh, M.P., Northurp, J.E. and Louie, S.G. (1988) *Phys. Rev. B* **38**, 5976–80.

[250] Surh, M.P., Chacham, H. and Louie, S.G. (1995) *Phys. Rev. B* **51**, 7464–70.

[251] Svane, A. and Gunnarsson, O. (1990) *Phys. Rev. Lett.* **65**, 1148–51.

[252] Svendsen, P.S. and von Barth, U. (1996) *Phys. Rev. B* **54**, 17402–13.

[253] Szotek, Z., Temmerman, W.M. and Winter, H. (1993) *Phys. Rev. B* **47**, 4029.

[254] Talman, J.D. and Shadwick, W.F. (1976) *Phys. Rev. A* **14**, 36.

[255] Terakura, K., Oguchi, T., Williams, A.R. and Kübler, J. (1984) *Phys. Rev. B* **30**, 4734–47.
[256] Ting, C.S., Lee, T.K. and Quinn, J.J. (1975) *Phys. Rev. Lett.* **34**, 870.
[257] Tjeng, L.H., Chen, C.T., Ghijsen, J., Rudolf, P. and Sette, F. (1991) *Phys. Rev. Lett.* **67**, 501.
[258] Treglia, G., Ducastelle, F. and Spanjaard, D. (1980) *Phys. Rev. B* **21**, 3729.
[259] Troullier, N. and Martins, J.L. (1992) *Phys. Rev. B* **46**, 1754.
[260] Tsui, D.C. (1967) *Phys. Rev.* **164**, 669.
[261] Tsui, D.C. and Stark, R.W. (1968) *J. Appl. Phys.* **39**, 1056.
[262] Verdozzi, C., Godby, R.W. and Holloway, S. (1995) *Phys. Rev. Lett.* **74**, 2327–30.
[262a] von Barth, U. and Aryasetiawan, F. (1990), unpublished.
[263] von Barth. (1995), private communication.
[264] von Barth, U. and Holm, B. (1996) *Phys. Rev. B* **54**, 8411–9.
[265] von der Linden, W. and Horsch, P. (1988) *Phys. Rev. B* **37**, 8351.
[266] Wang, C.S. and Pickett, W.E. (1983) *Phys. Rev. Lett.* **51**, 597–600.
[267] Ward, J.C. (1950) *Phys. Rev.* **78**, 182.
[268] Watson, R.E., Herbst, J.F., Hodges, L., Lundqvist, B.I. and Wilkins, J.W. (1976) *Phys. Rev. B* **13**, 1463.
[269] Whited, R.C., Flaten, C.J. and Walker, W.C. (1973) *Solid State Commun.* **13**, 1903.
[270] Zakharov, O., Rubio, A., Blase, X., Cohen, M.L. and Louie, S.G. (1994) *Phys. Rev. B* **50**, 10780-7.
[271] Zhang, S.B., Tomànek, D., Louie, S.G., Cohen, M.L. and Hybertsen, M.S. (1988) *Solid State Commun.* **66**, 585–8.
[272] Zhang, S.B., Wei, S.-H. and Zunger, A. (1995) *Phys. Rev. B* **52**, 13975.
[273] Zhu, X. and Louie, S.G. (1991) *Phys. Rev. B* **43**, 14142–56.
[274] Ziman, J.M. (1972) *Principles of the Theory of Solids*, (Cambridge University Press, Cambridge).
[275] Zunger, A., Perdew, J.P. and Oliver, G.L. (1980) *Solid State Commun.* **34**, 933 .
[276] Zunger, A. and Lindefelt, U. (1983) *Solid State Commun.* **45**, 343.

2. The LDA+U Method: Screened Coulomb Interaction in the Mean-field Approximation

VLADIMIR I. ANISIMOV and A.I. LICHTENSTEIN

1 Introduction

There are two ways to study the ground state properties and excitation spectrum of a many-electron system. The first one is to choose some model with one or more adjustable parameters, to calculate with this model some measurable property, for example the spectrum, and to fit the result to the experimental data to determine the parameters of the model. The second one is to find the eigenfunctions and eigenvalues of the Hamiltonian in a parameter-free approximation (first principles approach). Naturally, the first-principle approach is more appealing, since it has no adjustable parameters. Unfortunately, except for small molecules, it is impossible to solve the many-electron problem without severe approximations. The most successful first principles method is the density functional theory (DFT) within the Local (Spin-) Density Approximation (L(S)DA) [1], where the many-body problem is mapped into a non-interacting system with a one-electron exchange-correlation potential which is approximated by that of the homogeneous electron gas. LDA has proved to be very efficient for extended systems, such as large molecules and solids.

However, as an approximation, the LDA cannot be successful for all systems although the exact DFT should be capable of obtaining groundstate properties. Strongly correlated materials are examples where the deficiency of the LDA is seen most clearly. Such systems usually contain transition-metal or rare-earth metal ions with a partially filled d- (or f-) shell. When applying a one-electron method with an orbital-independent potential, like in the LDA, to transition-metal compounds, one has as a result a partially filled d-band with metallic type electronic structure and itinerant d-electrons. This is definitely a wrong answer for the late-transition-metal oxides and rare-earth metal compounds where d-(f-)electrons are well localized and there is a sizable energy separation between occupied and unoccupied subbands (the lower Hubbard band and upper Hubbard band in a model Hamiltonian approach).

There have been several attempts to improve on the LDA in order to take into account strong electron-electron correlations. One of the most popular approaches is the Self Interaction Correction (SIC) method [2] (see the contribution in the present book). It reproduces quite well the localized nature of the d-(or f-)electrons in transition metal (rare-earth metal) compounds, but the SIC one-electron energies are usually in strong disagreement with spectroscopy data (for example for transition metal oxides occupied d-bands are $\approx$ 1 Ry below the oxygen valence band).

The Hartree-Fock (HF) method [3] is appropriate for describing Mott insulators because it explicitly contains a term which cancels the self-interaction. The fact that the problem of self-interaction is treated in an average way in the LDA is the main reason why the LDA spectra are in such strong qualitative disagreement with experimental data. However, a serious problem of the Hartree-Fock approximation is the unscreened nature of the Coulomb interaction used in this method. The "bare" value of the Coulomb interaction parameter U is rather large (15–20 eV) while screening in a solid leads to much smaller values 8 eV or less [4, 5]. Due to the neglect of screening, the HF energy gap values are a factor of 2–3 larger than the experimental values [3].

The problem of screening is addressed in a rigorous way in the GW approximation (GWA) [6, 7] (see the contribution in the present book) which may be regarded as a Hartree-Fock theory with a frequency and orbital dependent screened Coulomb interaction. The GWA has been applied with success to real systems ranging from simple metals to transition metals, but applications to more complex systems have not been feasible up to now due to the large computational size. Another problem in using the GW approximation is that in its practical realization [8] a response function, needed to calculate the screened interaction, is computed with the help of the energy bands and wave functions obtained in the LDA calculation. While such a procedure is justified for the systems where correlation effects are small (such as semiconductors [9]), for strongly correlated systems one may need a better starting Hamiltonian than the LDA. This, for example, can be achieved by improving the LDA Hamiltonian using the calculated self-energy in a self-consistent procedure [10].

In the present review, the so-called LDA+U method [11–13] is described where the non-local and energy dependent self-energy is approximated by a frequency independent but non-local screened Coulomb potential. A similar approximation is used in a model Hamiltonian approach, which has proved to be successful in applications to strongly correlated systems [14, 15].

2 The LDA+U method: formulation and practical realization

As in the Anderson model [16] electrons are separated into two subsystems: localized d- or f- electrons for which Coulomb d-d interaction should be taken into account by a term $\frac{1}{2}U\sum_{i\neq j}n_in_j$ (n_i are d-orbital occupancies) as in a mean-field (Hartree-Fock) approximation and delocalized s-,p- electrons which could be described by using an orbital-independent one-electron potential (LDA). Let us consider a d-ion as an open system with a fluctuating number of d-electrons. If one assumes that the Coulomb energy of d-d interactions as a function of the total number of d-electrons $N = \sum n_i$ given by the LDA is a good approximation (but not the orbital energies (eigenvalues)!), then the correct formula for this energy is $E = UN(N-1)/2$. Let us subtract this expression from the LDA total energy functional and add a Hubbard like term (neglecting for a while exchange and nonsphericity). As a result we would have the following functional:

$$E = E_{LDA} - UN(N-1)/2 + \frac{1}{2}U\sum_{i\neq j} n_i n_j \, . \tag{1}$$

The orbital energies ϵ_i are derivatives of (1) with respect to the orbital occupations n_i :

$$\epsilon_i = \partial E/\partial n_i = \epsilon_{LDA} + U\left(\frac{1}{2} - n_i\right) . \tag{2}$$

This simple formula shifts the LDA orbital energy by $-U/2$ for occupied orbitals (n_i=1) and by $+U/2$ for unoccupied orbitals (n_i=0). A similar formula is found for the orbital dependent potential ($V_i(\mathbf{r}) = \delta E/\delta n_i(\mathbf{r})$ where a variation is taken not on the total charge density $n(\mathbf{r})$ but on the charge density of a particular i-th orbital $n_i(\mathbf{r})$):

$$V_i(\mathbf{r}) = V_{LDA}(\mathbf{r}) + U\left(\frac{1}{2} - n_i\right) . \tag{3}$$

The LDA+U orbital-dependent potential (3) gives upper- and lower Hubbard bands with energy separation between them equal to the Coulomb parameter U, thus reproducing qualitatively the correct physics for Mott-Hubbard insulators. To construct a quantitatively sound calculational scheme, one needs to define in a more general way an orbital basis set and to take into account properly the direct and exchange Coulomb interactions inside a partially filled d-(or f-) atomic shell.

All one needs physically is the identification of regions in space where the atomic characteristics of the electronic states have largely survived ('atomic spheres'), which is not a problem for at least d- or f electrons. Within these atomic spheres one can expand in a localized orthonormal basis $| \, inlm\sigma\rangle$ (i denotes the site, n the main quantum number, l- orbital quantum number, m- magnetic number and σ-spin index). Although not strictly necessary, let us specialize to the usual situation where only a particular nl shell is partly filled. The density matrix is defined by:

$$n^{\sigma}_{mm'} = -\frac{1}{\pi}\int^{E_F} ImG^{\sigma}_{inlm,inlm'}(E)dE, \tag{4}$$

where $G^{\sigma}_{inlm,inlm'}(E) = \langle inlm\sigma \mid (E - \widehat{H})^{-1} \mid inlm'\sigma\rangle$ are the elements of the Green function matrix in this localized representation, while $\widehat{H}$ will be defined later on. In terms of the elements of this density matrix $\{n^\sigma\}$, the generalized LDA+U functional [13] is defined as follows:

$$E^{LDA+U}[\rho^\sigma(\mathbf{r}), \{n^\sigma\}] = E^{LSDA}[\rho^\sigma(\mathbf{r})] + E^U[\{n^\sigma\}] - E_{dc}[\{n^\sigma\}] \quad (5)$$

Where $\rho^\sigma(\mathbf{r})$ is the charge density for spin-σ electrons and $E^{LSDA}[\rho^\sigma(\mathbf{r})]$ is the standard LSDA (Local Spin-Density Approximation) functional. Eq. (5) asserts that LSDA suffices in the absence of orbital polarizations, while the latter are driven by,

$$E^U[\{n\}] = \frac{1}{2}\sum_{\{m\},\sigma}\{\langle m, m'' \mid V_{ee} \mid m', m'''\rangle n^\sigma_{mm'} n^{-\sigma}_{m''m'''} +$$
$$(\langle m, m'' \mid V_{ee} \mid m', m'''\rangle - \langle m, m'' \mid V_{ee} \mid m''', m'\rangle) n^\sigma_{mm'} n^\sigma_{m''m'''}\}, \quad (6)$$

where V_{ee} are the screened Coulomb interactions among the nl electrons. Finally, the last term in Eq. (5) corrects for double counting (in the absence of orbital polarizations, Eq. (5) should reduce to E^{LSDA}) and is given by

$$E_{dc}[\{n^\sigma\}] = \frac{1}{2}UN(N-1) - \frac{1}{2}J[N^\uparrow(N^\uparrow - 1) + N^\downarrow(N^\downarrow - 1)], \quad (7)$$

were $N^\sigma = Tr(n^\sigma_{mm'})$ and $N = N^\uparrow + N^\downarrow$. U and J are screened Coulomb and exchange parameters [4, 5].

In addition to the usual LDA potential, we find an effective single-particle potential to be used in the effective single-particle Hamiltonian :

$$\widehat{H} = \widehat{H}_{LSDA} + \sum_{mm'} \mid inlm\sigma\rangle V^\sigma_{mm'} \langle inlm'\sigma \mid \quad (8)$$

$$V^\sigma_{mm'} = \sum_{\{m\}}\{\langle m, m'' \mid V_{ee} \mid m', m'''\rangle n^{-\sigma}_{m''m'''} +$$
$$+(\langle m, m'' \mid V_{ee} \mid m', m'''\rangle - \langle m, m'' \mid V_{ee} \mid m''', m'\rangle) n^\sigma_{m''m'''}\} - \quad (9)$$
$$-U(N - \frac{1}{2}) + J(N^\sigma - \frac{1}{2}).$$

The V_{ee}'s remain to be determined. We again follow the spirit of the LDA+U by assuming that within the atomic spheres these interactions retain largely their atomic nature. Moreover, it is asserted that the LSDA itself suffices to determine their values, following the well-tested procedure of the so-called supercell LSDA approach [12, 18]: the elements of the density matrix $n^\sigma_{mm'}$ have to be constrained locally and the second derivative of the LSDA energy with respect to the variation of the density matrix yields the wanted interactions. In more detail, the matrix elements can be expressed in terms of complex spherical harmonics and effective Slater integrals F^k [17] as

$$\langle m, m'' \mid V_{ee} \mid m', m'''\rangle = \sum_k a_k(m, m', m'', m''')F^k, \quad (10)$$

where $0 \leq k \leq 2l$ and

$$a_k(m, m', m'', m''') = \frac{4\pi}{2k+1} \sum_{q=-k}^{k} \langle lm \mid Y_{kq} \mid lm' \rangle \langle lm'' \mid Y^*_{kq} \mid lm''' \rangle \quad (11)$$

For d-electrons one needs F^0, F^2 and F^4 and these can be linked to the Coulomb- and Stoner parameters U and J obtained from the LSDA-supercell procedures via $U = F^0$ and $J = (F^2 + F^4)/14$, while the ratio F^2/F^4 is to a good accuracy a constant ~ 0.625 for the 3d elements [12, 18]. (For f-electrons the corresponding expression is $J = (286F^2 + 195F^4 + 250F^6)/6435$). The Coulomb parameter U is calculated as a second derivative of the total energy (or the first derivative of the corresponding eigenvalue) with respect to the occupancy of the localized orbitals of the central atom in a supercell with fixed occupancies on all the other atoms [4].

The new Hamiltonian in Eq. (8) contains an orbital-dependent potential in Eq. (9) in the form of a projection operator. This means that the LDA+U method is essentially dependent on the choice of the set of the localized orbitals in this operator. That is a consequence of the basic Anderson-model-like ideology of the LDA+U approach: the separation of the total variational space into a localized d- (f-)orbitals subspace, with Coulomb interaction between them treated with a Hubbard type term in the Hamiltonian, and the subspace of all other states for which the local density approximation for Coulomb interaction is regarded as sufficient. The arbitrariness of choice of the localized orbitals is not as crucial as it might be expected. The d-(f-)orbitals for which Coulomb correlation effects are important are indeed well localized in space and retain their atomic character in a solid. The experience of using the LDA+U approximation in various electronic structure calculational schemes shows that the results are not sensitive to the particular form of the localized orbitals.

Due to the presence of the projection operator in the LDA+U Hamiltonian in Eq. (8), the most straightforward calculational schemes would be to use atomic orbitals type basis sets, such as LMTO (Linear Muffin-Tin Orbitals) [19]. However, as soon as localized d-(f-orbitals) are defined, the Hamiltonian in Eq. (8) could be realized even in schemes using plane waves as a basis set, such as pseudopotential methods.

3 Relationship between the LDA+U and the *GW* approximation

Although there is no theoretical justification, it is customary to interpret the Kohn-Sham (KS) eigenvalues in the DFT as the quasiparticle energies measured in photoemission experiments. A proper way of calculating quasiparticle energies is provided by the Green function theory. In this approach, the many-body effects are contained in the self-energy operator Σ which is non-local and energy dependent:

$$H_0(\mathbf{r})\Psi(\mathbf{r}) + \int d\mathbf{r}_1 \Sigma(\mathbf{r}, \mathbf{r}_1, E)\Psi(\mathbf{r}_1) = E\Psi(\mathbf{r}) \tag{11}$$

where H_0 contains the kinetic energy, the Hartree potential and a possible one-particle external potential. We can think of the DFT exchange-correlation potential V_{xc} as a local and energy independent approximation to the self-energy which gives the correct ground state density. However, in some cases, the non-locality (orbital dependence) and energy dependence are crucial for obtaining excitation spectra.

Unfortunately, the self-energy is very hard to calculate and we have to resort to approximations. The simplest approximation to the self-energy, derived from the many-body perturbation theory, is the GW approximation [6, 7] (see also the contribution in the present book). But even in this simplest approximation, the computational effort is already quite large.

It can be shown [20], that GWA and LDA+U are both Hartree-Fock-like theories and, at least for localized states, such as the d- or f-orbitals of transition metal or rare-earth metal ions, the LDA+U may be regarded as an approximation to GWA.

The GW approximation (GWA) is given by

$$\Sigma(\mathbf{r}, \mathbf{r}'; \omega) = \frac{i}{2\pi} \int_{-\infty}^{\infty} d\omega' G(\mathbf{r}, \mathbf{r}'; \omega + \omega') W(\mathbf{r}, \mathbf{r}'; \omega') e^{i\delta\omega'} \tag{12}$$

W is a screened Coulomb potential obtained from the inverse dielectric function

$$\epsilon^{-1}(\mathbf{r}, \mathbf{r}'; \omega) = \delta(\mathbf{r} - \mathbf{r}') + \int d^3r'' v(\mathbf{r} - \mathbf{r}'') P(\mathbf{r}'', \mathbf{r}'; \omega) \tag{13}$$

where P is the full response function. We then have

$$\begin{aligned} W(\mathbf{r}, \mathbf{r}'; \omega) &= \int d^3r'' \epsilon^{-1}(\mathbf{r}, \mathbf{r}''; \omega) v(\mathbf{r}'' - \mathbf{r}') \\ &= v(\mathbf{r} - \mathbf{r}') + W_c(\mathbf{r}, \mathbf{r}'; \omega) \end{aligned} \tag{14}$$

where

$$W_c(\mathbf{r}, \mathbf{r}'; \omega) = \int d^3r_1 d^3r_2 v(\mathbf{r} - \mathbf{r}_1) P(\mathbf{r}_1, \mathbf{r}_2; \omega) v(\mathbf{r}_2 - \mathbf{r}') \tag{15}$$

The time-ordered Green function may be written in the spectral representation:

$$G(\mathbf{r}, \mathbf{r}'; \omega) = \int_{-\infty}^{\mu} d\omega' \frac{A(\mathbf{r}, \mathbf{r}'; \omega')}{\omega - \omega' - i\delta} + \int_{\mu}^{\infty} d\omega' \frac{A(\mathbf{r}, \mathbf{r}'; \omega')}{\omega - \omega' + i\delta} \tag{16}$$

where

$$A(\mathbf{r}, \mathbf{r}'; \omega) = -\frac{1}{\pi} Im\ G(\mathbf{r}, \mathbf{r}'; \omega)\ sgn(\omega - \mu) \tag{17}$$

In practice we use a zeroth order Green function so that

$$A(\mathbf{r}, \mathbf{r}'; \omega) = \sum_{\mathbf{k}n} \psi_{\mathbf{k}n}(\mathbf{r}) \psi^{*}_{\mathbf{k}n}(\mathbf{r}') \delta(\omega - \epsilon_{\mathbf{k}n}) \tag{18}$$

The self-energy can now be evaluated and we obtain

$$\Sigma(\mathbf{r}, \mathbf{r}'; \omega) = \Sigma_x(\mathbf{r}, \mathbf{r}') + \Sigma_c(\mathbf{r}, \mathbf{r}'; \omega) \tag{19}$$

where Σ_x is the bare exchange potential

$$\Sigma_x(\mathbf{r}, \mathbf{r}') = -\sum_{\mathbf{k}n}^{occ} \psi_{\mathbf{k}n}(\mathbf{r}) \psi^{*}_{\mathbf{k}n}(\mathbf{r}') v(\mathbf{r} - \mathbf{r}') \tag{20}$$

and Σ_c is the correlated part of the self-energy given by

$$\begin{aligned} \Sigma_c(\mathbf{r}, \mathbf{r}'; \omega) = & \sum_{\mathbf{k}n}^{occ} \psi_{\mathbf{k}n}(\mathbf{r}) \psi^{*}_{\mathbf{k}n}(\mathbf{r}') W_c^{-}(\mathbf{r}, \mathbf{r}'; \omega - \epsilon_{\mathbf{k}n}) \\ & + \sum_{\mathbf{k}n}^{unocc} \psi_{\mathbf{k}n}(\mathbf{r}) \psi^{*}_{\mathbf{k}n}(\mathbf{r}') W_c^{+}(\mathbf{r}, \mathbf{r}'; \omega - \epsilon_{\mathbf{k}n}) \end{aligned} \tag{21}$$

where

$$W_c^{\pm}(\mathbf{r}, \mathbf{r}'; \omega) = \frac{i}{2\pi} \int_{-\infty}^{\infty} d\omega' \frac{W_c(\mathbf{r}, \mathbf{r}'; \omega')}{\omega + \omega' \pm i\delta} \tag{22}$$

In short we can write

$$\Sigma(\mathbf{r}, \mathbf{r}'; \omega) = -\sum_{\mathbf{k}n} \psi_{\mathbf{k}n}(\mathbf{r}) \psi^{*}_{\mathbf{k}n}(\mathbf{r}') W_0(\mathbf{r}, \mathbf{r}'; \omega - \epsilon_{\mathbf{k}n}) \tag{23}$$

where

$$\begin{aligned} W_0(\mathbf{r}, \mathbf{r}'; \omega - \epsilon_{\mathbf{k}n}) \equiv & [v(\mathbf{r} - r') - W_c^{-}(\mathbf{r}, \mathbf{r}'; \omega - \epsilon_{\mathbf{k}n})] \theta(\mu - \epsilon_{\mathbf{k}n}) \\ & - W_c^{+}(\mathbf{r}, \mathbf{r}'; \omega - \epsilon_{\mathbf{k}n}) \theta(\epsilon_{\mathbf{k}n} - \mu) \end{aligned} \tag{24}$$

The self-energy in the GWA has the same form as that in the HFA except that it depends on energy and contains a term which depends on unoccupied states as a consequence of correlation effects. Thus the GWA can be interpreted as a generalization of the Hartree-Fock approximation (HFA) with a potential W_0 which contains dynamical screening of the Coulomb potential. Note, however, that W_0 is not the same as the dynamically screened potential W.

The LDA+U is designed to give the self-energy correction to localized states embedded in delocalized states. The localized states have a large Coulomb correlation which is accounted for by the U-term whereas the delocalized states are well described by the LDA. To make a connection between the GWA and the LDA+U we consider the correlated part of the self-energy in the GWA for an occupied core-like state ψ_d:

$$\begin{aligned}\langle\psi_d|\Sigma_c(\epsilon_d)|\psi_d\rangle &= \langle\psi_d\psi_d|W_c^-(0)|\psi_d\psi_d\rangle \\ &+ \sum_{\mathbf{k}n\neq d}^{occ} \langle\psi_d\psi_{\mathbf{k}n}|W_c^-(\epsilon_d-\epsilon_{\mathbf{k}n})|\psi_{\mathbf{k}n}\psi_d\rangle \\ &+ \sum_{\mathbf{k}n}^{unocc} \langle\psi_d\psi_{\mathbf{k}n}|W_c^+(\epsilon_d-\epsilon_{\mathbf{k}n})|\psi_{\mathbf{k}n}\psi_d\rangle \end{aligned} \tag{25}$$

Strictly speaking, the self-energy should be evaluated at the new energy $E_d = \epsilon_d +$ self-energy correction and it is understood to be the case here. If ψ_d is localized and well separated in energy from other states, then the first term is evidently much larger than the rest. The last term contains unoccupied Ψ_d states but they are orthogonal to the occupied ones so that this term is much smaller than the first. Thus we may make the following approximation:

$$\begin{aligned}\langle\psi_d|\Sigma_c(\epsilon_d)|\psi_d\rangle &\approx \langle\psi_d\psi_d|W_c^-(0)|\psi_d\psi_d\rangle \\ &= -\frac{1}{2}\langle\psi_d\psi_d|W_c(0)|\psi_d\psi_d\rangle\end{aligned} \tag{26}$$

The last step can be shown as follows. The correlated part of the screened potential can be written in terms of its spectral representation:

$$W_c(\omega) = \int_{-\infty}^{0} d\omega' \frac{B(\omega')}{\omega-\omega'-i\delta} + \int_{0}^{\infty} d\omega' \frac{B(\omega')}{\omega-\omega'+i\delta} \tag{27}$$

where

$$B(\omega) = -\frac{1}{\pi} Im\ W_c(\omega)\ sgn(\omega) \tag{28}$$

W_c is an even function of ω so that $B(\omega)$ is odd. We can now calculate

$$\begin{aligned}W_c(0) &= \int_{-\infty}^{0} d\omega' \frac{B(\omega')}{-\omega'-i\delta} + \int_{0}^{\infty} d\omega' \frac{B(\omega')}{-\omega'+i\delta} \\ &= -2\int_{0}^{\infty} d\omega' \frac{B(\omega')}{\omega'-i\delta}\end{aligned} \tag{29}$$

using the fact that $B(\omega)$ is odd.

$$\begin{aligned} W_c^-(0) &= \frac{i}{2\pi}\int_{-\infty}^{\infty} d\omega' \frac{W_c(\omega')}{\omega' - i\delta} \\ &= \frac{i}{2\pi}\int_{-\infty}^{\infty} d\omega' \frac{1}{\omega' - i\delta} \\ &\quad \times \left\{ \int_{-\infty}^{0} d\omega'' \frac{B(\omega'')}{\omega' - \omega'' - i\delta} + \int_{0}^{\infty} d\omega'' \frac{B(\omega'')}{\omega' - \omega'' + i\delta} \right\} \\ &= \int_{0}^{\infty} d\omega'' \frac{B(\omega'')}{\omega'' - i\delta} \\ &= -\frac{1}{2} W_c(0) \end{aligned} \tag{30}$$

This is a correction due to the work done on the electron by the polarization field from zero to $W_c(0)$ [6]. A similar result,

$$+\frac{1}{2}\langle\psi_d\psi_d|W_c(0)|\psi_d\psi_d\rangle$$

is obtained for an unoccupied core-like state of the same character so that the energy separation of the states is

$$\begin{aligned} \Delta &= \epsilon_2^{HF} - \epsilon_1^{HF} + \langle\psi_d\psi_d|W_c(0)|\psi_d\psi_d\rangle \\ &= \langle\psi_d\psi_d|v|\psi_d\psi_d\rangle + \langle\psi_d\psi_d|W_c(0)|\psi_d\psi_d\rangle \\ &= \langle\psi_d\psi_d|W(0)|\psi_d\psi_d\rangle \end{aligned} \tag{31}$$

which agrees with the intuitive result that the "gap" is given by the screened Coulomb interaction: $\Delta = U \approx W(0)$.

Within the above approximation, the GW self-energy for a localized state is given by

$$\Sigma(\mathbf{r}, \mathbf{r}'; \epsilon_d) = \Sigma_x(\mathbf{r}, \mathbf{r}') + \sum_{\mathbf{k}n=d} \psi_{\mathbf{k}n}(\mathbf{r})\psi_{\mathbf{k}n}^*(\mathbf{r}')W_c^0(\mathbf{r}, \mathbf{r}'; \epsilon_d) \tag{32}$$

with

$$W_c^0(\mathbf{r}, \mathbf{r}'; \epsilon_d) = -\frac{1}{2}W_c(\mathbf{r}, \mathbf{r}'; 0) \times [\theta(\mu - \epsilon_d) - \theta(\epsilon_d - \mu)] \tag{33}$$

It is clear that the self-energy correction to the LDA,

$$\Delta\Sigma(\mathbf{r}, \mathbf{r}'; \epsilon_d) = \Sigma(\mathbf{r}, \mathbf{r}'; \epsilon_d) - V_{xc}^{LDA}(\mathbf{r})\delta(\mathbf{r} - \mathbf{r}') \tag{34}$$

should be equated to the U-term in the LDA+U scheme. In the spirit of the LDA+U and the Anderson impurity model, let us divide the space into localized states $\{\phi_m\}$, such as d or f states, and delocalized states $\{\psi_{\mathbf{k}n}\}$:

$$\delta(\mathbf{r} - \mathbf{r}') = \sum_m \phi_m(\mathbf{r})\phi_m^*(\mathbf{r}') + \sum_{\mathbf{k}n} \psi_{\mathbf{k}n}(\mathbf{r})\psi_{\mathbf{k}n}^*(\mathbf{r}') \tag{35}$$

The self-energy correction can be written as follows:

$$\Delta\Sigma(\mathbf{r},\mathbf{r}';\epsilon_d) = \sum_{mm'} \phi_m(\mathbf{r})\Delta\Sigma_{mm'}(\epsilon_d)\phi^*_{m'}(\mathbf{r}')$$
$$+\sum_{\mathbf{k}nn'} \psi_{\mathbf{k}n}(\mathbf{r})\Delta\Sigma_{nn'}(\epsilon_d)\psi^*_{\mathbf{k}n'}(\mathbf{r}')$$
$$+\sum_{\mathbf{k}nm} \psi_{\mathbf{k}n}(\mathbf{r})\Delta\Sigma_{nm}(\mathbf{k},\epsilon_d)\phi^*_m(\mathbf{r}')$$
$$+\sum_{kmn} \phi_m(\mathbf{r})\Delta\Sigma_{mn}(\mathbf{k},\epsilon_d)\psi^*_{\mathbf{k}n}(\mathbf{r}') \quad (36)$$

Since we are interested in the localized states $\{\phi_m\}$, the first term will dominate and therefore

$$\Delta\Sigma(\mathbf{r},\mathbf{r}';\epsilon_d) \approx \sum_{mm'} \phi_m(\mathbf{r})\Delta\Sigma_{mm'}(\epsilon_d)\phi^*_{m'}(\mathbf{r}') \quad (37)$$

where

$$\Delta\Sigma_{mm'}(\epsilon_d) = \langle\phi_m|\Sigma_x - V_{xc}|\phi_{m'}\rangle + \sum_{m''m'''} \langle m,m''|W_c^0|m''',m'\rangle n_{m'',m'''} \quad (38)$$

with

$$n_{m'',m'''} = \sum_{\mathbf{k}n=d} \langle\phi_{m''}|\psi_{\mathbf{k}n}\rangle\langle\psi_{\mathbf{k}n}|\phi_{m'''}\rangle \quad (39)$$

We note that the orbitals $\{\phi_m\}$ can always be chosen to be localized within the atomic spheres. The rest of the self-energy correction is small and if necessary it can be incorporated into the one-particle term.

Let us suppose that we have a d-ion with d-orbitals $\psi_{m\sigma}$ only. Within the above approximation, the GW self-energy for a localized state is given by

$$\Sigma(\mathbf{r},\mathbf{r}';\epsilon_{m\sigma}) = \Sigma_x(\mathbf{r},\mathbf{r}') + \sum_{m'\sigma'} \psi_{m'\sigma'}(\mathbf{r})\psi^*_{m'\sigma'}(\mathbf{r}')W_c^0(\mathbf{r},\mathbf{r}';\epsilon_{m\sigma}) \quad (40)$$

with

$$W_c^0(\mathbf{r},\mathbf{r}';\epsilon_{m\sigma}) = -\frac{1}{2}W_c(\mathbf{r},\mathbf{r}';0) \times [\theta(\mu-\epsilon_{m\sigma}) - \theta(\epsilon_{m\sigma}-\mu)] \quad (41)$$

What will the matrix element of the total potential of the electron-electron interaction in the GWA be?

$$\langle\psi_{m\sigma}|V_{Hartree}+\Sigma_x+\Sigma_c|\psi_{m\sigma}\rangle =$$

$$\sum_{m'\sigma'}^{occup}\int\int d\mathbf{r}d\mathbf{r}'\psi^*_{m\sigma}(\mathbf{r})\psi_{m\sigma}(\mathbf{r})v(\mathbf{r}-r')\psi^*_{m'\sigma'}(\mathbf{r}')\psi_{m'\sigma'}(\mathbf{r}')$$

$$-\sum_{m'}^{occup}\int\int d\mathbf{r}d\mathbf{r}'\psi^*_{m\sigma}(\mathbf{r})\psi_{m'\sigma'}(\mathbf{r})v(\mathbf{r}-r')\psi_{m\sigma}(\mathbf{r}')\psi^*_{m'\sigma'}(\mathbf{r}')$$

$$+(\frac{1}{2}-n_{m\sigma})\sum_{m'}\int\int d\mathbf{r}d\mathbf{r}'\psi^*_{m\sigma}(\mathbf{r})\psi_{m'\sigma'}(\mathbf{r})W_c(\mathbf{r},r',0)\psi_{m\sigma}(\mathbf{r}')\psi^*_{m'\sigma'}(\mathbf{r}') \tag{42}$$

Here $n_{m\sigma}$ is the occupancy of the $m\sigma$-orbital which is equal 1 if $\mu-\epsilon_{m\sigma}>0$ and 0 if $\mu-\epsilon_{m\sigma}<0$.

The above matrix element can be written in the form:

$$V^{GWA}_{m\sigma}=\sum_{m'\sigma'}U^0_{mm'}n_{m'\sigma'}-U^0_{mm}n_{m\sigma}-\sum_{m'\neq m}J_{mm'}n_{m'\sigma}+(\frac{1}{2}-n_{m\sigma})\sum_{m'}W_{mm'}$$

where $U^0_{mm'}$ is the *unscreened* Coulomb interaction matrix, $J_{mm'}$ is the exchange matrix and $W_{mm'}$ is the matrix elements of the correlation potential $W_c(\mathbf{r},r',0)$. If we define the screening parameter W as:

$$W=-\sum_{m'}W_{mm'}$$

then the final formula for the GWA-potential matrix element will be:

$$V^{GWA}_{m\sigma}=\sum_{m'\sigma'}U^0_{mm'}n_{m'\sigma'}-(U^0_{mm}-W)n_{m\sigma}-\sum_{m'\neq m}J_{mm'}n_{m'\sigma}-\frac{1}{2}W$$

To write down the GWA-type correction to the LSDA we must express the matrix element of the LSDA potential in the same form as above. As LSDA is not derived from the orbital-orbital interaction formalism but is given as some effective local orbital-independent potential using the electron density dependence of the Coulomb interaction energy as in the homogeneous electron gas, it is not possible to do it in a rigorous way. Let us look at the energy of the electron-electron interaction in the d-ion as a function of the total number of d-electrons N, $E_{LSDA}[\rho(\mathbf{r})]=E_{LSDA}[N|\psi_{m\sigma}(\mathbf{r})|^2]$. It is known that while one-electron eigenvalues are not so good in LSDA, the total energy values are in a much better agreement with more rigorous calculations. So we can suppose that Hartree-Fock formulae could be a good approximation:

$$\begin{aligned}E_{LSDA}[\rho_\sigma(\mathbf{r})]&=E_{LSDA}[N_\sigma|\psi_{m\sigma}(\mathbf{r})|^2]\\&=\frac{1}{2}F^0N(N-1)-\frac{1}{4}JN(N-2)-\frac{1}{4}J(N_\uparrow-N_\downarrow)^2\end{aligned} \tag{43}$$

where F^0 is a first Slater integral , J is exchange parameter and

$$N_\sigma = \sum_m n_{m\sigma}, \quad N = N_\uparrow + N_\downarrow$$

LSDA electron interaction potential is a variational derivative of the total energy as a functional of the electron density $\rho(\mathbf{r})$:

$$V^\sigma_{LSDA}(\rho(\mathbf{r})) = \frac{\delta E_{LSDA}[\rho(\mathbf{r})]}{\delta\rho^\sigma(\mathbf{r})}$$

The derivative of the interaction energy as a function of the total number of d-electrons N_σ:

$$\begin{aligned}\frac{\partial E_{LSDA}[N_\sigma|\psi_{m\sigma}(\mathbf{r})|^2]}{\partial N_\sigma} &= \int d\mathbf{r}\frac{\delta E_{LSDA}[\rho(\mathbf{r})]}{\delta\rho^\sigma(\mathbf{r})}\frac{\partial\rho^\sigma(\mathbf{r})}{\partial N_\sigma} \\ &= \int d\mathbf{r}V^\sigma_{LSDA}(\rho(\mathbf{r}))|\psi_{m\sigma}(\mathbf{r})|^2 \\ &= F^0 N - \frac{1}{2}(F^0 - J) - JN_\sigma \end{aligned} \quad (44)$$

That gives us the matrix element of the LSDA potential:

$$V^{LSDA}_{m\sigma} = F^0 N - \frac{1}{2}(F^0 - J) - JN_\sigma$$

The GWA correction to LSDA potential is:

$$\begin{aligned}\delta V_{m\sigma} &= V^{GWA}_{m\sigma} - V^{LSDA}_{m\sigma} \\ &= \sum_{m'\sigma'} U^0_{mm'}n_{m'\sigma'} - (U^0_{mm} - W)n_{m\sigma} - \sum_{m'\neq m} J_{mm'}n_{m'\sigma} - \frac{1}{2}W \\ &\quad - F^0\sum_{m'\sigma'} n_{m'\sigma'} + J\sum_m n_{m\sigma} + \frac{1}{2}(F^0 - J) \\ &= \sum_{m'\sigma'}(U^0_{mm'} - F^0)n_{m'\sigma'} - (U^0_{mm} - W)n_{m\sigma} - \sum_{m'\neq m} J_{mm'}n_{m'\sigma} \\ &\quad - \frac{1}{2}W + J\sum_m n_{m\sigma} + \frac{1}{2}(F^0 - J) \end{aligned} \quad (45)$$

The difference $(U^0_{mm'} - F^0)$ does not depend on the Slater integral F^0 (it depends only on the Slater integrals F^k with $k \neq 0$) and is equal to $(U_{mm'} - U)$ where $U = F^0 - W$ is a screened Coulomb parameter and $U_{mm'}$ is a screened Coulomb matrix.

$$\begin{aligned}\delta V_{m\sigma} &= V^{GWA}_{m\sigma} - V^{LSDA}_{m\sigma} \\ &= \sum_{m'} U_{mm'}n_{m'-\sigma} + \sum_{m'\neq m}(U_{mm'} - J_{mm'})n_{m'\sigma} \\ &\quad - U(N - \frac{1}{2}) + J(N_\sigma - \frac{1}{2}) \end{aligned} \quad (46)$$

The above formula is equivalent to the LDA+U potential correction Eq. (9) if the occupation matrix (4) is diagonal:

$$
\begin{aligned}
n^{\sigma}_{mm'} &= n_{m\sigma}\delta_{mm'} \\
U_{mm'} &= \langle m, m' \mid V_{ee} \mid m, m' \rangle \\
J_{mm'} &= \langle m, m' \mid V_{ee} \mid m', m \rangle
\end{aligned}
$$

The only difference is the procedure of calculating screened Coulomb parameter U. In the LDA+U method it was done by the constrained LSDA supercell calculation and in the GWA it requires hard computations of the response function.

We have established a relationship between the GWA and LDA+U by recognizing that both are Hartree-Fock-like theories, thus giving a theoretical justification for the latter. At least for localized states, such as d or f states where on-site Coulomb correlation is very important, the LDA+U may be regarded as an approximation to the GWA. Whether they give the same results depends on three main things:

1. How close is the value of U to the static screened potential $W(0)$?
2. How important is the energy dependence of W which is neglected in the LDA+U ?
3. How important is self-consistency which is performed in the LDA+U calculations but not normally in the GW calculations ?

4 Localized states: 3d and 4f orbitals

4.1 Rare-earth metals: Gd

Gd metal is a good test to check how physically sound the approximations are that were used in the derivation of the LDA+U formula presented above. The Gd ion has 7 f-electrons so that the majority-spin subshell is completely filled and the minority-spin subshell is empty. The hybridization of the localized f-orbitals with the conduction bands is small and the 4f-shell could be regarded with good accuracy as that of the Gd^{3+} ion in $^8S_{\frac{7}{2}}$ ground state, well separated from the all other excited terms. That means that the ground state is well described by a single Slater determinant wave function and the LDA+U as a one-electron theory is valid here. The final $N-1$ and $N+1$ states of the removal (XPS) and addition (BIS) spectra are also well described by a single Slater determinant (^7F, neglecting spin-orbit coupling which gives splitting of the order of less than 1 eV). As a result, the theoretical XPS and BIS are (again neglecting spin-orbit coupling) single lines separated by $U+6J$. The calculation for Gd [21] gives U=6.7 eV and J=0.7 eV thus resulting in a theoretical value of the splitting between occupied and unoccupied 4f-bands ≈11 eV in good agreement with the experimental value ≈12 eV (Fig. 1).

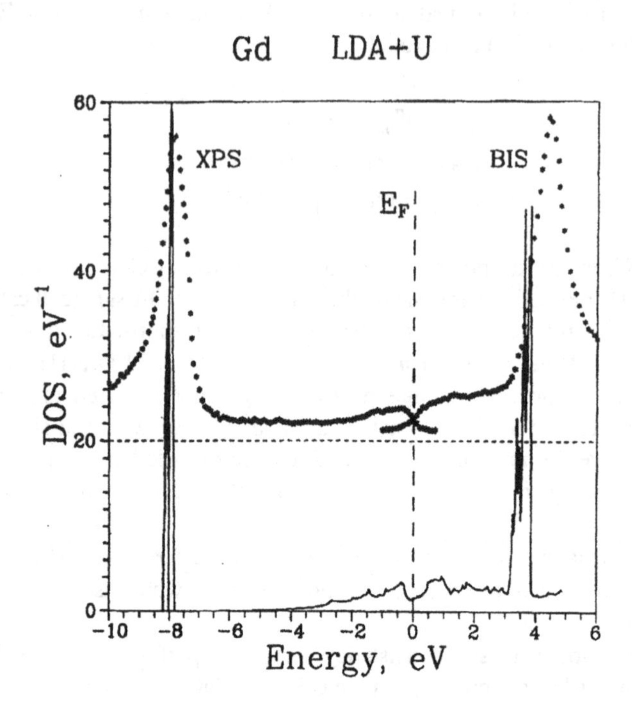

Figure 1: *Density of states for ferromagnetic Gd metal from LDA+U calculation and results of BIS (Bremstrahlung Isochromat spectroscopy) and XPS (X-ray Photoemission Spectroscopy) experiments.*

The advantage of the LDA+U method is the ability to treat *simultaneously* delocalized conduction band electrons and localized 4f-electrons in the same computational scheme. For such a method it is important to be sure that the relative energy positions of these two type of bands are reproduced correctly. The example of Gd gives us confidence in this (Fig. 1): there is a good agreement between calculated and experimental spectra not only for the separation between 4f-bands but also for the position of the 4f-peaks relative to the Fermi energy. Gd is usually presented as an example where the LSDA gives the correct electronic structure due to the spin polarization splitting of the occupied and unoccupied 4f-bands (in all other rare-earth metals LSDA gives unphysical 4f-peak on the Fermi energy). In the LSDA, the energy separation between 4f-bands is not only strongly underestimated (exchange splitting is only 5 eV instead of the experimental value 12 eV) but also the unoccupied 4f-band is very close to the Fermi energy thus strongly influencing the Fermi surface and magnetic ground state properties(in the LSDA calculation antiferromagnetic state is lower in total energy than ferromagnetic one in contradiction to the experiment). The LDA+U solves both of these problems [21].

4.2 Transition metal impurities in alkaline metals

The d-orbitals of transition metal ions are much less localized compared with the 4f-orbitals of the rare-earth metal ions. The hybridization of the d-states between themselves and the s- and p-states in transition metals results in d-band widths of few eV. However, there is one case where d-shells of transition metal atoms show properties typical for free ions: 3d and 4d impurities introduced into alkali metal hosts by ion implantation [22]. While it is typical for 3d ions in solids to have local magnetic moments, experimental observation of magnetic moments on 4d-impurities in Rb is very unusual. Moreover, there are indications that even the orbital moment is not quenched in these systems.

The reason for this is the large value of the Wigner-Seitz volume of Rb atoms compared with transition metal atoms. As a result, the energy of the d-states falls to the bottom (or even below the bottom) of the alkali metal conduction band and their hybridization with the hosts states becomes very small, leading to narrow resonant states, similar to the f-bands in the rare earth metals. The calculated impurity density of states [23] has a form of very narrow delta-function-like peaks, thus justifying discrete level free-ion description of the electronic structure of these impurities.

LDA+U calculations for d-impurities in Rb [23, 24] (Fig. 2) showed that the majority of elements have a T^{1+} valence state (where T is any transition metal), having one d-electron more than in a free atom. The Pd-impurity has unusual d^{10} configuration resulting in nonmagnetic state while all other impurities are magnetic, which agrees well with experimental data [22].

4.3 Rare-earth elements compounds: CeSb

While 4f-orbitals of rare-earth ions can usually be regarded as semi-core states , in some rare-earth compounds hybridization of 4f-orbitals with other states can be physically important. Examples of such systems are cerium monopnictides with their unusual magnetic properties. Among the pnictides CeSb has a special position. In addition to the anomalies inherent to all monopnictides, CeSb has a large magnetic anisotropy together with a small crystal-field splitting, an extremely complicated magnetic phase diagram, and the largest known Kerr angle [25].

The standard LSDA approach, where 4f-orbitals are treated as band states, fails to predict the ground-state properties: the value of the equilibrium magnetic moment, the additional orbits found in de Haas-van Alphen experiments, and the small density of states at the Fermi level [26]. But the exclusion of 4f-orbitals from the basis set (treating them as completely localized pseudo-core states) also did not

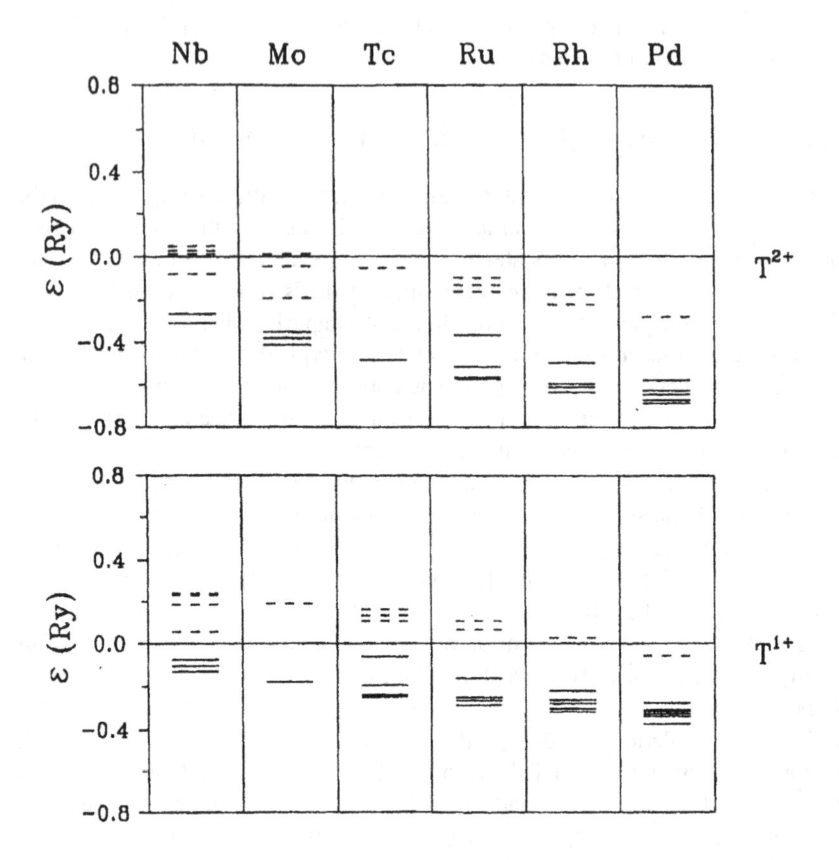

Figure 2: *Positions of the one-electron levels with respect to the Fermi energy obtained in the framework of LDA+U for two valence state (T^{2+} and T^{1+}) of 3d-impurities in Rb. The solid (dashed) lines were used for the states which were supposed to be occupied (empty) [24].*

lead to a satisfactory description of electronic and magnetic properties of CeSb. On the other hand, an empirical p-f model with a localized 4f -level 2–3 eV below E_F *hybridizing* with the conduction-band electrons, was able to describe CeSb qualitatively right [27].

The results of the LDA+U calculation for CeSb [28] gave some support to the above mentioned empirical p-f model. Without spin-orbital coupling, it was found that the Hartree-Fock-like one-electron 4f spin-up states with predominantly $m = 3$ and -3 character have the lowest total energy. This is in agreement with the first and second of Hund's rules. One should note that the symmetry of such orbitals is not cubic any more and Jahn-Teller tetragonal distortions give a lower

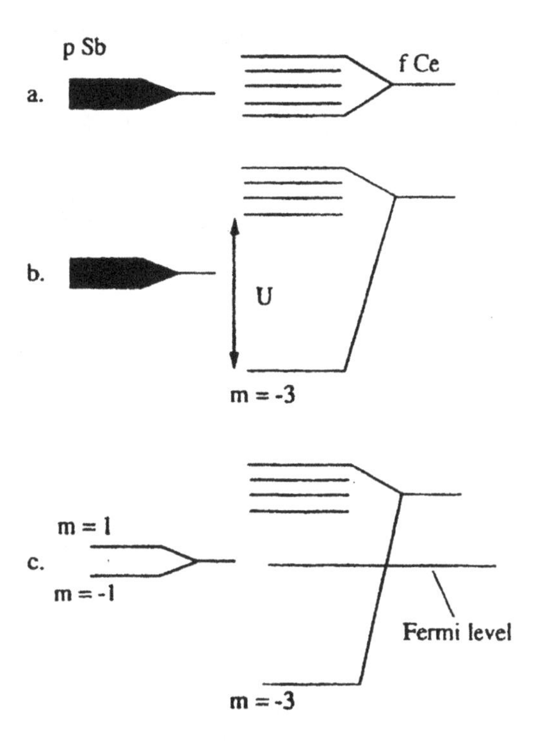

Figure 3: *Schematic picture of Hubbard-induced anisotropy formation in LDA+U scheme for CeSb: (a) normal LDA structure of p- and f-bands in CeSb; (b) a shift of one f-state of Ce (predominantly with m=-3 character); (c) the results of strong p-f hybridization and the creation of a highly polarized structure of p-states of Sb [21].*

energy. The spin-orbital coupling (~ 0.5 eV) lifts the degeneracy of the $m = \pm 3$ states according to the third Hund's rule and the lowest energy corresponds to the $| -3 \uparrow\rangle$ one-electron state (the spin and orbital moments are equal to -0.92μ_B and 2.86μ_B yielding a total magnetic moment of 1.94μ_B , which is close to the experimental value obtained for the antiferromagnetic ground state [(2.10±0.04) μ_B] [25] (Fig. 3).

The band structure of CeSb obtained in the LDA+U calculation (Fig. 4) has f bands split by approximately 6 eV, and the singly occupied f-band is located at 2 eV below the Fermi level. All unoccupied f bands are at approximately 4 eV above E_F and the broad bands which cross the Fermi level are formed by Sb p-states. There is a large reduction of the density of states at the Fermi level with LDA+U value of $N(E_F)$=6.5 states/Ry^{-1} compared to the LSDA value of about

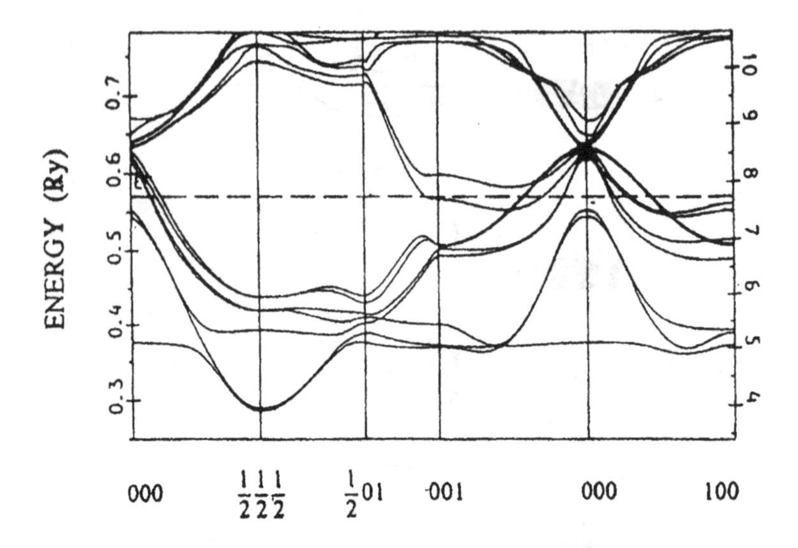

Figure 4: *Band structure of ferromagnetically ordered CeSb. Both Rydberg and electronvolts scales are shown [28].*

150 states/Ry^{-1}. The occupied f-band with mostly m =-3 character interacts in a very anisotropic way with Sb p-bands and even pushes one of the p-states (mostly of m =1 character) above the Fermi level along the $\Gamma - Z$ direction, but not in the $\Gamma - X$ direction. This electronic structure leads to an anisotropic Fermi surface (FS) which almost coincides with the FS in the p-f model [27]. The anisotropic p-f interaction helps explain the anomalous magnetic properties of CeSb with strong magnetic anisotropy in the ferromagnetically ordered phase (the calculated value of magnetic anisotropy as the total-energy difference with the magnetic field along [001] and [110] directions is 2.4 meV (standard LSDA calculations give 0.54 meV)). Such anisotropic p-f mixing for different m subbands near the Fermi level, together with the large spin-orbital coupling of Sb p states (~0.6 eV), leads to particularly strong magneto-optical effects.

4.4 $PrBa_2Cu_3O_7$

Another case where the hybridization of the nearly localized f-orbitals with the band states near the Fermi level leads to anomalous effects, is $PrBa_2Cu_3O_7$. Among all rare earths (RE) that form the $ReBaCu_3O_7$ structure only $PrBa_2Cu_3O_7$ is nonmetallic and nonsuperconducting. The only satisfactory model which explains qualitatively this puzzle is that of Fehrenbaher and Rice [29] which assumes a hole depletion in the CuO_2 planes due to the transfer of holes from the Cu-O pdσ band

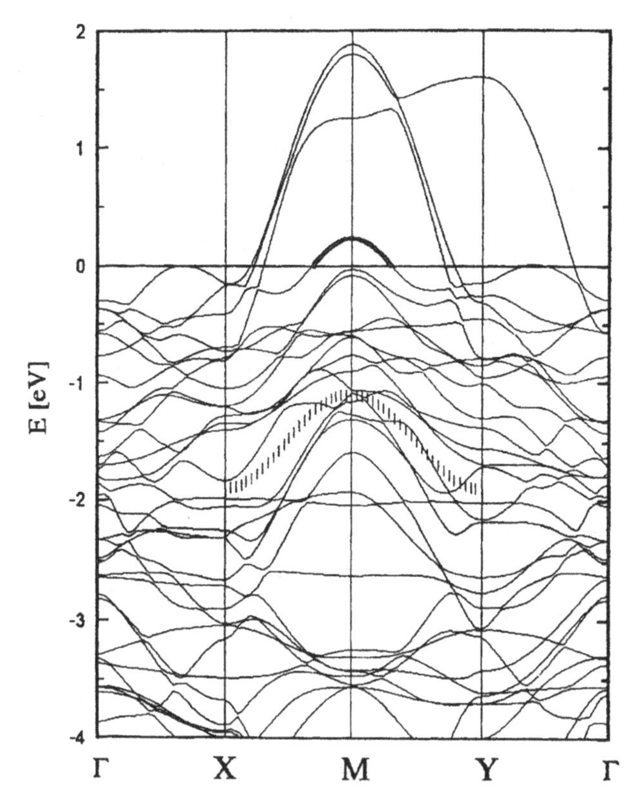

Figure 5: *LDA+U electron band structure of $PrBa_2Cu_3O_7$ for the majority spin. Fat line shows the depleted part of the Fehrenbacher-Rice band. Hatched strip indicates the position of this band in the spin-minority channel (or in $YBa_2Cu_3O_7$) [30].*

into p-f hybridized state which is a mixture of $4f^1$ and $4f^2L$ configurations (L is a ligand hole in the O2p orbital with $z(x^2 - y^2)$ symmetry around the Pr site and distributed over the eight nearest oxygen sites).

The results of the LDA+U calculation for $PrBa_2Cu_3O_7$ [30] has fully confirmed the Fehrenbaher and Rice model. In this calculation, the correlation correction (Eq. 9) was applied to the Pr4f-orbitals only but not to Cu3d-orbitals. The reason for this is that while LDA+U correction to potential acting on d-orbitals is necessary and sufficient for describing the insulating state of transition metal oxides with a partially filled d-shell [11], the transition from antiferromagnetic insulator to paramagnetic metal with doping by holes is beyond the scope of the mean-field approximation. This "Pr-U only" calculation was done to describe the transfer of holes from the Cu-O pdσ band into p-f hybridized state.

For $PrBa_2Cu_3O_7$ the LDA+U calculations yield occupied $f^{\uparrow}_{z(5z^2-3)}$ and $\sqrt{1-\alpha^2_{Pr}}f^{\uparrow}_{z(x^2-y^2)}+\alpha_{Pr}L^{\uparrow}$ bands ($\alpha_{Pr}\sim 0.2$), and a partly occupied antibonding $\alpha_{Pr}f^{\uparrow}_{z(x^2-y^2)}+\sqrt{1-\alpha^2_{Pr}}L^{\uparrow}$ band with a cylindrical hole pocket around the SR line $(\pi/a, \pi/b, k_z)$ (Fig 5). This partly occupied band grabs holes which normally would be in Cu-O pdσ band thus reducing its effective doping and making its occupation closer to half-filling where stable electronic state is known to be antiferromagnetic insulator. In $NdBaCu_3O_7$, the on-site $f^{\uparrow}_{z(x^2-y^2)}$ electron energy is so low that the top of the antibonding $\alpha_{Nd}f^{\uparrow}_{z(x^2-y^2)}+\sqrt{1-\alpha^2_{Nd}}L^{\uparrow}$ band ($\alpha_{Nd}<\alpha_{Pr}$) has fallen to $\sim$ 0.3 eV below the Fermi energy E_F. A relatively small doping with Pr could, however, push the top of this band partially above E_F.

5 Mott-Hubbard insulators

5.1 3d-transition metal oxides

The electronic structure of the rare-earth metal compounds is a relatively simple problem due to the weak hybridization of 4f-orbitals with the conduction band states. The advantages of the simultaneous treatment of the localized and delocalized electrons in the LDA+U method are seen most clearly for the transition metal compounds, where 3d-electrons, while remaining localized, hybridize quite strongly with other orbitals. Late-transition metal oxides, for which LSDA results strongly underestimate the energy gap and magnetic moment values (or even giving qualitatively wrong metallic ground state for the insulators CoO and $CaCuO_2$), are well described by the LDA+U [11] (see Table 1).

In Figs. 6-8 the partial densities of states (DOS) of $CaCuO_2$, NiO and CoO are shown. The DOS of FeO and MnO are similar to that of CoO, except for the growing number of unoccupied t_{2g} bands. First, focusing on the unoccupied density of states of NiO (Fig. 7), one can see that all the weight is concentrated in the narrow $e_{g\downarrow}$ peak, in agreement with the experimentally observed $d^8 \longrightarrow d^9$ peak [31]. In CoO (Fig. 8) the $3d_{xy\downarrow}$ orbital is emptied too, and this band is located at $\sim$ 0.5 eV lower energy. This crystal field splitting of the unoccupied d band is also found experimentally [32]. Comparing now the unoccupied DOS of CoO or NiO with that of $CaCuO_2$, we found that the width of the $3d_{x^2-y^2\downarrow}$ band of the cuprate is larger by a factor of 4–5 compared to that of the rocksalt oxides. As a result, a sharp $d^9 \longrightarrow d^{10}$ peak is missing, which is in striking agreement with experiment [32]. This is obviously related to the formation of a broad Cu $3d_{x^2-y^2}$$-O2p$ band caused by the relatively small in-plane Cu-O bond length and a Cu-O-Cu bond angle of 180°. In CuO, on the other hand, the bond angles are much smaller (between 96° and 146°) so that two neighboring Cu $3d_{x^2-y^2}$ orbitals hardly couple via the same ($2p_x$ or $2p_y$) O orbital. One expects thus a strong decrease

Table 1: *Experimental (exp) and calculated (LDA+U and LSDA) spin moment (m in μ_B) and energy gap (E in eV) values of the late-3d-transition metal oxides.*

	E_{LSDA}	E_{LDA+U}	E_{exp}	m_{LSDA}	m_{LDA+U}	m_{exp}
$CaCuO_2$	0.0	2.10	1.5	0.0	0.66	0.51
CuO	0.0	1.9	1.4	0.0	0.74	0.65
NiO	0.2	3.1	4.3,4.0	1.0	1.59	1.77,1.64,1.90
CoO	0.0	3.2	2.4	2.3	2.63	3.35,3.8
FeO	0.0	3.2	2.4	3.4	3.62	3.32
MnO	0.8	3.5	3.6-3.8	4.61	1.67	4.79,4.58

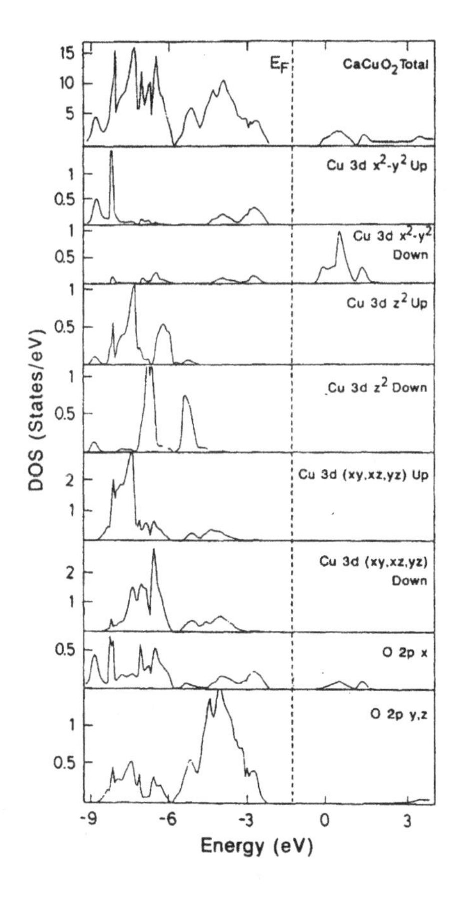

Figure 6: *The total and partial densities of states of Cu and O in $CaCuO_2$. O $2p_x$ and O $2p_{y,z}$ refer to the oxygen orbitals pointing towards and perpendicular to the Cu $3d_{x^2-y^2}$ orbitals, respectively [11].*

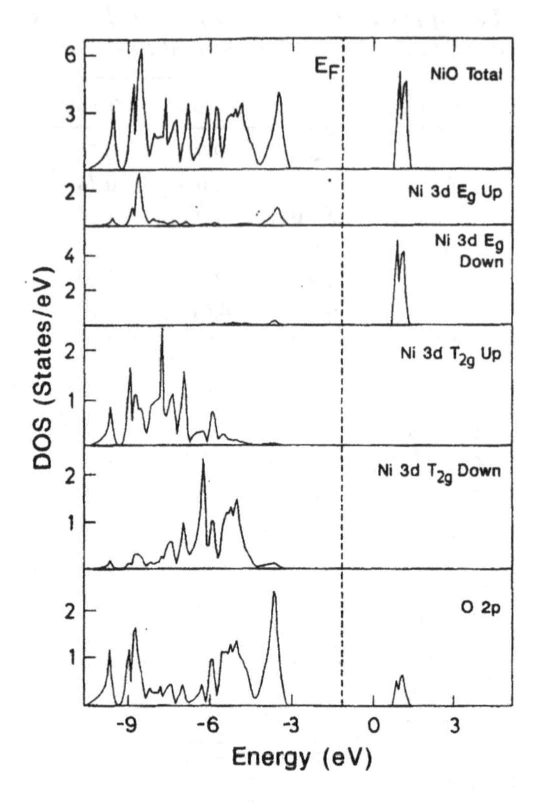

Figure 7: *The total and partial densities of states of NiO [11].*

of the bandwidth in going from $CaCuO_2$ to CuO [34], despite the similarity of the two systems on a local level [35]. In Fig. 9, the results of the LDA+U calculation for the DOS of CuO are shown. The CuO crystal structure has four equivalent Cu atoms per unit cell. Experimentally, the unoccupied DOS of CuO is characterized by a relatively sharp peak corresponding to the unoccupied *d* band, which is in strong contrast to the "blurred" unoccupied DOS of the high-T_c cuprates [36]. The LDA+U results suggest that this difference comes from the smaller bandwidth in the former (Fig. 9).

In the case of NiO, the LDA+U results are less conventional. In the past the peak at the top of the occupied valence band (corresponding to the lowest binding energy peak (BE) in the photoemission spectrum) has been ascribed to the high-spin d^{n-1} state (additional photoelectron hole in $t_{2g\downarrow}$ orbital) and the higher BE shoulder to the low-spin state (photoelectron hole in $e_{g\uparrow}$ orbital) [37], and this is also the outcome of several many-body-model calculations [14, 38]. According to LDA+U

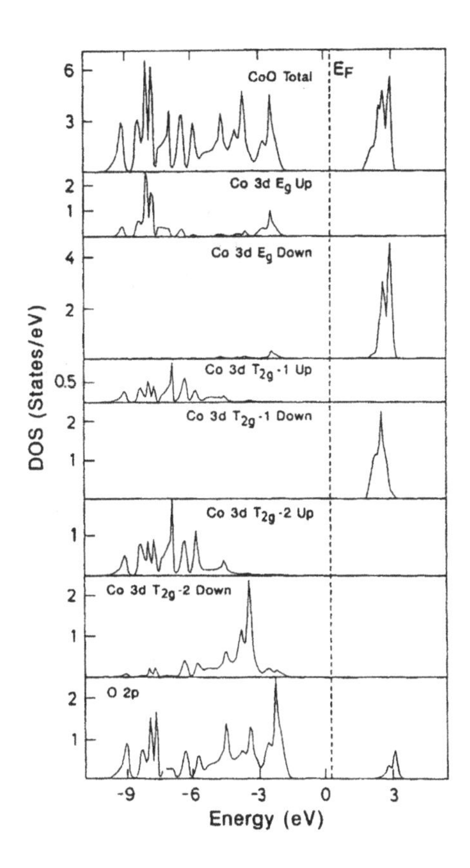

Figure 8: *The total and partial densities of states of CoO. Notice the inequivalence of the t_{2g} orbitals, due to the orbital dependence of the LDA+U potentials [11].*

calculations, this high-spin-low-spin identification has to be reversed. The peak on the top of the valence band in NiO is clearly of $e_{g\uparrow}$ character and of the same sort as the $3d_{x^2-y^2\uparrow}$ in the cuprates. This low-spin nature of the lowest ionization state of NiO is in agreement with experiment. This follows unambiguously from doping experiments. NiO can be doped with a large concentration of Li, and the Ni(III) compound $LiNiO_2$ is especially well characterized [39]. In this compound, every second (111) plane of Ni is replaced by Li and the local environment of Ni ions barely changes. $LiNiO_2$ is thus from a local perspective representative for NiO. According to x-ray absorption spectroscopy (XAS) data, the additional holes (introduced by Li doping) have $O2p$ character [40]. Further, $LiNiO_2$ is a low-spin ($S=\frac{1}{2}$) material [41]. The many-body interpretation is as follows: the added hole

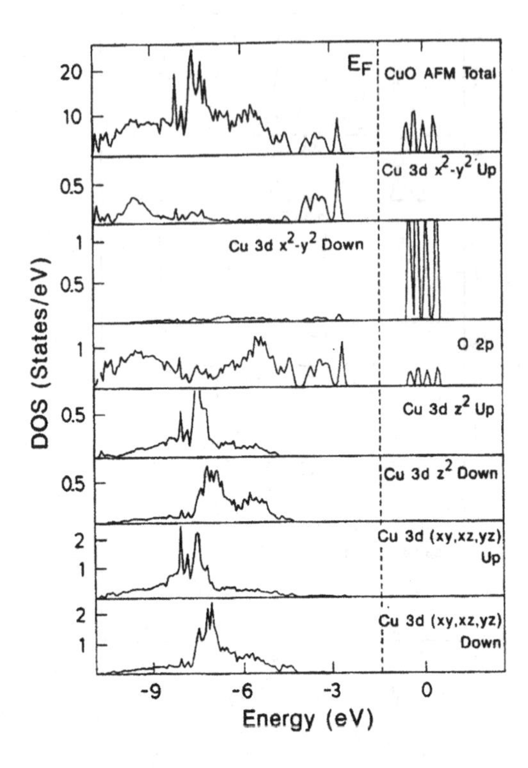

Figure 9: *The total and partial densities of states of CuO. Compared to $CaCuO_2$, the unoccupied,as well as the occupied local singlet, bandwidth is decreased [11].*

goes predominantly in the oxygen band and it gets antiferromagnetically bound to the Ni spin.

The results of LDA+U calculations for $LiNiO_2$ [11] are shown on Fig. 10 for the most stable (ferromagnetic, ferro-orbital-ordered) ground-state configuration. Compared to NiO, there are some similarities. One still can see a rather narrow $e_{g\downarrow}$ unoccupied $3d$ band at roughly the same position as in NiO, relative to the first occupied state. The new aspect is that a new unoccupied band of predominantly $O2p$ character is found inside the "NiO" gap, which is centered just above E_F. This is the same pattern as found by Kuiper *et al.* [11] in their XAS data. The magnetic moments values are 1.30 μ_B and -0.15 μ_B at the Ni and O site, respectively, and the net moment per NiO_2 unit is therefore exactly 1 μ_B (S=$\frac{1}{2}$).

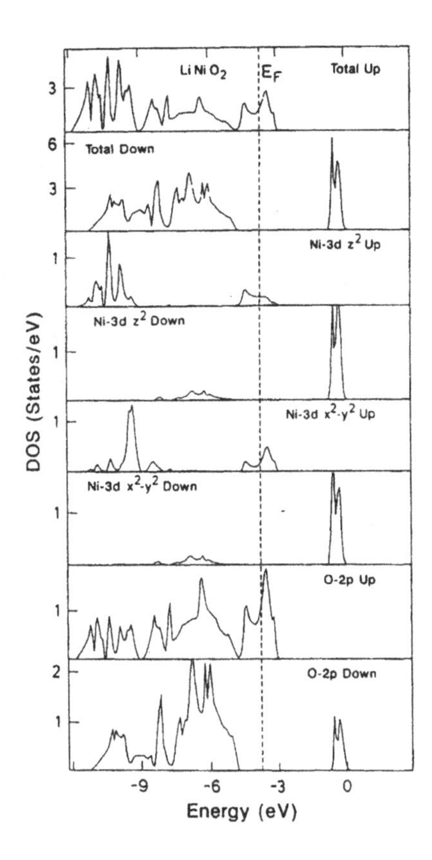

Figure 10: *The total and partial densities of states of* $LiNiO_2$ *[11].*

5.2 Transition metal impurities in insulators: Fe-impurity in MgO

The description of the electronic state of 3d-impurities in insulators is another example where Coulomb interactions inside the d-shell must be properly taken into account in order to cure the deficiency of the LSDA. The ground state of an Fe impurity in MgO [42] is $^5T_2$ (configuration $t_{2g}^4 e_g^2$) thus showing a high-spin magnetic Jahn-Teller ion. However an LSDA supercell calculation results in a nonmagnetic solution with configuration $t_{2g}^6 e_g^0$. The reason for this is that in a magnetic configuration the Fermi level of the LSDA solution crosses two narrow peaks (spin-down t_{2g} and spin-up e_g) and such a solution would therefore be energetically unfavorable. The orbital-dependent potential of the LDA+U method

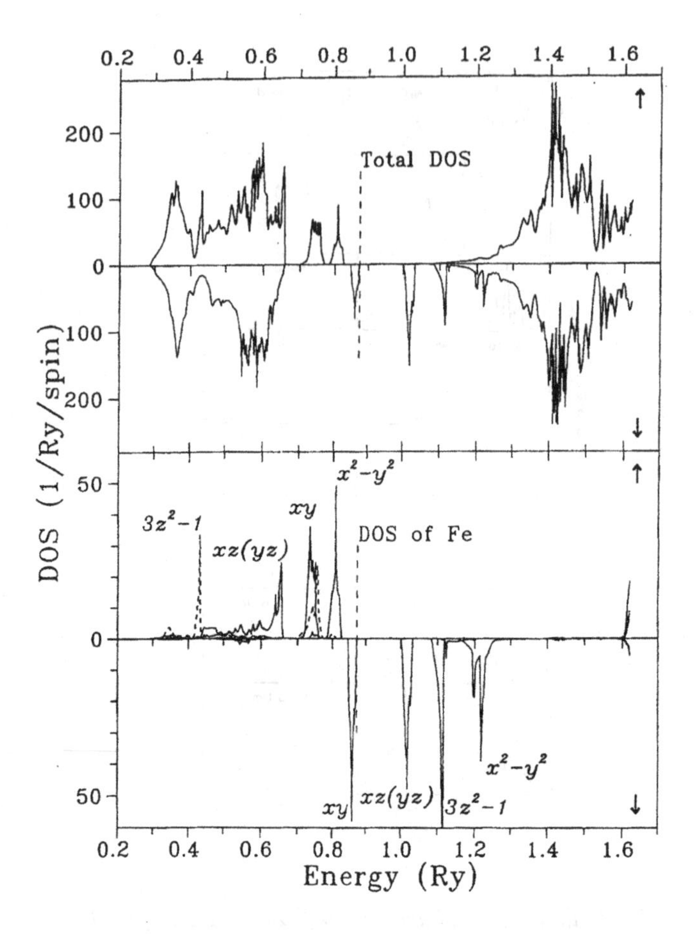

Figure 11: *Spin-resolved total density of states (DOS) and local orbital-resolved DOS at the Fe site in $FeMg_7O_8$ for the optimized geometry (LDA+U result) [42].*

splits partially occupied bands and results in a stable insulating magnetic solution with configuration $t_{2g}^4 e_g^2$ (Fig. 11).

The configuration $t_{2g}^4 e_g^2$ has a partially filled spin-down t_{2g} band and it is known that Fe-impurity in MgO exhibits a dynamic Jahn-Teller behavior. The optimization of the lattice around an impurity atom in the LDA+U calculation agreed quite well with this fact: the total energy has a minimum as a function of tetragonal distortion of the O_6 octahedron for the value of 0.5% of the lattice constant.

Although substitutional Fe in MgO exists mostly as a 2+ ion, the Fe^{3+} configuration is also known to exist. Most probably Fe^{3+} ions are formed due to the trapping of holes at Fe^{2+} sites. A supercell LDA+U calculation with one electron less than stoichiometric value indeed led to the hole localized on impurity site with Fe ion in Fe^{3+} state (high-spin $t_{2g}^3 e_g^2$ configuration).

5.3 Linear chain (MX) compounds

The halogen-bridged transition-metal linear-chain compounds, referred to as MX compounds because of their alternating transition-metal atoms M and halogen atoms X, form weakly coupled linear-chain-like structures and are of considerable interest due to a rich phase diagram. They exhibit a charge-density wave(CDW) with dimerization distortion as well as a spin-density wave (SDW) or even a peculiar mixture of them (spin-Peierls state) [43].

While LSDA successfully describes CDW systems (for example Pt-based MX-compounds) it fails to give the correct antiferromagnetic insulator solution for Ni-based MX predicting instead a nonmagnetic metal. This problem has the same origin as for the undoped cuprate superconductor materials: the fact that in LSDA the magnetic transition is driven by the spin polarization of a Stoner intra-atomic exchange interaction I (about 1 eV), instead of the much stronger Hubbard interaction U (about 8 eV). Again, as was the case with $CaCuO_2$ and La_2CuO_4, use of the LDA+U functional gave the correct antiferromagnetic insulating ground state Ni-based MX-compound [44], while dimerization distortion did not lead to the total energy lowering as it was with the Pt-based systems. It is interesting to mention that due to the large ligand-field splitting the Ni^{3+} ion in these compounds has a low-spin ground state (configuration $t_{2g}^6 e_g^1$) (Fig. 12).

6 Electron-lattice interactions: Jahn-Teller distortions and polarons

We must emphasize here that in spite of the model Hamiltonian spirit in the above derivation of the LDA+U formula, it remains a *first-principles*, *ab-initio* method preserving its ability to calculate the lattice properties such as the ground state crystal structure,equilibrium volume and even phonon frequencies. The orbital-dependent potential of the LDA+U method (9) makes it possible to treat properly the orbital polarization and the corresponding lattice Jahn-Teller distortions and polarons [13, 45].

6.1 Cooperative Jahn-Teller distortions in transition metal compounds: $KCuF_3$

This was demonstrated for example in the perovskite $KCuF_3$ [13]. This compound is subject to a collective Jahn-Teller like distortion which is more complicated than the simple tetragonal distortion of the cuprates, involving a staggering of

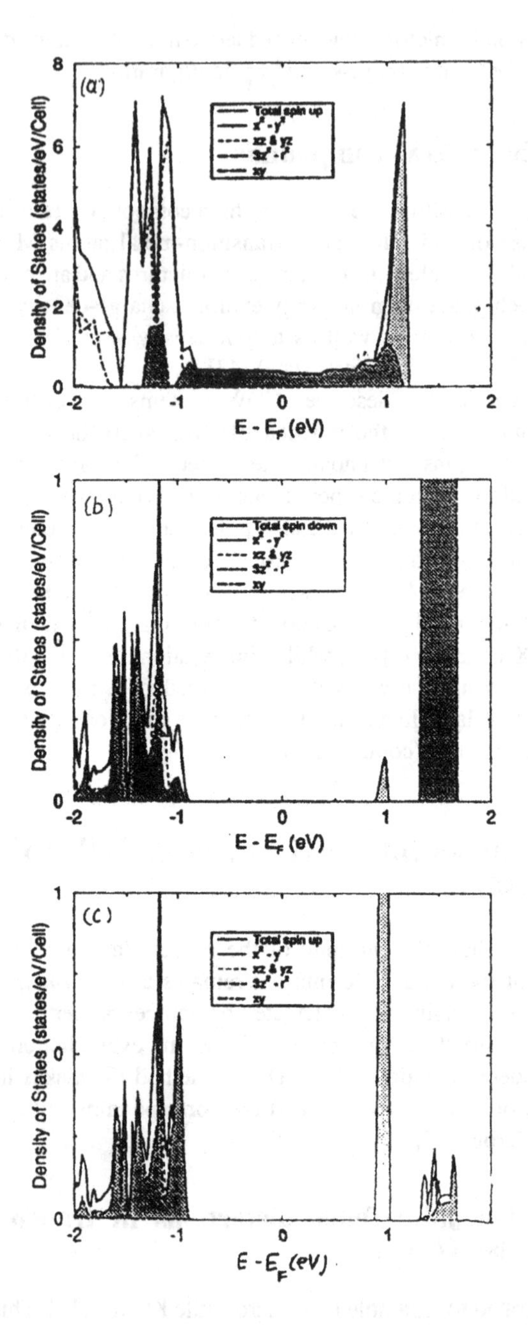

Figure 12: *The total and partial densities of states at Ni site for Ni based MX-chain compound. (a)- LSDA result; (b)-(c) LDA+U result [44].*

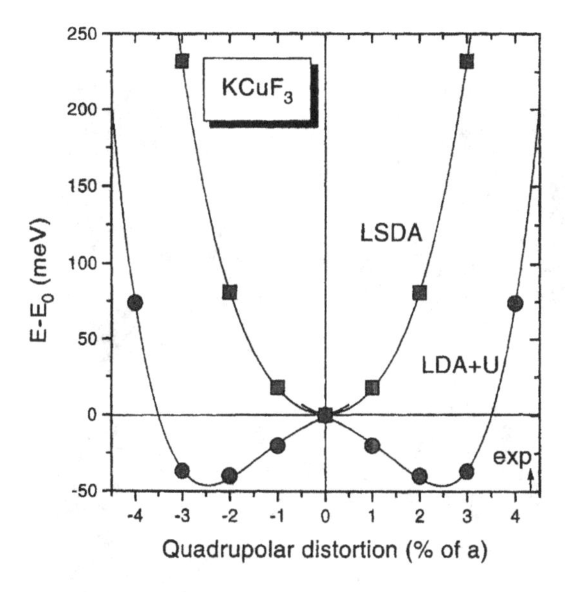

Figure 13: *The dependence of the total energy of $KCuF_3$ on the quadrupolar lattice distortion obtained in calculations with LSDA and LDA+U functionals [13].*

quadrupolar distorted CuF_4 units (2 short and 2 long Cu-F bonds) in the a-b planes. In a seminal work, Kugel and Khomskii [46] pointed out that this distortion is in the first instance electronically driven. They showed that the e_g $(x^2 - y^2, 3z^2 - 1)$ orbital degrees of freedom are, as the spins, subject to kinetic exchange interactions, while in addition, the spin- and orbital degrees of freedom are mutually coupled as well. Orbital ordering is found from these Hamiltonians involving a staggering of the orbitals in a-b directions ($\sim x^2 - z^2, y^2 - z^2$) and a ferromagnetic ordering of the spins. Because of this 'preexisting', electronically driven orbital polarization, any non-zero electron-phonon interaction then leads automatically to the observed lattice distortion [46].

$KCuF_3$ has the perovskite crystal structure with a slight tetragonal distortion (c/a ratio < 1) while the planes show the quadrupolar distortion. The spins are ferromagnetically ordered in the planes while the unit cell is doubled in the c-direction by antiferromagnetic spin ordering, so that the resulting unit cell contains four formula units. The prime subject of the investigation [13] was the quadrupolar distortion in planes which is directly connected with the peculiar orbital ordering. The total energy as a function of the shifts of the fluorine ions in the CuF_2 plane was calculated with the standard LSDA and LDA+U functionals (Fig. 13). The striking difference between the two calculation is that the LSDA solution has no instability against quadrupolar distortion while LDA+U curve has

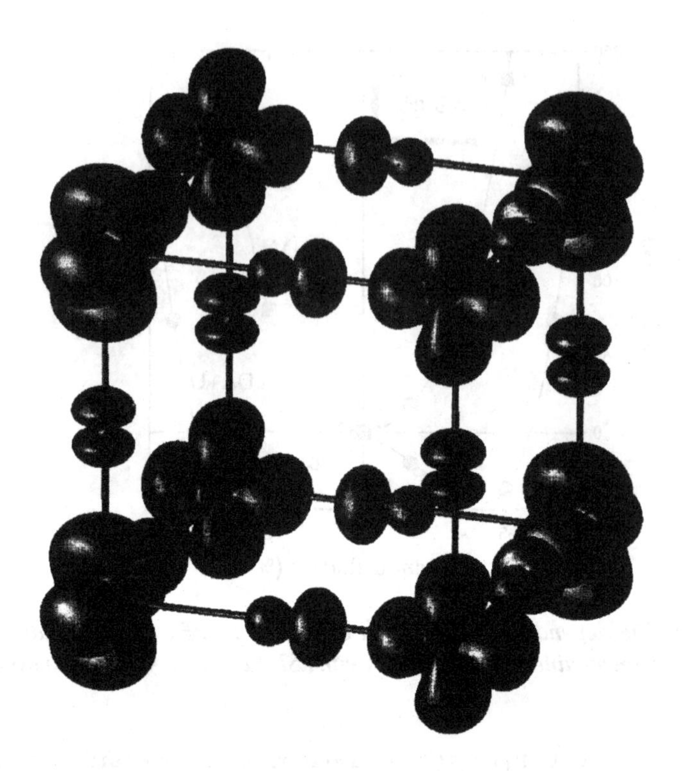

Figure 14: *The three-dimensional plot of the electron-spin-density distribution in* $KCuF_3$ *from the result of LDA+U calculation. Note that* x^2-y^2*- and* y^2-z^2*-like "d-orbitals" correspond to the spin density located at the copper atoms, while p-like density corresponds to fluorine ions [13].*

a minimum at Q=2.5% of a compared with the experimental value of 4.4%. It means that exchange-only and lattice-electron ("electron-phonon") interactions in LSDA are not enough to drive the observed orbital polarization and collective Jahn-Teller distortion. In order to be able to reproduce them, the orbital-dependent interaction terms must be included in the functional as is the case in the LDA+U. It is possible to directly observe the orbital ordering in $KCuF_3$ by plotting the 3-dimensional spin density obtained in the LDA+U calculation (Fig. 14). As there is only one hole in the d-shell of the Cu^{2+} ion, this spin density distribution gives the charge density of the holes. The picture agrees quite well with the orbital ordering of the alternating x^2-z^2 and y^2-z^2 Cu3d orbitals. We notice that the charge distribution changes only very little under the influence of the lattice distortion, emphasizing that this ordering is in the first instance of an electronic origin.

As expected, the electronic properties come out essentially correct. The 'Koopman theorem' gap in the LDA+U band structure should give an order of magnitude estimate for the single particle gap and was found to be similar to that in the cuprates (2 eV) [45]. Furthermore, the magnetic ordering is reproduced and to test the method more severely the magnetic exchange interactions were calculated as well. Using the Green function method to calculate the effective exchange interaction parameters as second derivatives of the ground state energy with respect to the magnetic moment rotation angle [47] in combination with Eq. (4–9), one obtains

$$J_{ij} = \sum_{\{m\}} I^{i}_{mm'} \chi^{ij}_{mm'm''m'''} I^{j}_{m''m'''}, \tag{47}$$

where the spin-dependent potentials I are expressed in terms of the potentials of Eq. (9),

$$I^{i}_{mm'} = V^{i\uparrow}_{mm'} - V^{i\downarrow}_{mm'}, \tag{48}$$

while the effective intersublattice susceptibilities are defined in terms of the LDA+U eigenfunctions ψ as,

$$\chi^{ij}_{mm'm''m'''} = \sum_{\mathbf{k}nn'} \frac{n_{n\mathbf{k}\uparrow} - n_{n'\mathbf{k}\downarrow}}{\epsilon_{n\mathbf{k}\uparrow} - \epsilon_{n'\mathbf{k}\downarrow}} \psi^{ilm^*}_{n\mathbf{k}\uparrow} \psi^{jlm''}_{n\mathbf{k}\uparrow} \psi^{ilm'}_{n'\mathbf{k}\downarrow} \psi^{jlm'''^*}_{n'\mathbf{k}\downarrow}. \tag{49}$$

It was found that the anti-ferromagnetic exchange in the CuF 'chains' amounts to $J_c = -20.7$ meV while the ferromagnetic exchange in the a-b planes is much smaller ($J_{ab} = 0.52$ meV), emphasizing the quasi 1D character of this S=1/2 spin system. This compares quite well with the neutron scattering measurements, showing the Luttinger liquid nature of the spin system, with the 1D exchange estimated to be $J_c = -17.5, -17.0$ meV and $J_{ab} = 0.17, 0.27$meV [48].

6.2 Polarons in doped Mott insulators: $La_{2-x}Sr_xCuO_4$ and $La_{2-x}Sr_xNiO_4$

LDA+U was designed to treat Mott-Hubbard insulators, but a real challenge is the problem of the theoretical description of the doped Mott insulator, the latter one being the model for high-T_c superconductors based on copper oxides. One important question in this field is the strength of the polaronic effects in these compounds.

In order to answer this question Sr doped (tetragonal) La_2CuO_4 was investigated [45]. The undoped system is quite well described by the LDA+U and a 2×2 supercell ($La_{8-x}Sr_xCu_4O_{16}$) was used to investigate the (self-) localization effects. At first, the problem of the hole in the undistorted lattice was considered to address the magnetic relaxation effects. Subsequently the interaction of the hole with both in-plane and out-of-plane lattice distortions was investigated. These calculations were then repeated for $La_{2-x}Sr_xNiO_4$.

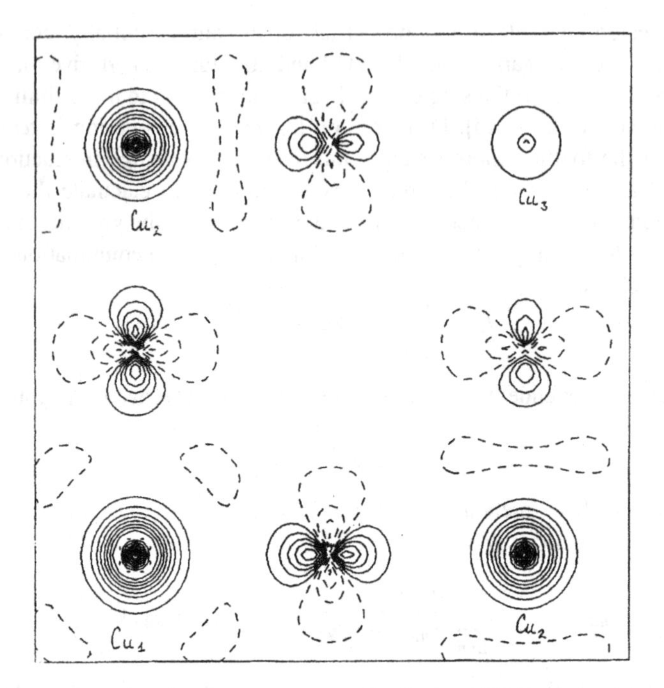

Figure 15: *The contour map of the hole density distribution in CuO_2 plane of La_2CuO_4 in supercell LDA+U calculation.The central copper atom (Cu_1) is in the left lower coner of the figure. The full lines correspond to the positive and dashed lines to the negative values of hole density.*

Except for the irrelevant La 4f states, the LDA+U electronic structure of undoped La_2CuO_4 looks very similar to that of $CaCuO_2$ (Fig. 6). According to these calculations, La_2CuO_4 is a charge-transfer insulator [49]. The lowest unoccupied states are Cu $3d_{x^2-y^2}$, while the lowest occupied states are more O $2p$ like. To see how well structural properties are handled, the frequency of the breathing-mode phonon was calculated. Using the frozen-phonon technique the calculated value was found to be $\omega_{br} = 660cm^{-1}$, quite close to the experimental value of 710 cm^{-1} [50].

It was found that the hole induces a parallel alignment of the moment of Cu_1 (central Cu atom in supercell) with those of Cu_2 (nearest neighbor). This local ferromagnetic "spin bag" in an antiferromagnetic background caused by the presence of a hole can be viewed as a mean-field analog of the Zhang-Rice singlet [51]: the hole is mostly localized on the four oxygens surrounding the central Cu_1, having its spin antiparallel to that of Cu_1 (Fig. 15 and Table 2). The hole induced states produce the peak in the former gap region which is close (0.15 eV) to the top of the valence band.

Table 2: *The dependence of the total energy (meV) (δE) and the magnetic moments (in μ_B) on the displacement (x) of either in plane- ('breathing', BR) or apical oxygens ('Jahn-Teller', JT) towards the central transition metal ion (TM_1) in the supercell (TM_2 is nearest- and TM_3 next nearest TM_1 neighbour), in the case of 'doped' La_2CuO_4 (LCO) and La_2NiO_4 (LNO).(For JT the total energy δE is measured relative to the energy of BR.)*

	x	δE	μ_{TM_1}	μ_{TM_2}	μ_{TM_3}
LCO	0%	0	−0.55	−0.59	0.72
	2% (BR)	−39	−0.43	−0.63	0.73
	11% (JT)	54	0.96	−0.64	0.73
LNO	0%	0	0.42	−1.58	1.67
	4% (BR)	−210	0.54	−1.58	1.67

In the study of lattice-polaronic effects two possible types of lattice deformations were investigated: 1) the 4 neighboring in-plane O-atoms were moved along the Cu-O bond-axis toward Cu_1, where the hole is localized ('breathing polaron'), 2) the motion of the 2 apical O's (O_{ap}) along the Cu_1-O_{ap} bond-axis was considered ('anti JT polaron'). As a function of the breathing distortion it was found that *the energy is a minimum for a finite distortion, corresponding to a 2% contraction of the Cu_1-O_1 bond.* The total energy is lowered by an amount of 39 meV, and the energy is lower than that of the undistorted lattice up to a distortion of 4%. The calculated polaron binding energy is small compared to the estimates for its kinetic energy [52] and small polaron effects are not expected.

It was found that the contraction of the Cu_1-O_{ap} bond increases the LDA+U total energy if the distortion is small. However, it was possible to find a local minimum in the total energy by allowing the additional hole to have $3z^2 - 1$ symmetry with respect to Cu_1, corresponding to a reduction of the Cu_1-O_{ap} bond-length of 0.26 A (11%), in remarkable agreement with the data of Egami et al. [53]. *The total energy of the anti-JT polaron is only 54 meV higher than that of the fully-relaxed breathing mode polaron.*

The above calculations were repeated for Sr doped La_2NiO_4. The undoped La_2NiO_4 was found to be a high-spin ($1.69\mu_B$) antiferromagnetic charge-transfer insulator with a p-d gap of 3.5 eV. The ground state is locally a $x^2 - y^2, 3z^2 - 1$ ($S = 1$) state, where the holes are rather strongly localized on Ni (10% $O2p$ admixture). Adding a hole to the supercell leads to an inhomogeneous state, which is qualitatively similar to the one in the cuprate. The additional hole has a large weight on the 4 in-plane O_1 atoms (nearest-neighbors to Ni_1), although the 3d admixture has increased compared to the cuprate. This state has $x^2 - y^2$ symmetry with respect to Ni_1 and the spin of the additional hole is anti-parallel to that of Ni_1, i.e. the hole is low spin. The $x^2 - y^2$ spins do not compensate exactly on Ni_1

($m_x = n_{x^2-y^2\uparrow} - n_{x^2-y^2\downarrow} = -0.30\mu_B$) and together with the larger polarization of the $3z^2-1$ hole ($0.70\mu_B$), give a net moment of $0.42\mu_B$ (Table 2). The additional hole is nearly entirely localized on Ni_1 and O_1. Hence, *the magnetic confinement effects are much stronger in the Nickelate than in the Cuprate*. This is not surprising, considering the larger gap and moment in the former. Then the breathing type lattice relaxation was studied, and *the stabilization energy of the breathing polaron was found to be 210 meV, 5 times larger than in the Cuprate.* The total energy is at minimum if the Ni_1-O_1 bond is contracted by 4% (Table 2). This large polaron binding energy could help to explain why doped nickelates are nonmetallic in contrast to cuprates.

7 The metal-insulator transition

It is well known that the mean-field approximation is too crude to be able to describe the metal-insulator transition in strongly correlated systems. However there are a few cases where LDA+U calculations can be relevant to this problem. These are FeSi [54] and $LaCoO_3$ [55], both of them being nonmagnetic insulators at low temperature and becoming metals with significant local magnetic moments at higher temperatures.

7.1 FeSi

FeSi displays an unusual crossover from a singlet semiconducting ground state with a narrow bandgap to a metal with an enhanced spin susceptibility and a Curie-Weiss temperature dependence in the vicinity of room temperature [56]. Various theoretical models have been put forward to explain this behavior starting with the very narrow band description of Jaccarino et al. [57]. Takahashi and Moriya [58] proposed a nearly ferromagnetic semiconductor model which predicted thermally induced spin fluctuations which were subsequently confirmed experimentally [59]. Recently, models based on treating FeSi as a transition metal analog of the Kondo insulators found in heavy fermion rare earth systems have been much discussed [60, 61].

The LDA electronic structure calculation by Mattheiss and Hamann [62] correctly accounts for the narrow gap semiconducting ground state but more is required to explain the anomalous behavior.

If one sets U=0 then the potential correction (9) vanishes and the LDA+U method becomes equivalent to the standard LDA. In Fig. 16a, the density of state (DOS) obtained in calculation with U=0 is shown. It is quite close to the results of the previous LDA calculations [62]. The Stoner parameter, I, is not strong enough to produce a magnetic state and the only stable solution is nonmagnetic. As one increases the value of U above the critical value, U_c=3.2 eV, a stable magnetic

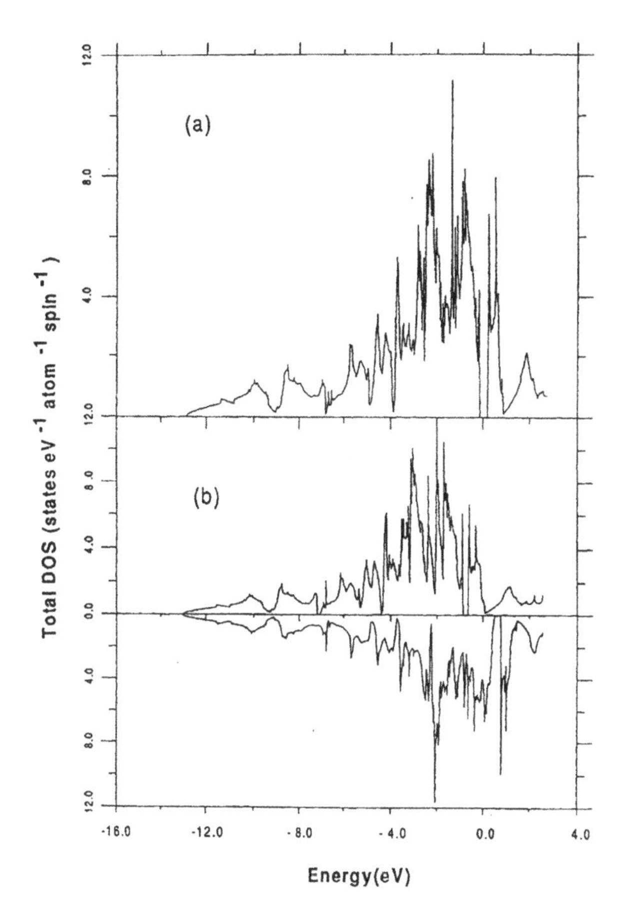

Figure 16: *Density of states (DOS) for FeSi from LDA+U calculations. The chemical potential is the zero energy. a)Nonmagnetic state with U=0; b)Majority and minority spin bands in a ferromagnetic state with a moment of 1 μ_B [54].*

solution appears with a magnetic moment on each Fe site, $\mu = 1\mu_B$ (Fig. 16b). The nonmagnetic solution is still present and has a total energy of $\approx$0.3 eV/Fe lower than the magnetic one. With further increase of U the total energy of the magnetic solution decreases relative to the nonmagnetic one and for $U > 4.6$ eV it is lower in energy and so becomes the ground state.

A so called "fixed spin moment" calculation was also performed where total energy $E_g(\mu)$ is calculated as a function of the magnetic moment, μ, on each Fe site (Fig. 17). One can see that for $U < U_c$ there is only one local minimum in the curve corresponding $\mu = 0$. Near $\mu \approx 1$ there is only a bend in the curve but

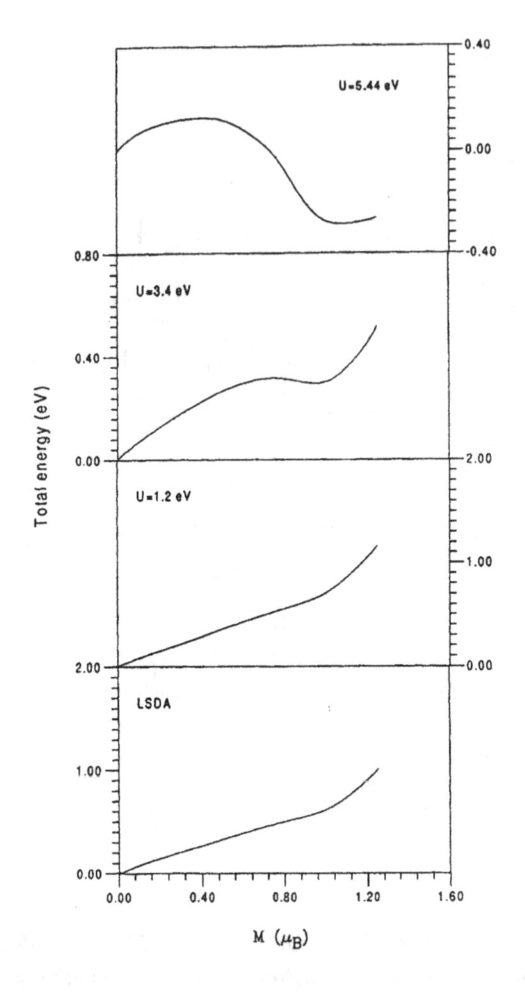

Figure 17: *Total energy of FeSi as a function of the spin moment M (μ_B/Fe) with various values of U [54].*

no minimum. For $U = 3.4$ eV a second local minimum appears in the curve but it lies higher in energy than the zero-moment minimum. However for $U = 5.4$ eV, the minimum corresponding to the magnetic solution is clearly lower than for the nonmagnetic one.

The existence of the second local minimum in $E_g(\mu)$ leads to a first order transition in an external magnetic field. Although the magnetic moment ($\mu = 1\mu_B/Fe$) of the ferromagnetic state is insensitive to the choice of U, the critical field, B_c, that determines the transition is very sensitive to U. For example for the choice $U = 3.4$ eV corresponding to Fig. 17, B_c is very large ($\sim 10^3 T$), but for $U \geq 4.6$ eV, $B_c = 0$ and the magnetic solution is the most stable. It is clearly not possible to make an accurate a priori estimate of B_c. The authors [54] resorted to simpler model calculations guided by the a priori calculations which included the effect of finite temperature and then adjusted the model parameters to obtain agreement with the measured spin susceptibility $\chi(T)$ and specific heat $C_p(T)$. They obtained the critical temperature value T_c=280K and critical field B_c=170T.

7.2 $LaCoO_3$

Among the systems showing a semiconductor to metal transition, $LaCoO_3$ is especially interesting due to the fact that it also displays very unusual magnetic behavior, often associated with a low-spin (LS) — high-spin (HS) transition [63]. Although a large number of investigations have been carried out since the early 1960's the character of the transition and the nature of the temperature dependence of the spin state is still unclear. For example, the temperature dependence of the magnetic susceptibility shows a strong maximum at around 90 K followed by a Curie-Weiss-like decrease at higher temperatures [64] which was interpreted by the authors as a LS to a HS transition. The semiconductor to metal-like transition occurs in a range of 400 – 600 K, well above this transition.

The electronic structure of $LaCoO_3$ was studied [55] in the LDA+U approach. In contrast to the standard LDA several stable solutions were found corresponding to different local minima of the LDA+U functional. The nonmagnetic insulating low-spin state (d-shell configuration of Co ion $t_{2g}^6 e_g^0$) is a ground state, in agreement with experiments. The unexpected result was that while the high-spin state (configuration $t_{2g}^4 e_g^2$) lies rather high in energy, two other orbitally-polarized magnetic solutions (configuration $t_{2g}^5 e_g^1$) corresponding to intermediate-spin (IS) states (one of them is a gapless semiconductor and the other is a metal) have total energy values only slightly higher than the LS ground state (for crystal structure parameters corresponding to the temperature 4 K). The total energy difference between the LS and IS states is very sensitive to the lattice constant. With increasing lattice constant the energy of the IS solution becomes smaller than the total energy of the LS solution thus giving the IS solution as a ground state. This crossover occurs at the lattice constant corresponding to $\simeq$150 K. The authors of [55] suggested the following interpretation of the transition in $LaCoO_3$. According to their scheme, with increasing temperature, at first a transition from a LS (nonmagnetic) insulating ground state to a state with an IS (configuration $t_{2g}^5 e_g^1$) occurs. Due to the strong Jahn-Teller nature of this configuration, this state may develop an orbital ordering. The orbitally-ordered state turns out to be nonmetallic (actually a nearly zero-gap semiconductor) in the LDA+U calculations. With the further increase of the temperature the orbital ordering may be gradually destroyed which can explain the transition to a metallic state observed in $LaCoO_3$ at 400 – 600 K.

8 Charge and orbital ordering

8.1 Magnetite Fe_3O_4

Magnetite Fe_3O_4 is a mixed valence $3d$ transition metal compound. It crystallizes in an inverted cubic spinel structure in which the tetrahedral A-sites contain one third of the Fe ions as Fe^{3+}, while the octahedral B-sites contain the remaining Fe ions, with equal numbers of Fe^{3+} and Fe^{2+} in B1- and B2-sites, respectively. Below 860 K, magnetite is ferrimagnetic with A-site magnetic moments aligned antiparallel to the B-site moments. At T_V=122K Fe_3O_4 undergoes a first order phase transition, the so-called Verwey transition [65], in which dc conductivity abruptly increases by two orders of magnitude on heating through T_V. Verwey interpreted the transition as an order-disorder transformation of Fe ions on the B-sites. Indeed, studies by electron and neutron diffraction, and nuclear magnetic resonance [66] show that below T_V the B1- and B2-sites are structurally distinguishable in a distorted crystal structure. Photoemission measurements clearly show a gap of $\simeq$0.14 eV in the spectra [67]. However a band structure calculation using the Local Spin-Density Approximation (LSDA) [68] gave only a metallic solution without charge ordering with partially filled band (containing one electron per two B-sites) originated from t_{2g} spin-down $3d$-orbitals of Fe ions in octahedral B-sites.

The problem of charge ordering cannot be treated by the standard LSDA. The reason for that is a spurious self-interaction which is present in the LSDA. In contrast to the Hartree-Fock approximation where self-interaction is explicitly excluded for every orbital, in the LSDA it is canceled only in the total energy integrals but not in one-electron potentials which are orbital-independent. The spurious self-interaction present in the LSDA leads to an increase in the Coulomb interaction when the distribution of the electron charge deviates from the uniform one. This effect can be illustrated in the following way. If one neglects inter-site Coulomb interaction then the electron under consideration feels the same potential on all sites independently of the occupancy of the particular site, as it does not interact with itself. However, as the LSDA potential is a functional of the electron density only, then, increasing the electron density on one site and decreasing it on another one, with a formation of the charge ordering, will lead to an increase of the potential on the first site and a decrease on the second one. As a result, in the self-consistency loops, the charge distribution will return to the uniform density.

In order to cure this deficiency, it is necessary to remove the spurious self-interaction. Formally LDA+U does it but this method was constructed for Mott insulators and for systems with a charge ordering it must be modified by taking into account intersite Coulomb interaction. In order to do it one must map the dependence of the Coulomb interaction energy on the number of t_{2g}-electrons in the LSDA on a model with on-site and inter-site terms, and then explicitly exclude the self-interaction on-site term [69].

If one defines n_i as a sum of the occupancies of t_{2g} orbitals $(n_{xy} + n_{xz} + n_{yz})$ for minority spin direction on B-site i then the model which imitates the LSDA is:

$$E[n_i] = \sum_i \{\frac{1}{2} U n_i (n_i - 1) + V \sum_j n_i n_j\} \tag{50}$$

(The index j numerates the neighbors of the site i.) In order to compute the value of the parameters U and V one must perform a constrained calculation for the two types of charge ordering where occupancies n_i on different sites are varied with the total number of t_{2g} electrons conserved. As a first type of charge ordering, the order suggested by Verwey [65] was chosen. It can be described as a lattice built from the neutral tetrahedra where every tetrahedron contains two atom from the B1 sublattice and two atoms from the B2 sublattice. The second type of charge ordering corresponds to charged tetrahedra where one of them contains only B1-type and another only B2-type.

Having determined U and V parameters one can now define new functional without a self-interaction by subtracting a $\frac{1}{2} U n_i (n_i - 1)$ term from the LSDA-functional. The real on-site Coulomb interaction energy is small due to the small probability to meet two t_{2g} electrons on the same site but it is nevertheless nonzero and it was taken into account by adding the corresponding term in the Hartree-Fock approximation:

$$E = E_{LSDA} - \frac{1}{2} \sum_i \{U n_i (n_i - 1) - \sum_{m,m' \neq m} U n_{i,m} n_{i,m'}\} \tag{51}$$

(here m, m' denotes different t_{2g}-orbitals for spin-down electrons of Fe ions on octahedral B-sites).The corresponding orbital dependent potential $V_{i,m}$ is given by the variation of the new functional (6) with respect to the occupancy of the particular t_{2g} -orbital $n_{i,m}$:

$$V_{i,m} = V_i^{LSDA} + U \left(\frac{1}{2} - n_{i,m} \right)$$

Constrained calculations with two types of charge ordering gave the following parameters: U=4.51 eV and V=0.18 eV.

Electronic structure calculations with a functional in Eq. (51) were performed for the Verwey type of charge ordering. In contrast to the LSDA, where the stable solution is a metal with a uniform distribution of the t_{2g}-spin-down electrons on the octahedral sites, self-interaction corrected functional (51) gave a charge ordered insulator with an energy gap of 0.34 eV (experimental value is 0.14 eV [67]).

According to the ionic model, charge ordering means Fe^{+3} and Fe^{+2} ions on octahedral sites sublattices B1 and B2 with configurations d^5 $(t_{2g\uparrow}^3 e_{g\uparrow}^2)$ and d^6 $(t_{2g\uparrow}^3 e_{g\uparrow}^2 t_{2g\downarrow}^1)$ respectively. In the actual calculations due to the strong covalency effects the number of d-electrons in atomic spheres were 5.91 and 6.23 with the charge difference 0.32 instead of the pure ionic value of 1.0 (however the difference in the occupancy of $t_{2g\downarrow}$, orbital for two sublattices is larger: 0.70).

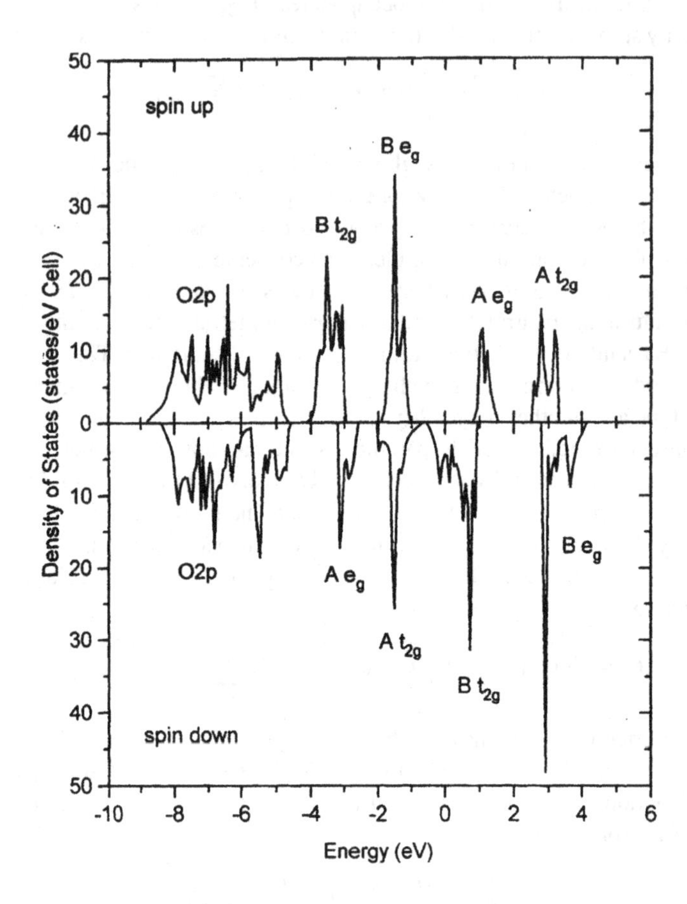

Figure 18: *The density of states for Fe_3O_4 in LSDA calculation. A - tetrahedral coordinated Fe-ions, B - octahedral Fe-ions [69].*

In Fig. 18, the densities of states (DOS) for Fe_3O_4 obtained in standard LSDA calculations are presented and in Fig. 19 the DOS calculated with the use of the functional in Eq. (51). For the LSDA one can see an oxygen band between -8 eV and -4 eV and above it (for both spin-up and spin-down DOS) four bands of Fe3d origin. For spin-up states the sequence is t_{2g} and e_g bands of iron in octahedral B-sites and above it e_g and t_{2g} bands of iron in tetrahedral A-sites. For spin-down states the order of the octahedral and tetrahedral sites bands is reversed so that the first two d-bands are e_g and t_{2g} bands of iron in tetrahedral sites (A) and above them are t_{2g} and e_g bands of iron in octahedral sites (B) with the Fermi energy lying in t_{2g} band.

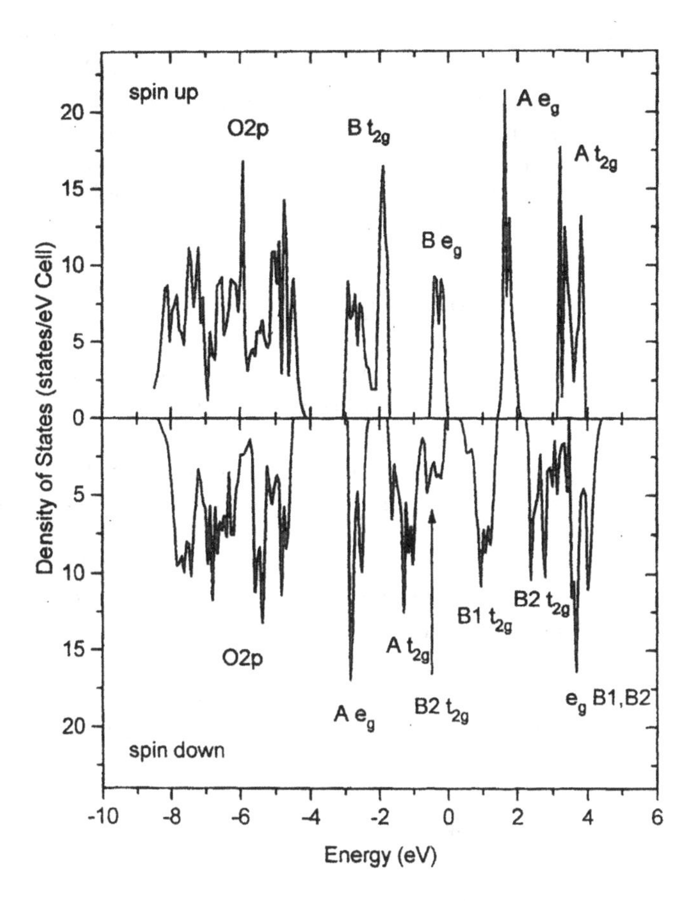

Figure 19: *The density of states for Fe_3O_4 in LDA+U calculation. A - Tetrahedral coordinated Fe-ions, B - octahedral Fe-ions (B1 correspoinds to Fe^{+3} and B2 to Fe^{+2} ions) [69].*

In the charge-ordered state the partially-filled t_{2g} spin-down band of the octahedral (B) ions is split into three parts (Fig. 19): just below the Fermi energy is subband corresponding to the occupied t_{2g} orbital of the B2(Fe^{+2}) sublattice, then immediately above the Fermi energy t_{2g} band of the B1(Fe^{+3}) sublattice and above it the band formed by the empty orbitals of B2(Fe^{+2}) ions and empty e_g -bands of octahedral (B1,B2) ions.

This result shows that,indeed, after subtracting the spurious self-interaction present in LSDA it is possible to obtain an insulating charge-ordered solution for Fe_3O_4. However what about the metal-insulator transition, can this method describe it? Knowing the value of the intersite Coulomb interaction parameter V it is possible to estimate the change in the potential acting on the t_{2g} electrons in going from Verway type charge order to a completely disordered state. The difference of the electrostatic potential for two sublattices in Verwey type charge order is equal to $\delta V = 4V\delta n_{t_{2g}}$. The calculation of [69] gave V=0.18 eV and $\delta n_{t_{2g}}$=0.70 which results in δV=0.50 eV and that is definitely larger than the calculated energy gap value 0.34 eV. That means that completely destroying the charge order would close an energy gap and lead to the metallic state.

8.2 Doped manganite $Pr_{1-x}Ca_xMnO_3$

The doped rare earth manganites $Re_{1-x}A_xMnO_3$ (Re is a rare earth such as La and A is a divalent element such as Sr or Ca) due to their peculiar correlation between magnetism and conductivity have been extensively studied during the 1950s and 60s [70]. The most thoroughly investigated was the $La_{1-x}Sr_xMnO_3$ system. Undoped (x=0), $LaMnO_3$ is an antiferromagnetic insulator. Upon doping with Sr, this perovskite oxide becomes a ferromagnetic metal; the connection between metallicity and ferromagnetism was well explained by the double exchange hopping mechanism [71]. The discovery of colossal magnetoresistance phenomena in samples with Sr dopant densities in the $0.2 \leq x \leq 0.4$ regime [72] brought a revival of interest to these systems.

The Mn^{+3} ion in the hole-undoped compound $LaMnO_3$ has the high-spin d^4 electron configuration $t^3_{2g\uparrow}e^1_{g\uparrow}$. The t_{2g} orbitals hybridize with O2p orbitals much more weakly than the e_g orbitals and can be regarded as forming a localized spin ($S = 3/2$). In contrast, e_g orbitals, which have lobes directed to the neighboring oxygen atoms, hybridize strongly with O2p producing as a result rather broad bands. The strong exchange interaction with $t_{2g\uparrow}$ subshell leads to a splitting of the e_g band into unoccupied $e_{g\downarrow}$ and half-occupied $e_{g\uparrow}$ subbands. The half filled $e_{g\uparrow}$ subband is a typical example of the Jahn-Teller system, and, indeed, $LaMnO_3$ has an orthorhombic crystal structure [73] with distorted (elongated) MnO_6 octahedra. This cooperative Jahn-Teller effect is usually considered responsible for the opening of the gap in the half-filled $e_{g\uparrow}$ band and the insulator ground state of $LaMnO_3$.

The orthorhombic crystal structure of $LaMnO_3$ ($Pbnm$ space group) can be described as a perovskite with two types of distortion from a cubic structure: 1) Tilting (rotation) of the MnO_6 octahedra, so that Mn-O-Mn angles become less than 180°, and 2) Jahn-Teller distortion of the octahedra, with one long Mn-O bond and two short bonds, the long bonds alternately in the a and b directions (Fig. 20). With doping by Sr the orthorhombicity (the deviation of the values of the lattice parameters b and $c/\sqrt{2}$ from a) becomes smaller and for $x \geq 0.175$ the stable crystal structure at room temperature becomes rhombohedral, where only the tilting of oxygen octahedra is present but all three Mn-O bonds are equal. Around the same value of the doping x, $La_{1-x}Sr_xMnO_3$ becomes metallic.

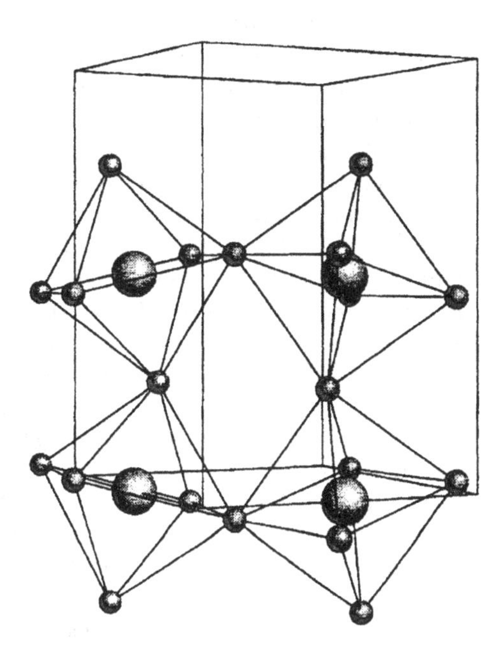

Figure 20: *The orthorombic crystal structure of* $PrMnO_3$. *Large spheres are Mn atoms, small spheres are oxygen atoms. Pr atoms are not shown.*

For Sr doping with $0.2 \leq x \leq 0.4$, $La_{1-x}Sr_xMnO_3$ is a ferromagnetic metal at low temperatures. However, as the temperature increases and approaches the Curie temperature T_c a sudden increase of the resistivity is observed [72]. As the temperature dependence of the resistivity $\rho(T)$ has typically semiconducting behavior ($d\rho/dT < 0$) above T_c, this increase in resistivity at around T_c is usually described as a metal-insulator transition. The colossal magnetoresistance effect happens when the temperature is close to T_c: the external magnetic field leads to suppression of the resistivity increase or to shifting the metal-insulator transition to higher temperatures resulting in a very large negative value of $[\rho(H)-\rho(0)]/\rho(0)$.

The crucial point in the understanding of the effect of the colossal magnetoresistance is the nature of the metal-insulator transition with a temperature variation across T_c. It was shown [74] that the double exchange mechanism alone is not enough to explain such transitions. Millis *et al.* [75] suggested local Jahn-Teller distortions (Jahn-Teller polarons) as a main mechanism causing localization. Their idea is that a E_{JT}/W ratio (E_{JT} is the self-trapping Jahn-Teller polaron energy and W is the effective band width) is close to the critical value in $La_{1-x}Sr_xMnO_3$ and as W decreases with destroying the ferromagnetic order above T_c (the effective hopping takes the maximum value for a parallel spin alignment), this ratio becomes larger than the critical value to result in polaron localization. Varma [76] argues that $LaMnO_3$ is not an insulator due to the Jahn-Teller distortion but is a Mott insulator and the Jahn-Teller distortion occurs parasitically. He explains localization of the holes in the paramagnetic phase by a spin polaron mechanism.

By replacing La with other trivalent ions with smaller ionic radius, the Mn-O-Mn bond angle becomes smaller and the e_g band width W is reduced to enhance the tendency to carrier localization and lattice distortion. A variety of dramatic phenomena have been observed in (Pr, Nd, Sm)(Sr, Ca)MnO_3 systems. As one such example, the results of the study of $Pr_{1/2}Ca_{1/2}MnO_3$ are described in this section [77]. This system has a very peculiar phase diagram [78, 79]. At low temperature, it is an antiferromagnetic insulator with a charge ordering of Mn^{3+} and Mn^{4+} accompanied by an orbital ordering: the orbital ordering and spin ordering is of CE-type. Between T_N ($\simeq$ 180 K) and T_{co} ($\simeq$ 240 K), there is no long-range magnetic ordering but the system remains insulating because of the persistence of the charge ordering. Above T_{co}, it is a paramagnetic insulator.

The local spin density approximation (LSDA) was used by several groups for the theoretical investigation of the electronic structure of $La_{1-x}Sr_xMnO_3$ [80–82]. Those calculations have confirmed the importance of the Jahn-Teller distortion for correct description by the LSDA of the insulating antiferromagnetic ground state of undoped $LaMnO_3$, because the calculations with an undistorted cubic perovskite crystal structure produce a half-filled metallic $e_{g\uparrow}$ band. However, in order to address the charge and orbital order observed in doped manganites it is necessary to go beyond LSDA by including the intra d-shell Coulomb interaction (LDA+U method).

As a first step the electronic structure of undoped $PrMnO_3$ was calculated. This compound has practically the same properties as $LaMnO_3$. The only difference is that due to the smaller ionic radius of the Pr ion compared with that of the La ion, the tilting of the oxygen octahedra in the orthorhombic $Pbnm$ crystal structure is stronger, resulting in a smaller effective $e_g - e_g$ hopping between Mn atoms and narrower band width.

The crystal structure of the doped $Pr_{1-x}Ca_xMnO_3$ does not show a transition to the rhombohedral symmetry, as is the case for $La_{1-x}Sr_xMnO_3$, preserving the orthorhombic $Pbnm$ space group [78]. However the values of the lattice parameters a,b and $c/\sqrt{2}$ for x=0.5 are so close (a=5.395Å, b=5.403Å, $c/\sqrt{2}$=5.382Å), that it can be called a pseudocubic structure. In this structure only the tilting of the oxygen octahedra is present (see Fig. 20), but octahedra themselves are not Jahn-Teller distorted as in the undoped crystal structure. In order to study the effects of the orbital polarization purely due to the intrashell d-d Coulomb interaction without any influence of the Jahn-Teller lattice distortion the calculation for both undoped $PrMnO_3$ and 50% doped $Pr_{1/2}Ca_{1/2}MnO_3$ was performed with the experimental crystal structure parameters corresponding to $Pr_{1/2}Ca_{1/2}MnO_3$ [78].

The LSDA+U calculation with the experimental A-type antiferromagnetic spin alignment resulted in a non-diagonal occupation matrix for spin density

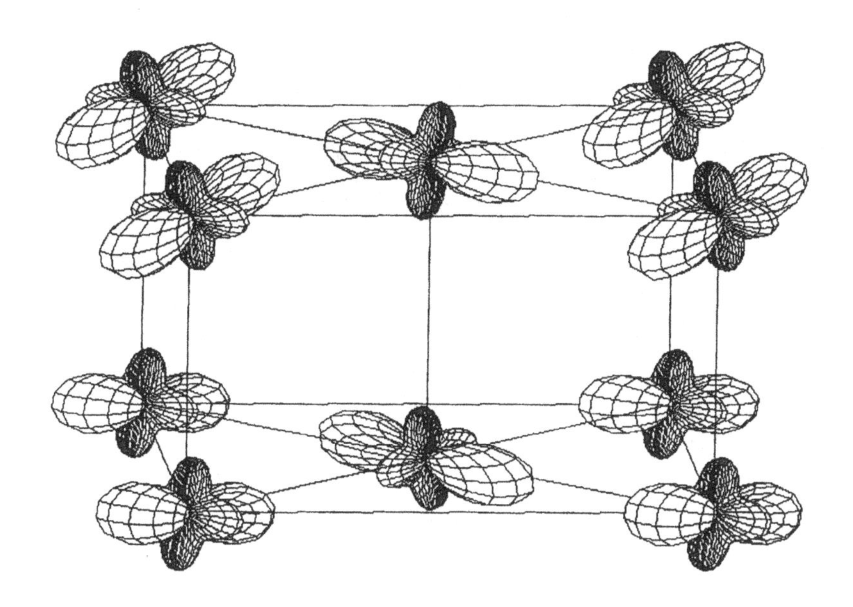

Figure 21: *The calculated angle distribution of the $e_{g\uparrow}$-electron spin density in $PrMnO_3$ [77].*

(for the e_g subspace ($3z^2 - r^2$ and $x^2 - y^2$ orbitals) of one particular Mn atom):

$$n^{\uparrow}_{mm'} - n^{\downarrow}_{mm'} = \begin{pmatrix} 0.31 & 0.22 \\ 0.22 & 0.58 \end{pmatrix} \tag{52}$$

The diagonalization of this matrix gives two new $e_{g\uparrow}$ orbitals: $\phi_1 = 3y^2 - r^2$ with occupancy 0.71 and $\phi_2 = z^2 - x^2$ with occupancy 0.18. For the second type of Mn atom, ϕ_1 and ϕ_2 orbitals can be obtained by the transposition of x and y.

The resulting orbital order can be presented graphically by plotting the angle distribution of the e_g-electron spin density:

$$\rho(\theta, \phi) = \sum_{mm'} (n^{\uparrow}_{mm'} - n^{\downarrow}_{mm'}) Y_m(\theta, \phi) Y_{m'}(\theta, \phi) \tag{53}$$

As a Mn^{+3} ion has formally only one electron in the partially filled $e_{g\uparrow}$ subshell, the e_g-spin density must correspond to the density of this electron (Fig. 21).

The plot of Fig. 21 and the form of the ϕ_1 and ϕ_2 orbitals diagonalizing the spin density occupation matrix show that the orbital order obtained in LDA+U calculations for $PrMnO_3$ is of the same symmetry as the one in $KCuF_3$. The "parasitical" lattice distortion, which one could expect from such orbital ordering: elongation of the MnO_6 octahedra in the direction of the lobes of the ϕ_1 orbitals, will reproduce the experimentally observed Jahn-Teller lattice distortion.

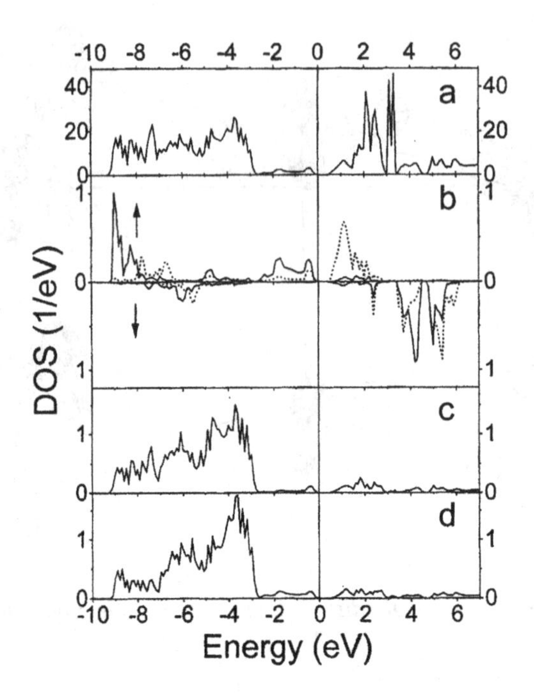

Figure 22: *The total and partial densities of states (DOS) from the result of LSDA+U calculation of* $PrMnO_3$. *a)Total DOS per unit cell (4 formula units); b)Partial* e_g *DOS of Mn. Solid lines are for the* ϕ_1 *-projected DOS, and dashed lines for the* ϕ_2*-projected DOS with* ϕ_1 *and* ϕ_2 *-orbitals diagonalizing the occupation matrix.; c) and d) Partial DOS for two types of oxygen orbitals [77].*

The orbital polarization is far from being 100%, the difference in the occupancy of ϕ_1 and ϕ_2 orbitals is only 0.53. The reason for this is the strong hybridization of $e_{g\uparrow}$-orbitals with oxygen 2p-orbitals. In Fig. 22, the total and partial densities of states (DOS) for $PrMnO_3$ are shown (for e_g-electrons DOS projected on ϕ_1 and ϕ_2 orbitals are also presented).

The electronic structure obtained in the LDA+U calculation is semiconducting with a band gap of 0.5 eV. Note that the standard LSDA needs the Jahn-Teller distorted crystal structure to reproduce the non-metallic ground state. The $e_{g\uparrow}$-band is split by this gap into two subbands: the occupied one with the predominantly ϕ_1 character, and the empty one with the ϕ_2 character. However there is significant admixture of ϕ_2 in the occupied subband and ϕ_1 in the empty one.

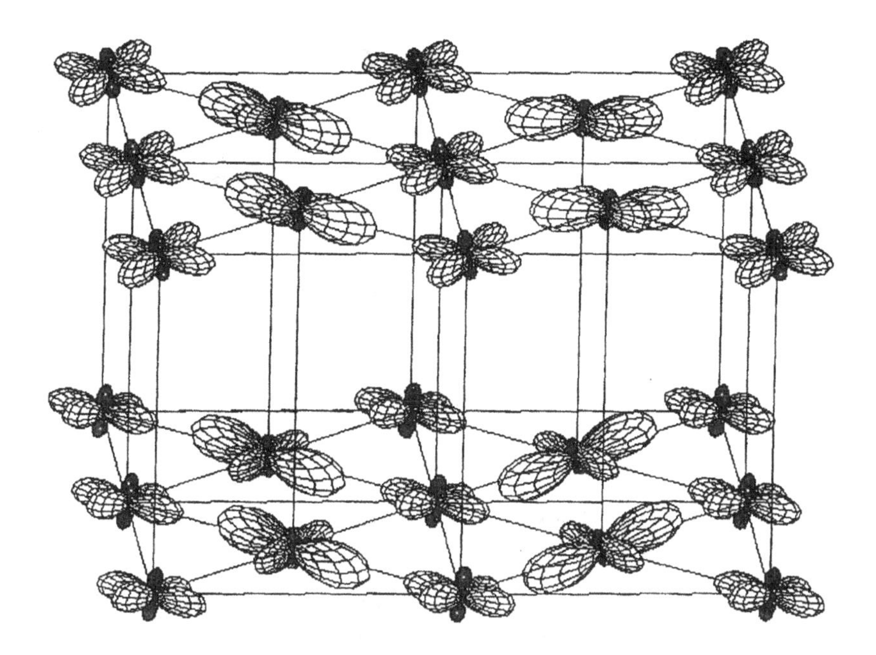

Figure 23: *The calculated angle distribution of the $e_{g\uparrow}$-electron spin density in $Pr_{1/2}Ca_{1/2}MnO_3$ [77].*

After checking that the calculation scheme gives a proper orbital order in undoped $PrMnO_3$, it was applied to $Pr_{1/2}Ca_{1/2}MnO_3$. In order to treat the experimentally observed CE-type of antiferromagnetism it was necessary to quadruple the $Pbnm$-type unit cell and the supercell had 16 formula units. The distribution of the Pr and Ca atoms in the lattice was chosen according to a model where all Pr ions have only Ca as their nearest neighbors in the lanthanide sublattice. No symmetry restrictions were imposed in the calculation, the integration in the reciprocal space being performed for the whole Brillouin zone, and the self-consistency iteration was started from a uniform distribution of e_g-electrons over both e_g-orbitals of all Mn atoms.

The result of the self-consistent calculation was a charge and orbital ordered insulator. The e_g-electron spin-density-plot calculated by Eq. 53 is presented in Fig. 23. This is in striking agreement with the orbital order derived in [78] from the neutron diffraction measurements (Fig. 24). The unexpected result is that the total number of d-electrons at all types of Mn sites are nearly equal (4.99 and 5.01), so that formally Mn^{4+} ions (in the Fig. 23 the ones with symmetric in-plane density distribution) and Mn^{3+} ions (the ones with the density strongly anisotropic) have nearly the same number of 3d-electrons. However the difference in the magnetic moment values is more pronounced: $3.34\mu_B$ for Mn^{4+} and $3.44\ \mu_B$ for Mn^{3+}.

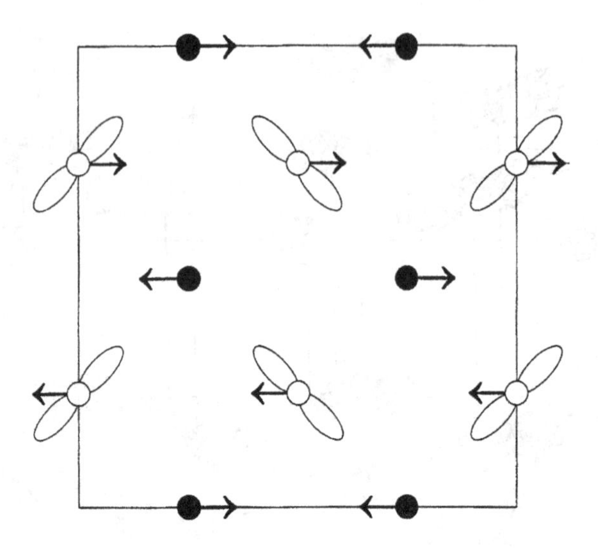

Figure 24: *The scheme of the spin, charge and orbital order in $Pr_{1/2}Ca_{1/2}MnO_3$ deduced from the neutron difraction data [78]. The open circles with the lobe of the e_g electron density distribution denote Mn^{3+} and the filled circles Mn^{4+}. The arrows denote the magnetic moments.*

9 Beyond the mean-field approximation: spectral properties and quasiparticle bands

In spite of its many successes LDA+U has obvious limitations as a one-electron method with a single Slater determinant as a trial function (a mean-field approximation). There are various ways to go beyond the mean-field approximation. Some are based on the perturbation theory in interaction strength U/W, where W is the effective bandwidth. Others start from the atomic limit and treat hybridization as a perturbation. Several attempts to do this are described below. In all of them the main idea of LDA+U method, the mapping of the LDA hamitonian in an atomic-type orbital basis set onto an extended Hubbard model, is used.

9.1 Spectral properties from the Anderson impurity model

It is well known that the spectra of the transition metal compounds are well described in the Configuration Interaction approximation (CI) with a trial function which is a linear combination of Slater determinants [14, 15].

The LDA+U can give all needed parameters for such calculations because the diagonal matrix elements for every single Slater determinant can be calculated in the framework of LDA+U and the off-diagonal matrix elements can be expressed through the one-electron parameters. The following [83] is the realization of this idea to the case of NiO but the procedure is general and could be applied to any material.

The ground state of NiO (with N particles) in the Anderson impurity model is:

$$\Psi_{GS}^{N} = \alpha_0 \mid d^8 > + \beta_0 \mid d^9 L > \tag{54}$$

where L is a hole in the ligand states (oxygen continuum) and the high-energy $\mid d^{10} L^2 >$ configuration is neglected. The final states of the the removal spectrum (with $N - 1$ particles) are (neglecting $\mid d^9 L^2 >$ configuration)

$$\Psi_{m}^{N-1} = \alpha_m \mid d^7 > + \beta_m \mid d^8 L > \tag{55}$$

The one-electron removal Green function is:

$$G(\omega) = \sum_m \frac{A_m}{\omega - E_m^{N-1} + E_{GS}^{N} + i\eta} \tag{56}$$

The pole strengths A_m are the square of the overlap between the eigenstates Ψ_m^{N-1} and the state obtained by suddenly removing an electron from the ground state Ψ_{GS}^{N-1}. The form of the function Ψ_{GS}^{N-1} depends on from which orbital an electron will be removed, from Ni $3d$ or from O $2p$. In the first case:

$$\Psi_{GS,d}^{N-1} = \alpha_0 \mid d^7 > + \beta_0 \mid d^8 L > \tag{57}$$

In the second case:

$$\Psi_{GS,p}^{N-1} = \alpha_0 \mid d^8 L > + \beta_0 \mid d^9 L^2 > \tag{58}$$

The corresponding pole strengths are:

$$A_m^d = \mid < \Psi_{GS,d}^{N-1} \mid \Psi_m^{N-1} > \mid^2 = \mid \alpha_0 \alpha_m + \beta_0 \beta_m \mid^2 \tag{59}$$

$$A_m^p = \mid < \Psi_{GS,p}^{N-1} \mid \Psi_m^{N-1} > \mid^2 = \mid \alpha_0 \beta_m \mid^2 \tag{60}$$

The corresponding one-electron Green's functions for removal of d- and p-electrons are:

$$G_d(\omega) = \alpha_0^2 G_{dd} + \alpha_0 \beta_0 (G_{dp} + G_{pd}) + \beta_0^2 G_{pp} \tag{61}$$

$$G_p(\omega) = \alpha_0^2 G_{pp} \tag{62}$$

where :

$$G_{\alpha\beta} = \sum_m \frac{\alpha_m \beta_m^*}{\omega - E_m^{N-1} + E_{GS}^N + i\eta} \tag{63}$$

Let us consider first the d-removal spectrum. The final state with three d-holes $| d^7 >$ can have any of three symmetries: ${}^2E$ ($e^1_{g\uparrow}e^2_{g\downarrow}$ hole configuration), ${}^4T_1$ ($t^1_{2g\downarrow}e^2_{g\downarrow}$) and ${}^2T_1$ ($t^1_{2g\uparrow}e^2_{g\downarrow}$). They can mix with the configurations $| d^8L >$ of the corresponding symmetries. The non-zero off-diagonal matrix elements can be expressed through the one-electron hopping parameters (properly given in LDA or LDA+U calculations) and the corresponding coefficients are tabulated in [14, 84].

How do we calculate the diagonal matrix elements, for example $< e^3 | e^3 >$ and $< e^2({}^3A_2)L_e | e^2({}^3A_2)L_e >$? If the energy of the ground state of $| d^8 >$ configuration ($| e^2({}^3A_2) >$) is zero then the former is the removal energy of the $e_{g\uparrow}$ electron and the latter is the removal energy of the ligand (oxygen $2p$) electron. In the LDA+U formalism the one electron energies of the occupied states have the meaning of the removal energies [12] so one can use the results of the self-consistent LDA+U calculation for the calculation of the diagonal matrix elements of the many-electron Hamiltonian. If the particular $| d^8L >$ state has a d^8 configuration not in the ground state ${}^3A_2$ (for example $| e^2({}^1E)L_e >$) then the diagonal matrix element will be the energy of L_e minus the energy difference $(E({}^1E) - E({}^3A_2))$.

In practice, the calculations were performed in the following way. At first the many electron Hamiltonian matrix, whose eigenfunctions are Ψ_m^{N-1} (55), was constructed from the parts of the one-electron Hamiltonian and then the Green functions (63) were calculated. Let us illustrate it for ${}^2E$ symmetry.

In the one-electron Hamiltonian matrix there is one block describing continuum states (O$2p$), two rows (columns) for d-orbitals of e_g symmetry and three rows (columns) for t_{2g}. In the many-electron Hamiltonian matrix for ${}^2E$ symmetry there are three blocks for continuum states (describing three types of the ligand holes in the final states $| d^8L >$) with the shifts of the diagonal elements of this blocks if the corresponding d^8 configuration is not in the ground state ${}^3A_2$. There is also one row (column) for the $| d^7 >$ (e^3) configuration with the diagonal matrix element equal to the energy of the $e_{g\uparrow}$ orbital in LDA+U calculations and with the off-diagonal matrix elements according to [14, 84].

To perform impurity calculations for NiO with only one d-shell in an oxygen continuum, the following procedure was used. We first push up the d-orbitals by adding a potential acting only on them:

$$H_0 = H_{LDA+U} + \sum_i |d_i\rangle \Delta \langle d_i| \tag{64}$$

Δ is a constant and i labels the site. The valence band is then mainly formed by the Ni$4s$, $4p$ and O$2s$, $2p$ orbitals with a small but non-zero hybridization with the Ni d decaying as $\sim 1/\Delta$. Using the Hamiltonian H_0 the Green function G^0_{dd}

was calculated, which has a small imaginary part around the valence band due to the small hybridization. The impurity Green function G_{dd} is then defined to be the one corresponding to the following Hamiltonian:

$$H = H_0 - |d_0\rangle\Delta\langle d_0| \tag{65}$$

i.e. at the origin where the impurity is located, the d-orbital is at its original position but at any other site it is high up in energy. Consequently, this Green function contains the hybridization with the oxygen continuum through H_0. The impurity Green function can be calculated from the Dyson equation:

$$G = G_0 + G_0 \Delta V G \tag{66}$$

where for the d-orbital, $V = |d_0\rangle\Delta\langle d_0|$ giving well known formulas [85]:

$$G_{dd} = \frac{G^0_{dd}}{1 - \Delta G^0_{dd}} \tag{67}$$

$$G_{dp} = \frac{G^0_{dp}}{1 - \Delta G^0_{dd}} \tag{68}$$

$$G_{pp} = \frac{G^0_{pd}\Delta G^0_{dp}}{1 - \Delta G^0_{dd}} + G^0_{pp} \tag{69}$$

Δ is the upward energy shift of the d-orbitals for calculating Green's functions G^0:

$$G^0_{ij}(\omega) = \int d\mathbf{k} \sum_n \frac{c^n_i(\mathbf{k})c^{n*}_j(\mathbf{k})}{\omega - E^n(\mathbf{k}) + i\eta} \tag{70}$$

The integration is over the Brillouin zone and $c^n_i(\mathbf{k})$, $E^n(\mathbf{k})$ are eigenvectors and eigenvalues of the band-structure calculation with the new Hamiltonian matrix.

Equations (67–70 and 59) were used to calculate the one-electron Green's function for a removal of a d-electron $G_d(\omega)$. (The α_0 and β_0 were extracted from the ground state self-consistent LDA+U calculation: $\alpha_0^2 \approx 0.9$ and $\beta_0^2 \approx 0.1$).

The total d-removal theoretical spectrum was calculated from the imaginary part of the $G_d(\omega)$ broadened by 0.5 eV (the summation over the three symmetries of the final state (^2E,^2T and ^4T) with appropriate weights was performed). On Fig. 25 this spectrum is plotted together with experimental XPS for NiO. One can see that two major features of the experimental spectra: the main line at the top of the valence band and a satellite at $\approx$ 8 eV lower in energy with smaller intensity, are well reproduced not only in energy separation but also in intensity ratio.

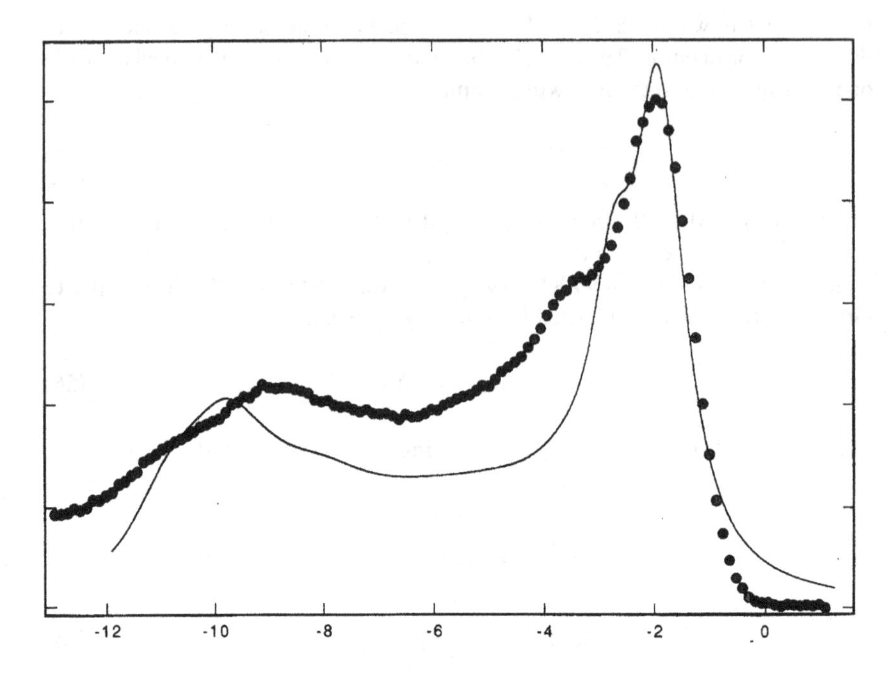

Figure 25: *The calculated (Green-function LDA+U) d-emission spectrum (line) and experimental XPS for NiO (points) [83].*

By using expression (61) one can calculate oxygen p-removal spectra. On Fig. 26 it is plotted together with the experimental $OK\alpha$ X-ray emission spectrum (XES) of NiO. The impurity approximation, which was used in the calculations, is a d-ion centered approach and not so well suited for the oxygen states, but it reproduced two major features of the experimental spectrum: the main peak of the oxygen bands and the high energy shoulder corresponding to the admixture of the oxygen states to the main line of the d-removal spectrum.

While XPS gives the distribution of the d-states and $OK\alpha$ XES of the p-states, the photoemission spectrum taken at a photon energy 40.8 eV, shown on Fig. 27, corresponds to nearly equal cross-sections of Ni 3d and O 2p states. Good agreement with the theoretical spectrum, obtained by adding the Ni 3d and the O 2p removal spectra, proves that the calculation gives not only the main line and satellite of the d-origin but also the relative position of the oxygen bands.

The final state of the addition spectrum for NiO has only one d-hole and the corresponding wave function could be chosen as a single Slater determinant. Hence the straightforward LDA+U calculation of the density of states for unoccupied bands must be compared with "Bremstrahlung isohromat spectra" (BIS). Such comparison is presented in Fig. 28 and shows quite good agreement.

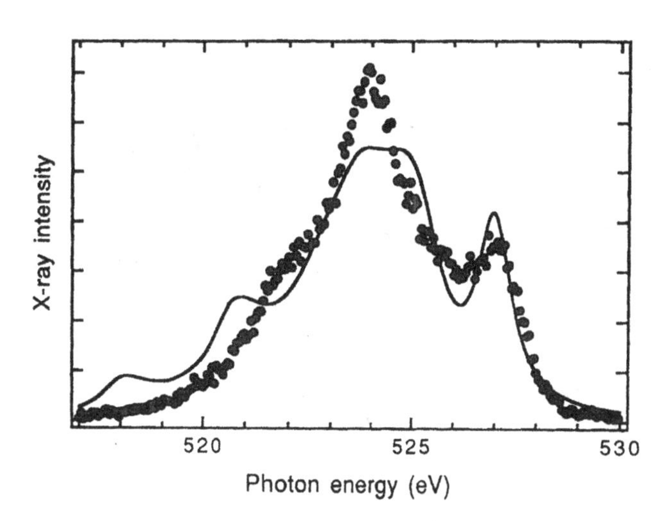

Figure 26: *The calculated (Green-function LDA+U) p-emission spectrum for NiO (line) and experimental O Kα X-ray emission spectrum for NiO [83].*

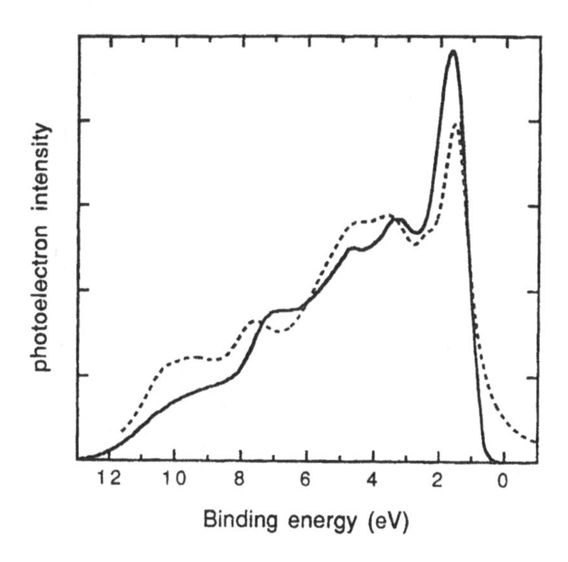

Figure 27: *The calculated (Green-function LDA+U) photoemission spectrum with equal weights of Ni 3d and O 2p states (dashed) compared to the experimental He II UPS (solid line) [83].*

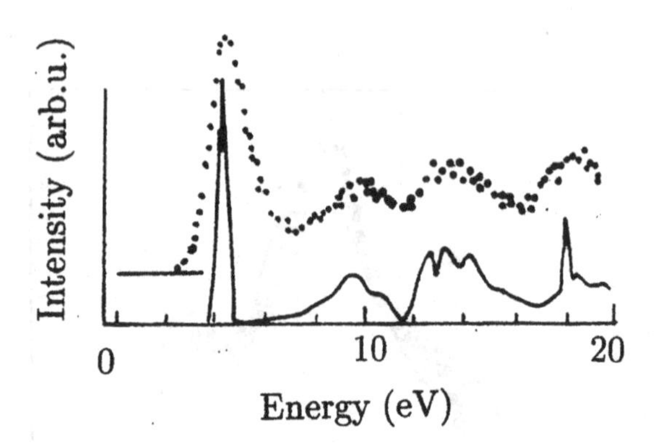

Figure 28: *The experimental (dots) and the calculated (solid line) BIS (Bremstrahlung isochromat spectrum) for NiO [12].*

9.2 Quasiparticle bands in FLEX approximation

If the Hubbard parameter U is smaller than the band-width, it is possible to use one of the "standard conserving approximations" - the fluctuation exchange approximation (FLEX) [87] and investigate the general $\mathbf{q}-$ and $\omega-$dependence of the self-energy. We briefly illustrate this approach to the problem of quasiparticle band structure for bi-layer superconducting cuprates, such as $YBa_2Cu_3O_7$.

In this case the total Green function matrix in the orbital and spin-space determined through the self-consistent solution of the FLEX-equations is :

$$\begin{aligned} G^{-1}(\mathbf{k},\omega_n) &= G_0^{-1}(\mathbf{k},\omega_n) - \Sigma(\mathbf{k},\omega_n) \\ \Sigma(\mathbf{k},\omega_n) &= \frac{T}{N}\sum_{\mathbf{q},\omega_m} V(\mathbf{q},\omega_m)G(\mathbf{k}-q,\omega_n-\omega_m) \end{aligned} \tag{71}$$

where N is a total number of momentum $\mathbf{k}$-points, $\omega_n = (2n+1)\pi T$ are the fermion Matsubara frequencies with n an integer and T the system temperature. The spin-fluctuation contribution to the effective interaction matrix V in the paramagnetic state is $V^{sf}(\mathbf{q},\omega_m) = 3/2U\chi(\mathbf{q},\omega_m)[(1-U\chi(\mathbf{q},\omega_m))^{-1}-1]U$, defined through the full particle-hole susceptibilities:

$$\chi^{ph}_{mm'm''m'''}(\mathbf{q},\omega_m) = -\frac{T}{N}\sum_{\mathbf{q},\omega_m} G_{mm''}(\mathbf{k},\omega_n)G_{m'm'''}(\mathbf{k}+q,\omega_n+\omega_m) \tag{72}$$

Here the U-matrix corresponds to antisymmetric coulomb interaction from Eq. (10): $U_{mm'm''m'''} =< mm''|V_{ee}|m'm''' > - < mm''|V_{ee}|m'''m' >$.

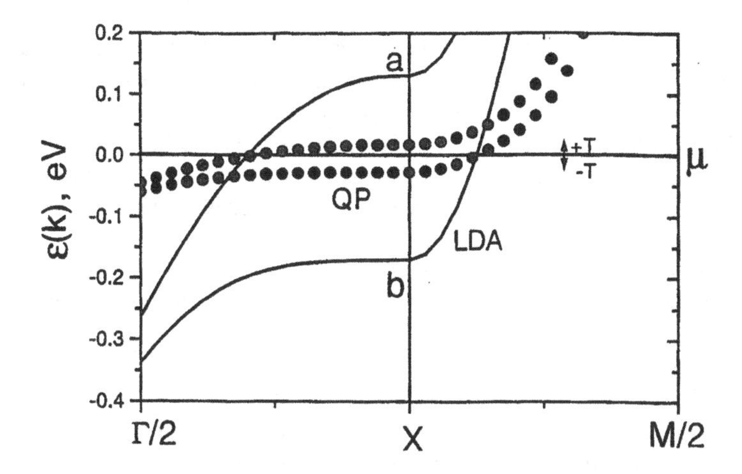

Figure 29: *LDA and QP bands near X-point for bi-layer cuprates. Arrows indicate the temperature scale (T=150 K),* $\Gamma/2 \equiv (\pi/2, 0)$, $M/2 \equiv (\pi, \pi/2)$.

For $YBa_2Cu_3O_7$ only two antibonding Cu-d$_{x^2-y^2}$ - O-p$_x$,p$_y$ LDA-bands were included, which cross the Fermi level for this bi-layer CuO_2 system [88]. The band width is approximately 4 eV, and the effective Hubbard parameter is 3 eV. The latter number takes into account the 65% weight Cu-d$_{x^2-y^2}$ orbital in these antibonding bands and an LDA screened electron-electron interaction parameter for the Cu$_d$ orbital in cuprates of about 7–8 eV. We solve the nonlinear integral FLEX-equations using the fast Fourier transform method on the discreet mesh of 64×64 momentum in the two dimensioned Brillouin zone and with 700–800 Matsubara frequencies with a cutoff of 20-30 eV in the energy (which corresponds to the temperature range of 80–200 K). Analytical continuation on the real axes was done by Padé approximation.

The resulting quasiparticle (QP) bands which correspond to the maximum in the spectral functions ($A(\mathbf{k}, \omega) = -1/\pi ImG(\mathbf{k}, \omega)$) are presented on Fig. 29, in comparison with the bare LDA bands. Due to the large spin-fluctuation for the $\mathbf{q} = (\pi, \pi)$ point in the two dimensioned Brillouin zone (BZ), there is a large renormalization of the bi-layer splitting near the X-point. While the Fermi-level crossing points are nearly the same for LDA-bands and (QP) bands due to the Luttinger theorem, the bonding (b) and antibonding (a) QP-bands are "pinned" to the Fermi level in the big portion of the BZ, forming so-called "extended van-Hove singularities". The important properties of such a QP band structure of HTC-compounds is the "vanishing" of bi-layer splitting in agreement with a recent angle-resolved photoemission study [89].

9.3 Dynamical mean-field theory

Recently the dynamical mean-field theory was developed [90] which is based on the mapping of lattice models onto quantum impurity models subject to a self-consistency condition. The resulting impurity model can be solved by various approaches (Quantum Monte Carlo, exact diagonalization) but the most promising for the possible use in a "realistic" calculation scheme is the Iterated Perturbation Theory (IPT) approximation, which was proved to give results in a good agreement with more rigorous methods.

In this section the dynamical mean-field theory in the iterated perturbation theory approximation is described in its application to the band structure calculations using a LMTO basis. The method was applied [91] to $La_{1-x}Sr_xTiO_3$ which is a classical example of a strongly correlated metal.

In order to be able to implement the achievements of the Hubbard model theory to LDA one needs a method which could be mapped onto a tight-binding model. The Linearized Muffin-Tin Orbitals (LMTO) method in the orthogonal approximation [92] can be naturally presented as a tight-binding calculation scheme (in real space representation):

$$H_{LMTO} = \sum_{ilm,jl'm',\sigma} (\delta_{ilm,jl'm'} \epsilon_{il} \hat{n}_{ilm\sigma} + t_{ilm,jl'm'} \hat{c}^{\dagger}_{ilm\sigma} \hat{c}_{jl'm'\sigma}) \tag{73}$$

(i - site index, lm - orbital indexes).

As we have mentioned above, the LDA one-electron potential is orbital independent and the Coulomb interaction between d-electrons is taken into account in this potential in an averaged way. In order to generalize this Hamiltonian by including Coulomb correlations, one must add the interaction term:

$$H_{int} = \frac{1}{2} \sum_{ilmm'\sigma\sigma' m\sigma \neq m'\sigma'} U_{il} \hat{n}_{ilm\sigma} \hat{n}_{ilm'\sigma'} \tag{74}$$

We neglected for a while the exchange terms and the dependence of the Coulomb parameter U on the particular pair of orbitals mm'. Through the following we will assume that only for one shell l_d of one type of atoms i_d (for example d-orbitals of the transition metal ions) the Coulomb interaction will be taken into account ($U_{il} = U\delta_{il,i_d l_d}$), and in the following indices il will be omitted. All other orbitals will be considered as resulting in the itinerant bands and well described by LDA. Such separation of the electronic states into localized and itinerant is close in spirit to the Anderson model.

To avoid double-counting one must at the same time subtract the averaged Coulomb interaction energy term, which we assume is present in LDA. Unfortunately there is no direct connection between the Hubbard model and LDA (because LDA is based on the homogeneous electron gas theory and not on the localized atomic type orbitals representation) and it is impossible to express rigorously the LDA energy through the d-d Coulomb interaction parameter U. However it is known that the LDA total energy as a function of the total number of electrons is a good approximation and the value of the Coulomb parameter U obtained in the LDA calculation agrees well with experimental data and the results of more rigorous calculations [4, 5]. That leads us to the suggestion that a good approximation for the LDA part of the Coulomb interaction energy will be:

$$E_{Coul} = \frac{1}{2} U n_d (n_d - 1) \tag{75}$$

($n_d = \sum_{m\sigma} n_{m\sigma}$ total number of d-electrons).

In the LDA Hamiltonian ϵ_d has the meaning of the LDA-one-electron eigenvalue for d-orbitals. It is known that in LDA the eigenvalue is the derivative of the total energy over the occupancy of the orbital:

$$\epsilon_d = \frac{d}{dn_d} E_{LDA} \tag{76}$$

If we want to introduce new ϵ_d^0 where d-d Coulomb interaction is excluded we must define them as:

$$\epsilon_d^0 = \frac{d}{dn_d}(E_{LDA} - E_{Coul}) = \epsilon_d - U\left(n_d - \frac{1}{2}\right) \tag{77}$$

Then the new Hamiltonian will have the form:

$$\begin{aligned} H &= H^0 + H_{int} \\ H^0 &= \sum_{ilm,jl'm',\sigma} (\delta_{ilm,jl'm'}\, \epsilon_{il}^0 \widehat{n}_{ilm\sigma} + t_{ilm,jl'm'} \widehat{c}_{ilm\sigma}^{\dagger} \widehat{c}_{jl'm'\sigma}) \end{aligned} \tag{78}$$

In reciprocal space the matrix elements of the operator H^0 are:

$$H^0_{qlm,q'l'm'}(\mathbf{k}) = H^{LDA}_{qlm,q'l'm'}(\mathbf{k}) - \delta_{qlm,q'l'm'} \delta_{ql,i_d l_d} U \left(n_d - \frac{1}{2}\right) \tag{79}$$

(q is an index of the atom in the elementary unit cell).

In the dynamical mean-field theory the effect of Coulomb correlation is described by a self-energy operator in the local approximation. The Green function is:

$$G_{qlm,q'l'm'}(i\omega) = \frac{1}{V_B} \int d\mathbf{k}\, [i\omega + \mu - H^0_{qlm,q'l'm'}(\mathbf{k}) - \delta_{qlm,q'l'm'} \delta_{ql,i_d l_d} \Sigma(i\omega)]^{-1} \tag{80}$$

($[...]^{-1}$ means inversion of the matrix, integration is over Brillouin zone, μ is chemical potential, V_B is the volume of the Brillouin zone).

In the following we will consider the paramagnetic case, orbital and spin degenerate system, so that the self-energy $\Sigma(i\omega)$ does not depend on orbital and spin indexes. One can define the effective Anderson model Green function through:

$$G(i\omega) = G_{i_d l_d m, i_d l_d m}(i\omega) = (i\omega + \mu - \Delta(i\omega) - \Sigma(i\omega))^{-1} \tag{81}$$

where $\Delta(i\omega)$ is an effective impurity hybridization function. The effective medium "bath" Green function G^0 is defined as:

$$G^0(i\omega) = (i\omega + \widetilde{\mu} - \Delta(i\omega))^{-1} = (G^{-1}(i\omega) + \Sigma(i\omega) + \widetilde{\mu} - \mu)^{-1}$$

($\widetilde{\mu}$ is chemical potential of the effective medium).

The chemical potential of the effective medium $\widetilde{\mu}$ is varied to satisfy the Luttinger theorem condition:

$$\frac{1}{\beta}\sum_{i\omega_n} e^{i\omega_n 0^+} G(i\omega_n)\frac{d}{d(i\omega_n)}\Sigma(i\omega_n) = 0 \tag{83}$$

In the iterated perturbation theory approximation the *anzatz* for the self-energy is based on the second order perturbation theory term calculated with the "bath" Green function G^0:

$$\Sigma^0(i\omega_s) = -(N-1)U^2\frac{1}{\beta^2}\sum_{i\omega_n}\sum_{ip_m} G^0(i\omega_m + ip_n)G^0(i\omega_m)G^0(i\omega_s - ip_n) \tag{84}$$

N is a degeneracy of orbitals including spin, $\beta = \frac{1}{kT}$, Matsubara frequencies $\omega_s = \frac{(2s+1)\pi}{\beta}$; $p_n = \frac{2n\pi}{\beta}$; s, n integer numbers.

The term Σ^0 is renormalized to insure the correct atomic limit:

$$\Sigma(i\omega) = UN(N-1) + \frac{A\Sigma^0(i\omega)}{1 - B\Sigma^0(i\omega)} \tag{85}$$

(n is orbital occupation $n = \frac{1}{\beta}\sum_{i\omega_n} e^{i\omega_n 0^+} G(i\omega_n)$),

$$B = \frac{U[1-(N-1)n] - \mu + \widetilde{\mu}}{U^2(N-1)n_0(1-n_0)} \tag{86}$$

$$A = \frac{n[1-(N-1)n] + (N-2)D[n]}{n_0(1-n_0)} \tag{87}$$

$$n_0 = \frac{1}{\beta}\sum_{i\omega_n} e^{i\omega_n 0^+} G^0(i\omega_n) \tag{88}$$

the correlation function $D[n] \equiv < \widehat{n}\widehat{n} >_{CPA}$ is calculated using the Coherent Potential Approximation (CPA) for the Green function with parameter $\delta\mu$ chosen to preserve orbital occupation n:

$$G_{CPA}(i\omega) = \frac{[1 - n(N-1)]}{i\omega + \mu - \Delta(i\omega) + \delta\mu} + \frac{n(N-1)}{i\omega + \mu - \Delta(i\omega) - U + \delta\mu} \tag{89}$$

$$n = \frac{1}{\beta} \sum_{i\omega_n} e^{i\omega_n 0^+} G_{CPA}(i\omega_n) \tag{90}$$

$$D[n] = n \sum_{i\omega_n} e^{i\omega_n 0^+} \frac{1}{i\omega + \mu - \Delta(i\omega_n) - U + \delta\mu} \tag{91}$$

The Matsubara frequency convolution in (84) was performed with time variables representation using the Fast Fourier Transform algorithm for the transition from energy to time variables and back:

$$G^0(\tau) = \frac{1}{\beta} \sum_{i\omega_n} e^{-i\omega_n \tau} G^0(i\omega_n) \tag{92}$$

$$\Sigma(\tau) = -(N-1)U^2 G^0(\tau) G^0(\tau) G^0(-\tau) \tag{93}$$

$$\Sigma^0(i\omega_n) = \int_0^\beta d\tau\, e^{i\omega_n \tau}\, \Sigma(\tau) \tag{94}$$

The serious problem is to perform the integration in **k**-space over the Brillouin zone. For this the generalized Lambin-Vigneron algorithm [93] was used. We define the new matrix $H(\mathbf{k}, z)$ as:

$$H(\mathbf{k}, z) = H^0(\mathbf{k}) + \Sigma(z) \tag{95}$$

where z is the complex energy and the term $\Sigma(z)$ is added only to the diagonal elements of the H-matrix corresponding to d-orbitals. In this matrix notation the Green function is :

$$G(z) = \frac{1}{V_B} \int d\mathbf{k}[z - H(\mathbf{k}, z)]^{-1} \tag{96}$$

After diagonalization, the $H(\mathbf{k}, z)$ matrix can be expressed through the diagonal matrix of its eigenvalues $D(\mathbf{k}, z)$ and the eigenvectors matrix $U(\mathbf{k}, z)$:

$$H(\mathbf{k}, z) = U(\mathbf{k}, z) D(\mathbf{k}, z) U^{-1}(\mathbf{k}, z) \tag{97}$$

and the Green function is:

$$G(z) = \frac{1}{V_B} \int d\mathbf{k} U(\mathbf{k}, z)[z - D(\mathbf{k}, z)]^{-1} U^{-1}(\mathbf{k}, z) \tag{98}$$

A particular matrix element of the Green function is calculated as:

$$G_{ij}(z) = \sum_n \frac{1}{V_B} \int d\mathbf{k} \frac{U_{in}(\mathbf{k}, z) U_{nj}^{-1}(\mathbf{k}, z)}{z - D_n(\mathbf{k}, z)} \tag{99}$$

In the analytical tetrahedron method the irreducible wedge of the Brillouin zone is divided into a set of tetrahedra and the total integral is calculated as a sum over the tetrahedra. To perform the integration over a given tetrahedron with four corners at vectors $\mathbf{k}_i$ ($i = 1, 2, 3, 4$) the denominator of the fraction in equation (99) is interpolated as a linear function in $\mathbf{k}$-space. In the result the integral over one tetrahedron is expressed through the values of the numerator and denominator at the corners of the tetrahedron:

$$\sum_n \frac{1}{V_B} \int_v d\mathbf{k} \frac{U_{in}(\mathbf{k}, z) U_{nj}^{-1}(\mathbf{k}, z)}{z - D_n(\mathbf{k}, z)} = \sum_n \sum_{i=1}^{4} r_i^n U_{in}(\mathbf{k}_i, z) U_{nj}^{-1}(\mathbf{k}_i, z) \frac{v}{V_B} \tag{100}$$

v is tetrahedron volume

$$r_i^n = \frac{(z - D_n(\mathbf{k}_i, z))^2}{\prod_{k(\neq i)} (D_n(\mathbf{k}_k, z) - D_n(\mathbf{k}_i, z))} + \sum_{j(\neq i)} \frac{(z - D_n(\mathbf{k}_j, z))^3}{\prod_{k(\neq j)} (D_n(\mathbf{k}_k, z) - D_n(\mathbf{k}_j, z))} \frac{\ln[(z - D_n(\mathbf{k}_j, z))/(z - D_n(\mathbf{k}_i, z)]}{(D_n(\mathbf{k}_i, z) - D_n(\mathbf{k}_j, z))} \tag{101}$$

The self-energy $\Sigma(i\omega_n)$ and Green function $G(i\omega_n)$ are calculated at the imaginary Matsubara frequencies $i\omega_n = i\pi(2n+1)/\beta$. It is enough to calculate expectation values, such as orbital occupancies n, but in order to calculate the spectral properties one needs to know the Green function on the real axis. The real axis equivalent of equations (84) is much more complicated and hard to implement numerically than the Matsubara frequencies version. It is much more convenient to perform analytical continuation from imaginary energy values to the real ones. For such continuation the Padé approximant algorithm [94] was used. If one has a set of the complex energies z_i ($i = 1, ..., M$) and the set of values of the analytical function u_i, then the approximant is defined as a continued fraction:

$$C_M(z) = \frac{a_1}{1+} \frac{a_2(z - z_2)}{1+} \cdots \frac{a_M(z - z_{M-1})}{1} \tag{102}$$

where the coefficients a_i are to be determined so that:

$$C_M(z_i) = u_i, \; i = 1, ..., M \tag{103}$$

The coefficients a_i are then given by the recursion:

$$a_i = g_i(z_i),\ g_1(z_i) = u_i,\ i = 1, ..., M \quad (104)$$

$$g_p(z) = \frac{g_{p-1}(z_{p-1}) - g_{p-1}(z)}{(z - z_{p-1})g_{p-1}(z)},\ p \geq 2 \quad (105)$$

The recursion formula for the continued fraction finally yields:

$$C_M(z) = A_M(z)/B_M(z) \quad (106)$$

where

$$A_{n+1}(z) = A_n(z) + (z - z_n)a_{n+1}A_{n-1}(z)$$
$$B_{n+1}(z) = B_n(z) + (z - z_n)a_{n+1}B_{n-1}(z) \quad (107)$$

and

$$A_0 = 0,\ A_1 = a_1,\ B_0 = B_1 = 1$$

We have found that the most convenient way is to use analytical continuation not for the Green function G but only for the self-energy Σ, and then to calculate G directly on the real axis through the Brillouin zone integration (100).

The above described calculation scheme was applied to the doped Mott insulator $La_{1-x}Sr_xTiO_3$. $LaTiO_3$ is a Pauli-paramagnetic metal at room temperature and below T_N=125 K an antiferromagnetic insulator with a very small gap value (0.2 eV). Doping by a very small value of Sr (few percent) leads to the transition to a paramagnetic metal with a large effective mass. As photoemission spectra of this system also show a strong deviation from the noninteracting electron picture, $La_{1-x}Sr_xTiO_3$ is regarded as an example of a strongly correlated metal.

The crystal structure of $LaTiO_3$ is slightly distorted cubic perovskite. The Ti ions have an octahedral coordination of oxygen ions and the t_{2g}-e_g crystal field splitting of d-shell is strong enough to survive in the solid. On Fig. 30 the total and partial DOS of paramagnetic $LaTiO_3$ are presented as obtained in LDA calculations (LMTO method). On 3 eV above O2p-band there is Ti-3d-band split on the t_{2g} and e_g subbands which are well separated from each other. Ti^{4+}-ions have a d^1 configuration and the t_{2g} band is 1/6 filled.

As only the t_{2g} band is partially filled and the e_g band is completely empty, it is reasonable to consider Coulomb correlations between $t_{2g}-$ electrons only and the degeneracy factor N in Eq. (84) is equal to 6. The value of the Coulomb parameter U was calculated by the supercell procedure [4, 5] regarding only the $t_{2g}-$electrons as localized ones and allowing the e_g-electrons to participate in the screening. This calculation resulted in a value of 3 eV. As the localization must lead to an energy gap between electrons with the same spin, the effective Coulomb interaction will be reduced by the value of exchange parameter J=1 eV. So the

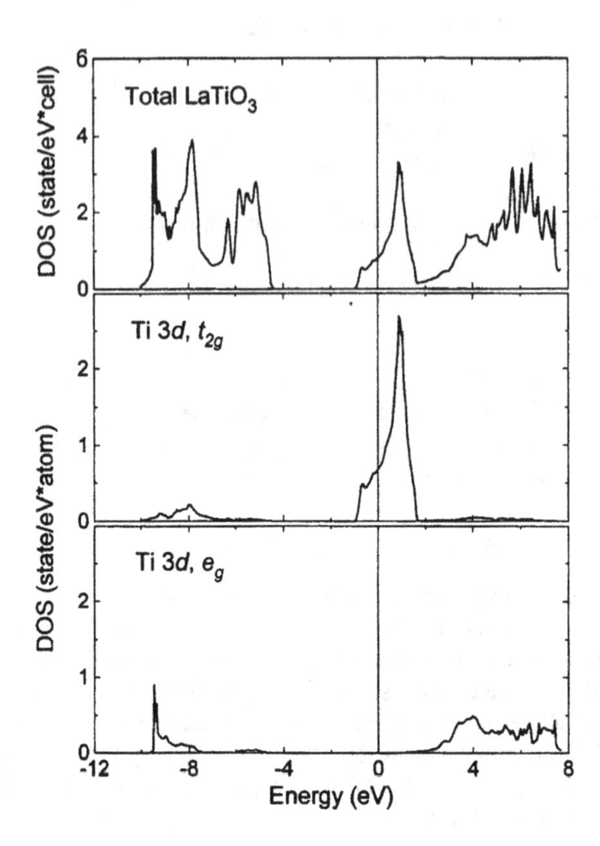

Figure 30: *Noninteracting (U=0) total and partial density of states (DOS) for $LaTiO_3$. [91].*

effective Coulomb parameter U_{eff}=2 eV was used. The results of the calculation for x=0.06 (doping by Sr was imitated by the decreasing on x the total number of electrons) are presented in the form of the t_{2g}-DOS on Fig. 31. Its general form is the same as for model calculations: a strong quasiparticle peak on the Fermi energy and incoherent subbands below and above it corresponding to the lower and upper Hubbard bands.

The appearance of the incoherent lower Hubbard band in the calculated DOS leads to qualitatively better agreement with photoemission spectra. In Fig. 32 the experimental XPS for $La_{1-x}Sr_xTiO_3$ (x=0.06) [95] is presented with the noninteracting (LDA) and interacting (IPT) occupied DOS broadened to imitate experimental resolution. The main correlation effects, the simultaneous presence of a coherent and incoherent band in XPS is successfully reproduced in the

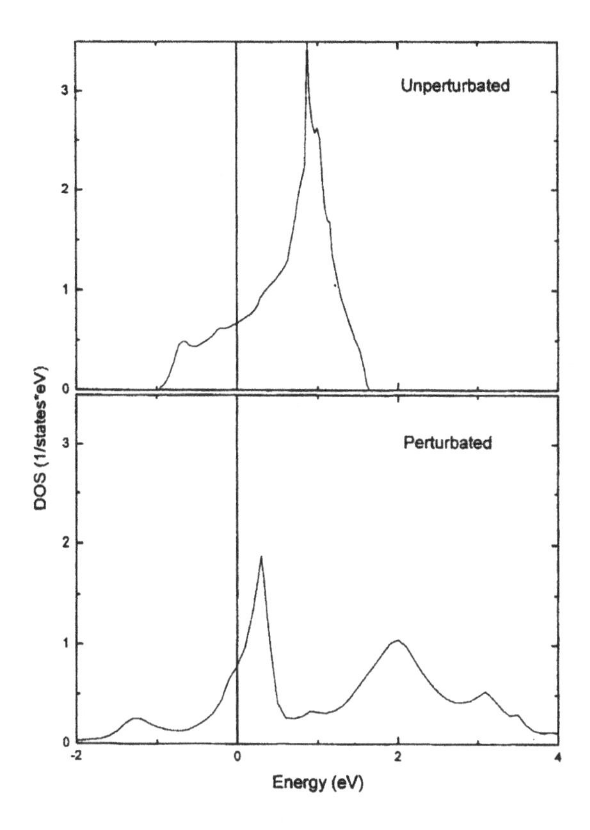

Figure 31: *Partial (t_{2g}) DOS obtained in IPT calculations in comparison with noninteracting DOS [91].*

IPT calculation. However, as one can see, IPT overestimates the strength of the coherent subband.

10 Conclusion

The LDA+U method was proven to be a very efficient and reliable tool in calculating the electronic structure of systems where the Coulomb interaction is strong enough to cause localization of the electrons. It works not only for nearly core-like 4f-orbitals of rare-earth ions, where the separation of the electronic states on the subspaces of the infinitely slow localized orbitals and infinitely fast itinerant ones is valid, but also for such systems as transition metal oxides, where 3d-orbitals hybridize quite strongly with oxygen 2p-orbitals. In spite of the fact

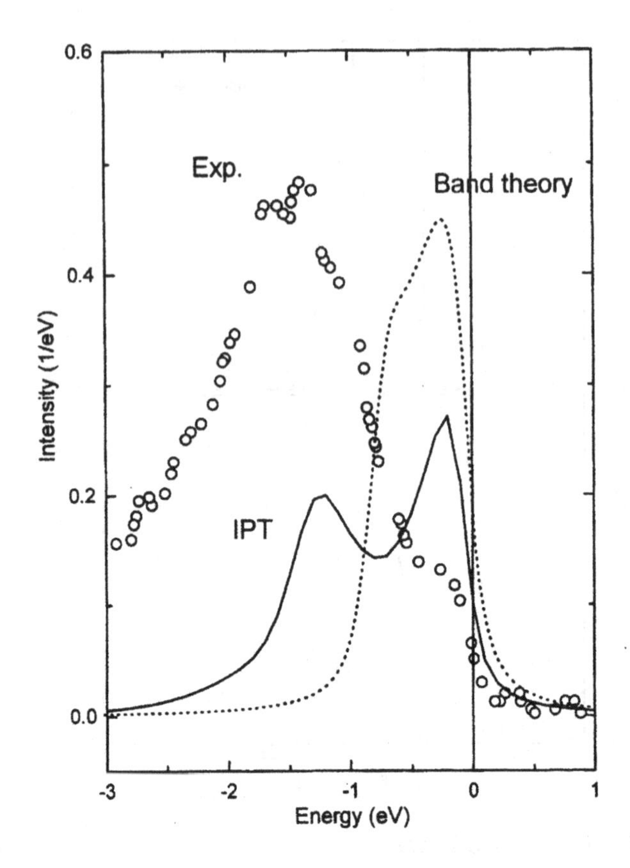

Figure 32: *Experimental and theoretical photoemission spectra of* $La_{1-x}Sr_xTiO_3$ *(x=0.06) [91].*

that the LDA+U is a mean-field approximation which is in general insufficient for the description of the metal-insulator transition and strongly correlated metals, in some cases, such as the metal-insulator transition in FeSi and $LaCoO_3$, LDA+U calculations gave valuable information by giving insight into the nature of these transitions. However in general LDA+U overestimates the tendency to localization as is well know for Hartree-Fock type methods. The main advantage of the LDA+U method over model approaches is its *first principle* nature with a complete absence of adjustable parameters. Another asset is the fully preserved ability of LDA-based methods to address the intricate interplay of the electronic and lattice degrees of freedom by computing the total energy as a function of lattice distortions. When the localized nature of the electronic states with Coulomb interaction between them is properly taken into account, this ability allows us to describe such effects as polaron formation and orbital polarization. As the spin and charge density of the electrons is calculated self-consistently in the LDA+U method, the resulting diagonal and off-diagonal matrix elements of the one-electron Hamiltonian could be used in more complicated calculations where many-electron effects are treated beyond the mean-field approximation. The main idea of the LDA+U method is that the mapping of the LDA hamiltonian on the multi-band lattice Anderson model can be used for constructing *ab-initio* calculating schemes based on the achievements of the Anderson and Hubbard model studies.

References

[1] P. Hohenberg and W. Kohn, *Phys. Rev.* **136**, B864 (1964); W.Kohn and L.J.Sham, *ibid.* **140**, A1133 (1965).

[2] A. Svane and O. Gunnarsson, *Phys. Rev. Lett.* **65**, 1148 (1990).

[3] S. Massida, M. Posternak, A. Baldereschi, *Phys. Rev. B* **48**, 5058 (1993).

[4] O. Gunnarsson, O.K. Andersen, O. Jepsen, J. Zaanen, *Phys. Rev. B* **39**, 1708 (1989).

[5] V.I. Anisimov and O. Gunnarson, *Phys. Rev. B* **43**, 7570 (1991).

[6] L. Hedin, *Phys. Rev.* **139**, A796 (1965).

[7] L. Hedin and S. Lundqvist, in *Solid State Physics* **23** p. 1, eds. H. Ehrenreich, F. Seitz and D. Turnbull (Academic, New York, 1969).

[8] F. Aryasetiawan, *Phys. Rev. B* **46**, 13051 (1992).

[9] R.W. Godby, M. Schlüter and L.J. Sham, *Phys. Rev. B* **37**, 10159 (1988).

[10] F. Aryasetiawan and O. Gunnarsson, *Phys. Rev. Lett.* **74**, 3221 (1995).

[11] V.I. Anisimov, J. Zaanen and O.K. Andersen, *Phys. Rev. B* **44**, 943 (1991).

[12] V.I. Anisimov, I.V. Solovyev, M.A. Korotin, M.T. Czyzyk and G.A. Sawatzky, *Phys. Rev. B* **48**, 16929 (1993).

[13] A.I. Lichtenstein, J. Zaanen, V.I. Anisimov, *Phys. Rev. B* **52**, R5467, (1995).

[14] A. Fujimori and F. Minami, *Phys. Rev. B* **30**, 957 (1984).

[15] J. van Elp, R.H. Potze, H. Eskes, R. Berger and G.A. Sawatzky, *Phys. Rev. B* **44**, 1530 (1991).

[16] P.W. Anderson, *Phys. Rev.* **124**, 41 (1961).

[17] B.R. Judd, *Operator techniques in atomic spectroscopy*, McGraw-Hill, New York, 1963.

[18] F.M.F. de Groot, J.C. Fuggle, B.T. Thole and G.A. Sawatzky, *Phys. Rev. B* **42**, 5459 (1990).

[19] O.K. Andersen, *Phys. Rev. B* **12**, 3060 (1975).

[20] Vladimir I. Anisimov, F. Aryasetiawan, A.I. Lichtenstein, *J. Phys.: Condens. Matter* **9**, 767 (1997).

[21] B.N. Harmon, V.P. Antropov, A.I. Lichtenstein, I.V. Solovyev, V.I. Anisimov, *J. Phys. Chem. Solids* **56**, 1521 (1995).

[22] D. Riegel, K.D. Gross and M. Luszik-Bhadra, *Phys. Rev. Lett.* **59**, 1244 (1987); K.D. Gross, D. Riegel, R. Zeller, *Phys. Rev. Lett.* **63**, 1176 (1989); K.D. Gross and D. Riegel, *Phys. Rev. Lett.* **61**, 1249 (1988).
[23] V.I. Anisimov, P.H. Dederichs, *Solid State Comm.* **84**, 241 (1992).
[24] I.V. Solovyev, P.H. Dederichs, V.I. Anisimov, *Phys. Rev. B* **50**, 16861 (1994).
[25] P. Fisher, B. Lebech, G. Meier, B.D. Rainford, O. Vogt, *J. Phys. C* **11**, 345 (1978); G. Meier, P. Fisher, W. Haly, B. Lebech, B.D. Raiford, O. Vogt, *ibid.* **11**, 1173 (1978).
[26] M.R. Norman, D.D. Koelling, *Phys. Rev. B* **33**, 6730 (1986).
[27] T. Kasuya, O. Sakai, J. Tanaka, H. Kitazawa, T. Suzuki, *J. Magn. Magn. Mater.* **63–64**, 9 (1987).
[28] A.I. Lichtenstein, V.P. Antropov, B.N. Harmon, *Phys. Rev. B* **49**, 10770 (1994).
[29] R. Fehrenbacher, T.M. Rice, *Phys. Rev. Lett.* **70**, 3471 (1993).
[30] A.I. Lichtenstein, I.I. Mazin, *Phys. Rev. Lett.* **74**, 1000 (1995).
[31] G.A. Sawatzky, J.W. Allen, *Phys. Rev. Lett.* **53**, 2239 (1984).
[32] F.M.F. de Groot, M. Grioni, J.C. Fuggle, J. Ghijsen, G.A. Sawatzky, H. Petersen, *Phys. Rev. B* **40**, 5715 (1989).
[33] K.C. Haas, in *Solid State Physics*, edited by C.H. Eherenreinch and D. Turnbull (Academic, New York, 1989), Vol. 42.
[34] J. Zaanen, in *Proceedings of the International Symposium on High-T_c Superconductivity, Jaipur, India*, 1989, edited by K.B. Karg (Oxford, New Delhi).
[35] O. Gunnarsson, O. Jepsen, Z.-X. Shen, *Phys. Rev. B* **42**, 8707 (1990).
[36] J.C. Fuggle, P.J.W. Weiss, R. Schoorl, G.A. Sawatzky, J. Fink, N. Nucker, P.J. Durham, W.M. Temmerman, *Phys. Rev. B* **37**, 123 (1988).
[37] G.K. Wertheim, H.J. Guggenheim, S. Hüffner, *Phys. Rev. Lett.* **30**, 1050 (1973); D.E. Eastman, J.F. Freeouf, *ibid.* **34**, 395 (1975).
[38] J. Zaanen, G.A. Sawatzky, *Can. J. Phys.* **65**, 1262 (1987).
[39] K. Hirokawa, H. Kadowaki, K. Ubikoshi, *J. Phys. Soc. Jpn.* **54**, 3526 (1985).
[40] P. Kuiper, G. Kruizinga, J. Ghijsen, G.A. Sawatzky, H. Verwey, *Phys. Rev. Lett.* **62**, 221 (1989).
[41] W. Bronger., H. Bade, W. Klemm, Z. Anorg. *Algem. Chem.* **333**, 188 (1964); J.B.Goodenough, D.G. Wickham, W.J. Croff, *J. Phys. Chem. Sol.* **5**, 17 (1958).
[42] M.A. Korotin, A.V. Postnikov, T. Neumann, G. Borstel, V.I. Anisimov, M. Methfessel, *Phys. Rev. B* **49**, 6548 (1994).
[43] H.Okamoto *et al.*, *Phys. Rev. B* **42**, 10381 (1990).
[44] V.I. Anisimov, R.C. Albers, J.M. Wills, M. Allouani, J.W. Wilkins, *Phys. Rev. B* **52**, R6975 (1995).
[45] V.I. Anisimov, M.A. Korotin, J. Zaanen, O.K. Andersen, *Phys. Rev. Lett.* **68**, 345 (1992).
[46] K.I. Kugel and D.I. Khomskii, *Usp. Fiz. Nauk.* **136**, 621 (1982) *Sov.Phys.-Usp.* **25**, 231 (1982)].

[47] A.I. Lichtenstein, M.I. Katsnelson, V.P. Antropov, V.A. Gubanov, *J. Magn. Magn. Mater.* **67**, 65 (1987).
[48] S.K. Satija, J.D. Axe, G. Shirane, H. Yoshizawa, K. Hirokawa, *Phys. Rev. B* **21**, 2001 (1980); M.T. Hutchings, H. Ikeda, J.M. Milne, *J. Phys. C: Solid State Phys.* **12**, L739 (1979).
[49] J. Zaanen, G.A. Sawatzky, J.W. Allen, *Phys. Rev. Lett.* **55**, 418 (1985).
[50] W. Weber, *Phys. Rev. Lett.* **58**, 1371 (1987).
[51] F.C. Zhang, T.M. Rice, *Phys. Rev. B* **37**, 3759 (1988).
[52] M.S. Hybertsen *et al.*, *Phys. Rev. B* **41**, 11068 (1990).
[53] T. Egami *et al.*, in *Electronic Structure and Mechanisms of High Temperature Superconductivity*, edited by J. Ashkenazi and G. Vezzoli (Plenum, New York, 1991).
[54] V.I. Anisimov, S.Yu. Ezhov, I. Elfimov, I.V. Solovyev, T.M. Rice, *Phys. Rev. Lett.* **76**, 1735 (1996).
[55] M.A. Korotin, S.Yu. Ezhov, I.V. Solovyev, V.I. Anisimov, D.I. Khomskii, G.A. Sawatzky, *Phys. Rev. B* **54**, 5309 (1996).
[56] G. Foex, J. *Phys. Radium* **9**, 37 (1938).
[57] V. Jaccarino, G.K. Wertheim, J.H. Wernick, L.R. Walker and S. Arajs, *Phys. Rev.* **160**, 476 (1987).
[58] Y. Takahashi and T. Moriya, J. *Phys. Soc. Jpn.* **46**, 1451 (1979); S.N. Evangelou and D.M. Edwards, J. *Phys. C* **16**, 2121 (1983).
[59] G. Shirane, J.E. Fisher, Y. Endoh and K. Tajima, *Phys. Rev. Lett.* **59**, 351 (1987); K. Tajima, Y. Endoh, J.E. Fisher and G. Shirane, *Phys. Rev. B* **38**, 6954 (1988).
[60] T. Mason, *et al.*, *Phys. Lett.* **69**, 490 (1992); G. Aeppli in *Strongly Correlated Electronic Materials* ed. K.S. Bedell *et al.*, Addison-Wesley (1994) p. 3.
[61] D. Mandrus *et al.*, *Phys. Rev. B* **51**, 4763 1995, see also B.C. Sales *et al.*, *Phys. Rev. B* **50**, 8207 (1994) and C. Fu and S. Doniach *Phys. Rev. B* **51**, 17439 (1995).
[62] L.F. Mattheiss and D.R. Hamann, *Phys. Rev. B* **47**, 13114 (1993).
[63] P.M. Raccah and J.B. Goodenough, *Phys. Rev.* **155**, 932 (1967).
[64] K. Asai, O. Yokokura, N. Nishimori, H. Chou, J.M. Tranquada, G. Shirane, S. Higuchi, Y. Okajima and K. Kohn, *Phys. Rev. B* **50**, 3025 (1994).
[65] E.J.W. Verwey and P.W. Haayman, *Physica* **8**, 979 (1941); E.J. Verwey, P.W. Haayman and F.C. Romeijn, *J. Chem. Phys.* **15**, 181 (1947).
[66] Y. Fujii *et al.*, *Phys. Rev. B* **11**, 2036 (1975); M. Izumi and G. Shirane, *Solid State Com.* **17**, 433 (1975); S. Ida *et al.*, *J. Appl. Phys.* **49**, 1456 (1978).
[67] A. Chainani, T. Yokoya, T. Morimoto, T. Takahashi, S. Todo, *Phys. Rev. B* **51**, 17976 (1995); J.-H. Park, L. Tjeng, J.W. Allen, P. Metcalf, C.T. Chen, preprint.
[68] A. Yanase and K. Saratory, *J. Phys. Soc. Jpn.* bf 53, 312 (1984); Z. Zhang and S. Satpathy, *Phys. Rev. B* **44**, 13319 (1991).

[69] V.I. Anisimov, I.S. Elfimov, N. Hamada, K. Terakura, *Phys. Rev. B* **54**, 3847 (1996).
[70] G.H. Jonker and J.H. van Santen, *Physica (Amsterdam)* **16**, 337 (1950); J.H. van Santen and G.H. Jonker, *Physica (Amsterdam)* **16**, 599 (1950).
[71] C. Zener, *Phys. Rev.* **82**, 403 (1951); P.W. Anderson and H. Hasegava, *Phys. Rev.* **100**, 675 (1955); P.G. de Gennes, *Phys. Rev.* **118**, 141 (1960).
[72] R.M. Kusters, J. Singleton, D.A. Keen, R. McGreevy and W. Hayes, *Physica (Amsterdam)* **155B**, 362 (1989); A. Urushibara, Y. Moritomo, T. Arima, A. Asamitsu, G. Kido and Y. Tokura, *Phys. Rev. B* **51**, 14103 (1995).
[73] J.B.A.A. Elemans, B. van Laar, K.R. van der Veen and B. O. Loopstra, *J. Solid State Chem.* **3**, 238 (1971).
[74] A.J. Millis, P.B. Littlewood and B. Shraiman, *Phys. Rev. Lett.* **74**, 5144 (1995).
[75] A.J. Millis, B. Shraiman and R. Mueller, *Phys. Rev. Lett.* **77**, 175 (1996).
[76] C.M. Varma, preprint.
[77] V.I. Anisimov, I.S. Elfimov, M.A. Korotin, K. Terakura, *Phys. Rev. B* **55**, 15494 (1997).
[78] Z. Jirak, S. Krupicka, V. Nekvasil, E. Pollert, G. Villeneuve and F. Zounova, *J. Magn. Magn. Mater.* **15–18**, 519 (1980); Z. Jirak, S. Krupicka, Z. Simsa, M. Dlouha and Z. Vratislav, *ibid.* **53**, 153 (1985).
[79] Y. Tomioka, A. Asamitsu, H. Kuwahara, Y. Moritomo and Y. Tokura, *Phys. Rev. B* **53**, R1689 (1996).
[80] N. Hamada, H. Sawada and K. Terakura, in *Spectroscopy of Mott Insulators and Correlated Metals*, edited by A. Fujimori and Y. Tokura (Springer-Verlag, Berlin, 1995), pp. 95–105.
[81] W. Pickett and D. Singh, *Phys. Rev. B* **53**, 1146 (1996).
[82] S. Satpathy, Z. Popovic and Ph. Vukajlovic, *Phys. Rev. Lett.* **76**, 960 (1996).
[83] V.I. Anisimov, P. Kuiper, J. Nordgren, *Phys. Rev. B* **50**, 8257 (1994).
[84] J. Zaanen, Ph.D. thesis, University of Groningen (The Netherlands), 1986.
[85] V.I. Anisimov, V.P. Antropov, A.I. Liechtenstein, V.A. Gubanov, and A.V. Postnikov, *Phys. Rev. B* **37**, 5598 (1988).
[86] A. Georges, G. Kotliar, W. Krauth, and M.J. Rosenberg, *Rev. Mod. Phys.* **68**, 13 (1996).
[87] N.E. Bickers, D.J. Scalapino, *Annals of Physics,* **193**, 206 (1989).
[88] O.K. Andersen, O. Jepsen, A.I. Liechtenstein, and I.I. Mazin *Phys. Rev. B* **49**, 4145 (1994).
[89] H. Ding *et al.*, *Phys. Rev. Lett.* **74**, 2784 (1995).
[90] Georges A., Kotliar G., Krauth W. and Rozenberg M.J., *Reviews of Modern Physics* **v68**, n.1, 13 (1996).
[91] V.I. Anisimov, A.I. Poteryaev, M.A. Korotin, A.O. Anokhin, G. Kotliar, *J. Phys.: Condens. Matter* **9**, 7359 (1997).
[92] O. Gunnarsson, O. Jepsen, O.K. Andersen, *Phys. Rev. B* **27**, 7144 (1983).
[93] Ph. Lambin and J.P. Vigneron, *Phys. Rev.* **B 29**, 3430 (1984).

[94] H.J. Vidberg and J.W. Serene, *Journal of Low Temperature Physics* v **29**, 179 (1977).

[95] A. Fujimori, I. Hase, H.Namatame, Y.Fujishima, Y.Tokura, H.Eisaki, S.Uchida, K.Takegahara, F.M.F de Groot, *Phys. Rev. Lett.* **69**, 1796 (1992). (Actually in this article the chemical formula of the sample was $LaTiO_{3.03}$, but the excess of oxygen produce 6 holes which is equivalent to doping of 6 Sr).

3. LSDA and Self-Interaction Correction

TAKEO FUJIWARA, MASAO ARAI and YASUSHI ISHII

1 Introduction

Density Functional Theory (DFT) is now the fundamental tool for first principles electronic structure calculations [1–3] . In principle, the electronic structures of small finite systems can be calculated exactly by the Hartree-Fock (HF) – configuration interaction (CI) method. For infinite systems with many-body effects or correlation effects, the HF-CI calculation is actually impossible. The DFT theory is applicable in this case and any many-body system can be uniquely mapped into a virtual non-interacting system, where many-body effects are properly considered in a self-consistent effective one-body potential. One usually uses the local approximation for the exchange-correlation energy functional (Local Spin Density Approximation = LSDA).

Though the LSDA has successfully provided the insight for the ground state properties such as the cohesive properties and even the ground state magnetic moment in transition metals [4], the LSDA usually underestimates the band gaps in insulators and semiconductors and even the proper ground state for strongly correlated electron systems in, e.g. transition metal oxides. Perdew *et al.* [5] and Sham *et al.* [6] showed the discontinuity of the exchange-correlation potential with respect to an infinitesimal change of the electron number. It is now widely known that the underestimate of the band gaps in insulators and semiconductors is not due to the functional form of the exchange-correlation energy in the local approximation, but due to the DFT itself. Although there have been several attempts to implement the discontinuity of the exchange-correlation energy functionals [6, 7], it would lose the easy handling of the LSDA framework.

Perdew and Zunger [8] proposed another treatment for this problem, by introducing the self-interaction correction term, which may be guided by the physical picture for the many-body effect. This approach loses its mathematical rigor but is easily manipulated in the LSDA formalism and applicable to many problems. Actually the problem of underestimating band gaps is essentially related to the localization/delocalization of valence electrons and this kind of improvement for the LSDA may help greatly for strongly correlated systems.

In the present article, we review the general idea of the self-interaction correction and show several applications to atoms, molecules and solids, especially strongly correlated systems such as the transition metal and rare earth metal compounds. We attribute the reason for successful results by the self-interaction correction to the formation of the non-local exchange correlation hole. We stress the importance of concepts of *localized* orbitals and *canonical* orbitals which relate to the orbitals of the removed electrons. The self-interaction correction can have a possibility of localized orbitals for the ground states in transition metal monoxides and rare earth metals. In Ce metals, the self-interaction correction can allow a unique scenario of the unified picture of the localized and itinerant f-electrons. We will also attribute the reason for unsatisfactory results of overestimated band gaps in the LSDA-SIC to the unscreened Coulomb repulsive interaction in the proposed self-interaction correction.

2 Total Energy and Self-Interaction Correction

In this section, we will give an expression for the LSDA total energy and discuss its shortcomings. Incompleteness of the LSDA gives rise to a self-interaction term in the total energy expression. On the contrary, a correction term for the self-interaction, called the self-interaction correction (SIC), causes another difficulty, that is the non-invariance of the total energy by the unitary transformation of the one-electron wavefunctions, and requires the concept of *canonical* orbitals.

2.1 LSDA total energy

The total energy is given in atomic units as

$$\begin{aligned} &E^{LSDA}[n_\uparrow, n_\downarrow] \\ &= T_0[n_\uparrow, n_\downarrow] + \sum_\sigma \int d\mathbf{r} V^\sigma_{ext}(\mathbf{r}) n_\sigma(\mathbf{r}) \\ &+ E_H[n] + E^{LSDA}_{xc}[n_\uparrow, n_\downarrow] , \end{aligned} \tag{1}$$

where, $n(\mathbf{r}) = n_\uparrow(\mathbf{r}) + n_\downarrow(\mathbf{r})$ is the valence electron density and $n_{\uparrow/\downarrow}(\mathbf{r})$ is the up/down spin density. Terms T_0, $E_H[n]$ and E_{xc} are kinetic energy of a non-interacting system, electrostatic Coulomb energy and exchange-correlation energy functionals, respectively, and V^σ_{ext} is a spin-dependent external potential.

The kinetic energy T_0 can be written as

$$T_0[n_\uparrow, n_\downarrow] = \underset{\{\psi_{i\sigma}\}}{Min} \sum_{i\sigma} f_{i\sigma} \langle \psi_{i\sigma} | -\frac{1}{2}\nabla^2 | \psi_{i\sigma} \rangle, \tag{2}$$

where $f_{i\sigma}$ is the occupation (0 or 1) of an orbital $\psi_{i\sigma}$. The minimization is taken over for all one-electron wavefunctions $\{\psi_{i\sigma}\}$ with a fixed spin density $n_\sigma(\mathbf{r})$ given by

$$n_\sigma(\mathbf{r}) = \sum_i f_{i\sigma} |\psi_{i\sigma}(\mathbf{r})|^2 , \tag{3}$$

and a constraint of orthonormality

$$\langle \psi_{i\sigma} | \psi_{j\sigma} \rangle = \delta_{ij} . \tag{4}$$

As for the exchange-correlation energy functional, we usually adopt the local spin density approximation (LSDA) and a functional form of

$$E_{xc}^{LSDA}[n_\uparrow, n_\downarrow] = \sum_\sigma \int d\mathbf{r} n_\sigma(\mathbf{r}) \epsilon_{xc}^\sigma(n_\uparrow(\mathbf{r}), n_\downarrow(\mathbf{r})) , \tag{5}$$

where $\epsilon_{xc}^\sigma(n_\uparrow, n_\downarrow)$ is the exchange-correlation energy density for a homogeneous system.

Minimizing Eq. (1) with respect to one-electron wavefunction with a constraint Eq. (4), we obtain the Kohn-Sham equation as

$$\begin{aligned} H_{LSDA}^\sigma \psi_{i\sigma}(\mathbf{r}) &= \sum_j \varepsilon_{ji}^{(\sigma)} \psi_{j\sigma}(\mathbf{r}) \\ H_{LSDA}^\sigma &= -\frac{1}{2}\nabla^2 + V_{ext}^\sigma(\mathbf{r}) + \int d\mathbf{r}' \frac{n(\mathbf{r}')}{|\mathbf{r} - \mathbf{r}'|} \\ &+ \left[\epsilon_{xc}^\sigma(n_\uparrow, n_\downarrow) + n_\sigma \frac{\partial \epsilon_{xc}^\sigma(n_\uparrow, n_\downarrow)}{\partial n_\sigma} \right]_{n_\sigma = n_\sigma(\mathbf{r})} , \end{aligned} \tag{6}$$

where the last two terms represent the exchange-correlation potential in the local approximation. Since the total energy Eq. (1) is invariant under a unitary transformation for occupied orbitals $\psi_{i\sigma}(\mathbf{r})$, we may choose them as eigenstates of the Kohn-Sham Hamiltonian H_{LSDA}^σ and a standard form of the Kohn-Sham equation is obtained as

$$H_{LSDA}^\sigma \psi_{i\sigma}(\mathbf{r}) = \varepsilon_i^{(\sigma)} \psi_{i\sigma}(\mathbf{r}) . \tag{7}$$

This point is not the case for the SIC, in which the total energy depends on the choice of orbitals, as discussed later.

The most serious difficulty in the LSDA may be the fact that a band gap in insulators predicted by the LSDA calculation is always smaller than the experimental ones by about 50%. The band gap in an insulator is defined as

$$E_G = \{E_{M+1} - E_M\} - \{E_M - E_{M-1}\} ,$$

where E_M is the *true* total energy of a system of M electrons. This can be rewritten as

$$\begin{aligned} E_G =& \frac{\partial E}{\partial n}\bigg|_{N=M+0} - \frac{\partial E}{\partial n}\bigg|_{N=M-0} \\ =& \varepsilon_{M+1}(M+0) - \varepsilon_M(M-0) \\ =& \varepsilon_{M+1}^{K-S}(M) - \varepsilon_M^{K-S}(M) + \frac{\partial E}{\partial n}\bigg|_{N=M-0}^{N=M+0} . \end{aligned} \tag{8}$$

Here $\varepsilon_L(M+\delta)$ is the L-th *true* one-electron energy of a system of $M+\delta$ electrons, and $\varepsilon_L^{K-S}(M)$ is the L-th one-electron energy of the Kohn-Sham equation of a system of M electrons. Therefore, we may miss a contributing term from the discontinuity of the *true* exchange-correlation potential, once we estimate the band gap by the one-electron energies obtained by the LSDA calculation where the exchange-correlation energy functional is assumed to be continuous as a function of the electron number. A simple model calculation shows that, in the ground state of the Mott insulator, the discontinuity of the exchange-correlation potential contributes significantly and, on the contrary, that in semiconductors may be negligibly small [9].

2.2 LSDA Self-Interaction Correction

The on-site Coulomb repulsion U is crucial to understand the electronic structure in strongly correlated electron systems, which splits the one-particle energy by an amount U. Though the value of U with the screening effects due to surrounding conduction electrons can be estimated successfully within the LSDA [10], the LSDA fails to predict this splitting of one-electron energy. The LSDA assumes the same local potentials for both occupied and unoccupied orbitals. On the other hand, the one-electron potential in the Hartree-Fock (HF) approximation is not the same for occupied and unoccupied orbitals because it is non-local, and one can expect a much larger energy gap between occupied and unoccupied orbitals.

2.2.1 SIC Energy Functionals and Minimization Orbitals

The origin of the difference between the LSDA and the HF approximation lies in a different expression for the exchange-correlation energy functionals. The HF approximation uses the exact expression of the exchange energy but omits the correlation energy, and the resulting effective potential is non-local. The LSDA uses an approximate local form of the exchange-correlation energy and the Kohn-Sham effective potential is local. The LSDA treatment causes a non-vanishing effective electron-electron potential (self-interaction) even in a dilute limit such as a hydrogen atom, e.g. for a fully polarized one-electron system,

$$E_H[n_{i\sigma}] + E_{xc}^{LSDA}[n_{i\sigma}, 0] \neq 0 , \tag{9}$$

whereas cancelation of the electrostatic and the exchange energies is perfect in the HF. This unsatisfactory treatment gives rise to an unphysical effective potential for the occupied orbitals caused by themselves. To remove the failure of the self-interaction, the self-interaction correction is introduced by assuming a self-interaction free total energy functional

$$E^{LSDA-SIC} = E^{LSDA}[n_\uparrow, n_\downarrow] + \sum_{i\sigma} E^{SIC}[n_{i\sigma}] ,$$
$$E^{SIC}[n_{i\sigma}] = -\left\{E_H[n_{i\sigma}] + E_{xc}^{LSDA}[n_{i\sigma}, 0]\right\} . \tag{10}$$

It may be noteworthy that the correction term is exact, strictly speaking, only in a dilute density limit.

The self-interaction corrected functional may be used to introduce a corrected Kohn-Sham equation [8],

$$H_{LSDA-SIC}^{i\sigma}\phi_{i\sigma}(\mathbf{r}) = \sum_j \varepsilon_{ji}^{(\sigma)}\phi_{j\sigma}(\mathbf{r})$$
$$H_{LSDA-SIC}^{i\sigma} = H_{LSDA}^{\sigma} + V_{SIC}^{i\sigma}(\mathbf{r})$$
$$V_{SIC}^{i\sigma} = -\left\{\int d\mathbf{r}' \frac{n_{i\sigma}(\mathbf{r}')}{|\mathbf{r}-\mathbf{r}'|} + \frac{\delta E_{xc}^{LSD}[n_{i\sigma}, 0]}{\delta n_{i\sigma}}\right\}$$
$$n_{i\sigma}(\mathbf{r}) = f_{i\sigma}|\phi_{i\sigma}(\mathbf{r})|^2 . \tag{11}$$

Here orbitals $\phi_{i\sigma}(\mathbf{r})$ minimizing the SIC total energy Eq. (10) are required to satisfy the orthonormality.

For electrons in a crystal, the charge density and the effective potential are periodic functions of the spatial coordinates because of the translational symmetry. Hence solutions of the Kohn-Sham equation are extended Bloch orbitals within the LSDA. To obtain localized solutions, which are appropriate for one-particle states in strongly correlated electron systems, we should introduce a symmetry-breaking field. The SIC gives a physical origin of such an effective symmetry-breaking field through orbital dependence of $V_{SIC}^{i\sigma}$. However, the SIC also introduces more subtle features into the formalism. We will discuss about them in the next subsection.

2.2.2 Non-orthogonality and Non-unitarity of SIC Orbitals

The self-interaction correction (SIC) leads to serious difficulty in the Kohn-Sham formalism because it breaks the assumption of the existence of the kinetic energy functional independent of a choice of $E_{xc}^{LSDA}[n_\uparrow, n_\downarrow]$. As a result, two unusual features emerge: (1) if we approximate the orbitals $\phi_{i\sigma}(\mathbf{r})$ by the eigenstates of the corrected Kohn-Sham Hamiltonian Eq. (11), they are not, in general, orthogonal to each other, and (2) the SIC total energy is not invariant under a unitary transformation for the occupied orbitals and hence those orbitals minimizing the total energy break the symmetry of the system.

First, we should remember that the orthogonality of the eigenfunction cannot be guaranteed in the present formalism because the SIC potential is orbital-dependent. This means that the minimization of the total energy is not a naive eigenvalue problem any more.

To see the non-unitary property of the minimization orbitals more explicitly, we estimate the SIC energy as follows [8]:

$$E_H[n_{i\sigma}] \leq 1.092 \int d\mathbf{r} n_{i\sigma}^{4/3}(\mathbf{r})$$

$$E_x^{LSDA}[n_{i\sigma}, 0] = -0.9305 \int d\mathbf{r} n_{i\sigma}^{4/3}(\mathbf{r})$$

$$E^{SIC}[n_{i\sigma}] \geq -0.16 \int d\mathbf{r} n_{i\sigma}^{4/3}(\mathbf{r}) \,. \tag{12}$$

Then the estimated value of $E^{SIC}[n_{i\sigma}]$ may be of order 1 for the localized orbitals. On the contrary, for the delocalized orbitals, the $E^{SIC}[n_{i\sigma}]$ vanishes because $n_{i\sigma} \propto V^{-1}$ and $E^{SIC}[n_{i\sigma}] \propto V^{-1/3}$ where V is the volume of the system, and the SIC is more effective for the localized orbitals. This implies that the choice of orbitals is crucial for the minimization of the total energy and that the ground states should be uniquely determined by the choice of orbitals.

2.2.3 SIC Treatment of Isolated Atoms

Far from a nucleus in an isolated ion of a charge Q, the true one-electron potential for the bounded electron should be $-(Q+1)/r$, because an electron cannot be affected by its own charge. However, in the LSDA, an electron sees a tail of $-Q/r$ far from the nucleus. This causes several serious difficulties, e.g. ionization energy, stability of negative ions, surface properties of solids etc. The SIC treatment is a unique method to give a correct behavior of the one-electron potential in the long-range region from a nucleus.

In isolated atoms, the off-diagonal values of $\varepsilon_{ij}^{(\sigma)}$ in the LSDA-SIC treatment may be small and we can consider the diagonal element as a removal energy of unit electron charge. The large cancelation between the self-Coulomb energy and the self-exchange energy can happen and the LSDA-SIC treatment gives a fairly reasonable agreement with the HF approximation [8]. This solves the serious difficulty of ds-level conversion in $d^{n-1}s^1$ configuration of $3d$ transition metal atoms.

In Table 1, the outer-orbital energy which is the SIC corrected diagonal energy for the Schmidt orthogonalized orbital is compared with the first ionization potential for several atoms [8]. From these results, we can see the great benefit of the LSDA-SIC treatment in atomic and ionic systems. On average, the LSDA-SIC eigenvalues show better agreement with experimental first ionization potentials than HF eigenvalues.

Table 1: *Outer-orbital eigenenergies and measured first ionization potentials I_1. (in eV unit) This table is from Ref. 8.*

	$-\varepsilon^{\text{HF}*}$	$-\varepsilon^{\text{LSDA}}$ **	$-\varepsilon^{\text{LSDA-SIC}**}$	I_1***
H	13.6	7.3	13.6	13.6
He	25.0	15.5	25.8	24.6
Li	5.3	3.2	5.4	5.4
Na	5.0	3.1	5.1	5.1
N	15.4	8.3	14.9	14.5
Mn	6.7	4.6	7.1	7.4

*) C. F. Fischer, *The Hartree-Fock Method for Atoms* (Wiley, New York, 1977).
) J. P. Perdew and A. Zunger, *Phys. Rev.* **B15, 5048 (1981).
***) C. E. Moore, *Natl. Bur. Stand.* (U.S.) Ref. Data Ser, **34**, 1 (1970).

According to the Janak's theorem [11] for the LSDA, the electron removal energy is calculated as

$$E^{LSDA}(M) - E^{LSDA}(M-1) = \int_0^1 df \varepsilon(f), \tag{13}$$

where $E^{LSDA}(M)$ is the LSDA total energy of a system of M-electrons and $\varepsilon(f)$ is the orbital energy of the highest occupied state with the occupation number f. If the effective potential in the Kohn-Sham equation is insensitive to the change of the occupation number and $\varepsilon(f)$ is independent of f, $\varepsilon(f)$ gives a good estimate of the ionization energy. This is realized in the LSDA-SIC when we interpret the diagonal element $\varepsilon_{ii}^{(\sigma)}$ as $\varepsilon(f)$.

In Fig. 1, we show the orbital energy for the Ne $2p$ state as a function of its occupation number [12]. The orbital energy obtained by the LSDA varies significantly with the occupation number while that obtained by the LSDA-SIC is almost constant except for $f \approx 0$. The SIC in the exchange-correlation potential is responsible for the difference in the orbital energies. For the case of a more-than half-filled orbital, the SIC lowers the effective potential because the self-Coulomb correction dominates. For the less-than half-filled case, on the other hand, the self-exchange part raises the effective potential. Since the self-exchange part is proportional to $f^{1/3}$ for small f, the LSDA-SIC orbital energy varies in a singular way near $f = 0$. Although the SIC does not describe directly the discontinuity in the exchange-correlation potential, the behavior of the exchange-correlation potential with varying the electron number is improved in the LSDA-SIC

The "exchange-only" LSDA, in which the correlation effect is not considered, gives the exchange energy with an absolute value smaller by 5–15% than the exact one obtained by the HF calculation for isolated atoms. The LSDA-SIC certainly improves this point to reduce the error to 2–3% [8]. For the correlation energy,

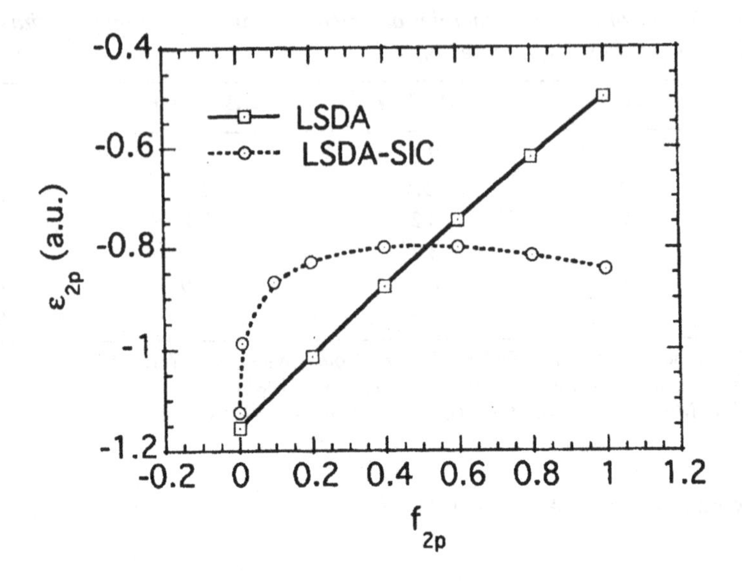

Figure 1: *Variation of the orbital energy of the Ne $2p$ state as a function of the occupation number.*

the LSDA typically overestimates by 100–200% in comparison with the difference in the total energies between the HF calculation and the experimental value. The LSDA-SIC totally improves an error in the LSDA for the exchange-correlation energy. If the exchange-correlation energy is evaluated more accurately as in the generalized gradient approximation [13], however, the self-Coulomb and self-exchange energies are almost cancelled [12].

The exchange-hole density of an isolated Ne atom is shown in Fig. 2. The LSDA-SIC clearly gives much realistic description of the spatial shape of the exchange-correlation hole and the density peak locates closer to the nucleus than the result of LSDA. The exchange-correlation hole in the LSDA-SIC treatment can be expressed as

$$\begin{aligned} n_{xc}(\mathbf{r}, \mathbf{r}+\mathbf{R}) = n_{xc}^{LSDA}&(\mathbf{R}; n_{\uparrow}(\mathbf{r}), n_{\downarrow}(\mathbf{r})) \\ &- \frac{1}{n(\mathbf{r})} \sum_{i\sigma} [n_{i\sigma}(\mathbf{r}) n_{xc}^{LSDA}(\mathbf{R}; n_{i\sigma}(\mathbf{r}), 0) \\ &+ n_{i\sigma}(\mathbf{r}) n_{i\sigma}(\mathbf{r}+\mathbf{R})] , \end{aligned} \tag{14}$$

where $n_{xc}^{LSDA}(\mathbf{R}; n_{\uparrow}(\mathbf{r}), n_{\downarrow}(\mathbf{r}))$ is the LSDA exchange-correlation hole. The LSDA exchange-correlation hole is spherical symmetric in the sense that it depends on the relative distance $|\mathbf{R}|$ and, on the contrary, the LSDA-SIC hole is not. Because of

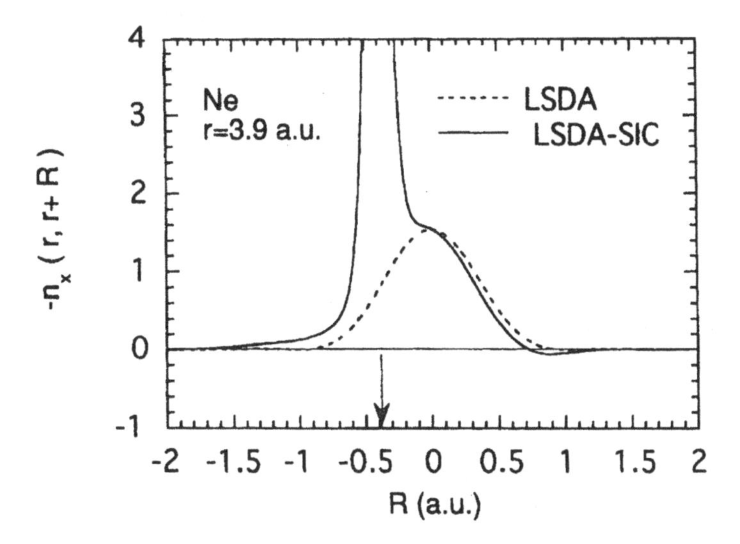

Figure 2: *The exchange-hole density for a Ne atom. The LSDA and LSDA-SIC results are shown by dotted and solid lines, respectively. An electron is located at 0.39 a.u. from the nucleus whose position is indicated by an arrow.*

this non-sphericity due to the non-locality of the last term in Eq. (14), the density of the LSDA-SIC exchange-correlation hole is very much improved.

2.2.4 Early SIC Treatment of Solids

A natural extension for the LSDA-SIC was reported for several bulk solids [14, 15]. Heaton *et al.* [14] formulated the SIC by using the LCAO scheme and got good results for large gaps in insulators. Hamada and Ohnishi [15] calculated band gaps of semiconductors in the LSDA-SIC by the LAPW framework, where an atomic SIC is included only in LAPW muffin-tin spheres.

We suppose a limiting case where isolated atoms are periodically arranged with a macroscopic lattice constant. Then the resulting LSDA band structures may show zero-width bands at atomic energy levels. The ionization energy or the band gap energy are underestimated, where the Bloch states are the eigenstates and the SIC vanishes. However the atomic Wannier representation is also the eigenstate and it can improve the one-particle energy in the LSDA-SIC. Next we introduce a small overlap between atomic orbitals. Then the resulting bands have small dispersion, and the Wannier representation cannot be an eigenstate any more. In this case the unique eigenstates satisfying the periodicity of the system are in the Bloch representation and the SIC vanishes. This is unphysical, because infinitesimally small overlap would cause a kind of phase transition.

Table 2: *Band gaps in wide gap insulators Ar and LiCl and semiconductors Si and GaAs.*

	Calculated (eV)		Experiment (eV)
	LSDA	LSDA-SIC	
Ar *	7.89	13.5	14.2
LiCl *	5.81	10.1	9.4–9.9
Si **	0.43	1.27	1.17
GaAs **	0.097	1.05	1.52

*) R. A. Heaton, J. G. Harrison and C. C. Lin, *Phys. Rev.* B**28**, 5992 (1983)
) Y. Hatsugai and T. Fujiwara, *Phys. Rev.* B37**, 1280 (1988).

One trial has been introduced [16] and the SIC was applied to the physically natural states for the one-particle density in semiconductors. The LSDA-SIC calculation was applied to Harrison's bond orbital model [17], which provides a physical nature of localized charge density between bonds. Then the SIC as calculated to localized bond charge density after the self-consistent LSDA calculation. The results are summarized in Table 2, together with the result by an atomic Wannier SIC calculation in wide gap insulators.

2.3 Canonical Orbitals

In the LSDA, Janak's theorem [11] described that the one-electron eigenvalue of the Kohn-Sham equation $\varepsilon_i^{(\sigma)}$ is an energy cost of a unit electron charge for an infinitesimally small amount of electron charge;

$$\frac{\partial E^{LSDA}}{\partial f_{i\sigma}} = \varepsilon_i^{(\sigma)} . \tag{15}$$

The SIC not only introduces non-orthogonality and non-unitarity, but requires interpretation of one-electron eigenvalues by Janak's theorem to be extended and modified.

The LSDA-SIC calculation was extended in molecules by Pederson *et al.* [18], and they showed the equivalence of *minimization* of the SIC total energy Eq. (10), with respect to individual one-electron orbital, to the *localization* condition

$$\langle \phi_{i\sigma} | V_{SIC}^{i\sigma} - V_{SIC}^{j\sigma'} | \phi_{j\sigma'} \rangle = 0 , \tag{16}$$

using a minimization condition by an infinitesimal orthogonality transformation. Since the SIC is effective for localized orbitals, the minimization orbitals $\phi_{i\sigma}$ may be localized. The Lagrange multiplier $\varepsilon_{ji}^{(\sigma)}$ can be calculated as

$$\varepsilon_{ji}^{(\sigma)} = \langle \phi_{j\sigma} | H_{LSDA-SIC}^{i\sigma} | \phi_{i\sigma} \rangle ,$$

and one can simply obtain, together with Eq. (16), the hermiticity of the Lagrange multiplier

$$\varepsilon_{ji}^{(\sigma)} = \varepsilon_{ij}^{(\sigma)*} . \tag{17}$$

One may expect that Janak's theorem preserves again for states diagonalizing the Lagrange multiplier $\varepsilon_{ji}^{(\sigma)}$. We then interpret the eigenvalues and the eigenstates of $\varepsilon_{ji}^{(\sigma)}$ as the ionization potential and the orbitals where one can remove an electron. The orbitals diagonalizing $\varepsilon_{ji}^{(\sigma)}$ are called *canonical orbitals* whereas the orbitals $\{\phi_{i\sigma}\}$ are called *localized orbitals*. The minimized total energy can be calculated in terms of *localized orbitals* and the diagonalized energy spectrum spanned by the canonical orbitals may be understood as the energy band for hopping electrons over the localized orbitals.

More explicitly, we could discuss as follows. We can diagonalize the hermitian Lagrange multiplier matrix with a unitary matrix $\{M_{\kappa i}^{(\sigma)}\}$;

$$\sum_{ij} M_{\kappa i}^{(\sigma)} \varepsilon_{ij}^{(\sigma)*} M_{\kappa' j}^{(\sigma)*} = \tilde{\varepsilon}_{\kappa}^{(\sigma)} \delta_{\kappa\kappa'} \tag{18}$$

Here, i and j run over the occupied states. The canonical orbitals are the corresponding eigenvectors, defined as

$$|\varphi_{\kappa\sigma}\rangle = \sum_{i} M_{\kappa i}^{(\sigma)} |\phi_{i\sigma}\rangle . \tag{19}$$

Let us consider a system with a spin density of occupation number $q_{\kappa\sigma}$ of canonical orbitals

$$n_{\sigma}(\mathbf{r}) = \sum_{\kappa} q_{\kappa\sigma} |\varphi_{\kappa\sigma}(\mathbf{r})|^2 \tag{20}$$

and define the "localized" orbitals as

$$|\tilde{\phi}_{i\sigma}\rangle = \sum_{\kappa} (M_{\kappa i}^{(\sigma)})^* \sqrt{q_{\kappa\sigma}} |\varphi_{\kappa\sigma}\rangle . \tag{21}$$

The "localized" orbitals $\{\tilde{\phi}_{i\sigma}\}$ are neither normalized nor orthogonalized for the fractional occupation $q_{\kappa\sigma} \neq 0,\ 1$. Once we assume the SIC total energy Eq. (11) where $f_{i\sigma}|\phi_{i\sigma}(\mathbf{r})|^2$ is replaced by $|\varphi_{\kappa\sigma}(\mathbf{r})|^2$, the derivative of the LSDA-SIC total energy by the occupation number $q_{\kappa\sigma}$ is just the energy eigenvalue $\tilde{\varepsilon}_{\kappa}^{(\sigma)}$ at the integral occupation configuration;

$$\tilde{\varepsilon}_{\kappa}^{(\sigma)} = \left.\frac{\partial E^{LSDA-SIC}}{\partial q_{\kappa\sigma}}\right|_{q_{\kappa\sigma}=1} . \tag{22}$$

To illustrate the idea of the canonical orbitals more precisely, we consider the one-dimensional Hubbard Hamiltonian [19, 20]

$$H_{Hubbard} = \sum_{ij\sigma} t_{ij} c_{i\sigma}^{\dagger} c_{j\sigma} + U \sum_{i} \hat{n}_{i\uparrow} \hat{n}_{i\downarrow} \tag{23}$$

and the corresponding LSDA total energy functional

$$E_{Hubbard}^{LSDA} = \sum_{ij\sigma} t_{ij} \langle c_{i\sigma}^{\dagger} c_{j\sigma} \rangle + \frac{U}{2} \sum_{i} [n_{i\uparrow} + n_{i\downarrow}]^2 + E_{xc}^{LSDA}[n_{\uparrow}, n_{\downarrow}] , \tag{24}$$

where $\hat{n}_{i\sigma} = c_{i\sigma}^{\dagger} c_{j\sigma}$ and $n_{i\sigma} = \langle \hat{n}_{i\sigma} \rangle$. The LSDA exchange-correlation energy is assumed to be

$$E_{xc}^{LSDA}[n_{\uparrow}, n_{\downarrow}] = \sum_{i} [n_{i\uparrow} + n_{i\downarrow}]^{4/3} [-a - bf(\xi(i))] \tag{25}$$

where

$$f(x) = \frac{[1+x]^{4/3} + [1-x]^{4/3} - 2}{2^{4/3} - 2}$$

with

$$\xi(i) = \frac{n_{i\uparrow} - n_{i\downarrow}}{n_{i\uparrow} + n_{i\downarrow}}$$

and parameters $a(= 0.3840)$ and $b(= 0.0705)$ are determined so that the atomic results ($t_{ij} = 0$) are consistent with the LSDA results of a free hydrogen atom. The canonical orbitals minimizing the SIC total energy are obtained by using the simulated annealing method, which is explained later.

The ionization energy I and the electron affinity A are shown in Fig. 3 [19], compared with exact results for the one-dimensional Hubbard model. The ionization energy and electron affinity should be defined, using the total energy E_M of the M-electron system, by

$$I(M) = E_{M-1} - E_M , \quad A(M) = E_M - E_{M+1} . \tag{26}$$

The ionization energy and the electron affinity explicitly depend on the electron number M and the results shown here are the ones extrapolated to infinite system size, $M \to \infty$. A doped single hole localizes at one site and, on the other hand, a doped single electron delocalizes over the whole system.

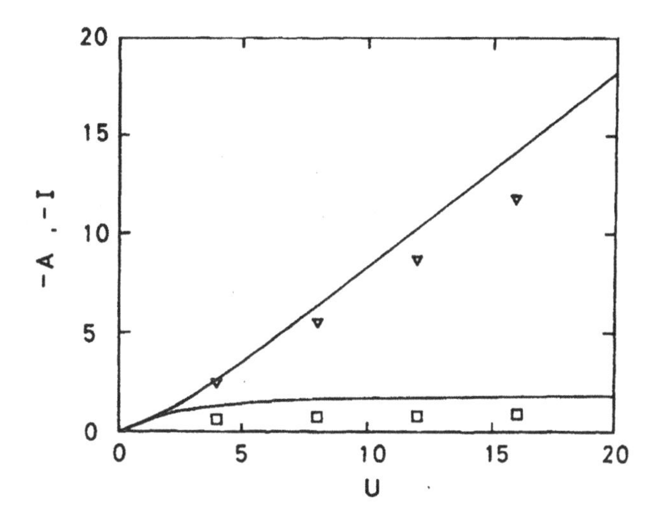

Figure 3: *The ionization energy I and electron affinity A as functions of the Coulomb repulsion U of the Hubbard Hamiltonian. (square (–I) and triangle (–A).)*

The energy spectra for the half-filled case, again in the one-dimensional Hubbard model, are shown in Fig. 4. The occupied band (the lower band) is obtained by diagonalizing the Lagrange multiplier matrix $\varepsilon_{ji}^{(\sigma)}$, while the unoccupied band (the upper band) is obtained by diagonalizing the LSDA Hamiltonian with the charge distribution from the LSDA-SIC calculation in the half-filled case. The band gap in LSDA is underestimated and the band width is overestimated, because of the less localized nature in the LSDA scheme compared with the cases of HF and LSDA-SIC approximation. The HF scheme tends to overestimate the band gap and the LSDA-SIC scheme gives reasonable results overall.

The SIC generally pushes down the energy eigenvalues of occupied levels. Therefore, we may expect to widen the energy gap between occupied and unoccupied levels. The ionization energy and the electron affinity calculated from the total energy difference are indicated in the figure by the arrows, which shows that the total energy difference can be approximated by the energy eigenvalue of canonical orbitals of the electron doped system.

2.4 Self-Interaction Correction in Metallic Systems

For the half-filled case discussed in the previous section, we assume a broken-symmetry solution, for which the lattice periodicity is doubled and the Brillouin zone is folded to give an energy gap in a natural way. This is essential to construct the localized orbitals. If we assume a solution with the original lattice periodicity

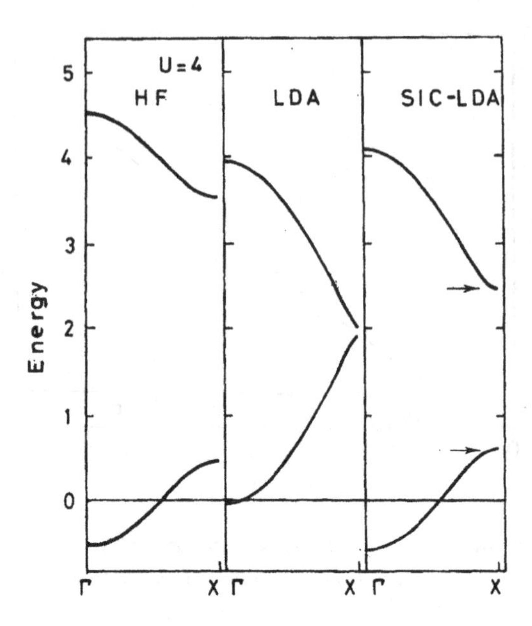

Figure 4: *The one particle excitation spectra for $U/t = 4$ in the HF, LDA and SIC-LDA. Arrows denote $-I(M)$ and $-A(M)$.*

or consider a non-half-filled case, on the other hand, we have a partially filled band and cannot construct any localized states from the occupied orbitals. This is because the occupied orbitals do not span the complete Hilbert space.

The tricky nature of the SIC for metallic cases has been discussed by one of the present authors and collaborators [21]. Consider a single impurity in a metal. If an impurity atomic level is below the bottom of the host-metal band, a true bound state is formed and the SIC is effective for such a bound state. When the impurity level is in the host-metal band, an electron is bound *virtually* near the impurity atom. Miyazaki *et al.* [21] have found that a description of such a virtual bound state in metals is also improved by considering the localized orbital.

2.5 Canonical Orbitals and Many-Electron Wavefunctions

Now we discuss how the canonical orbitals relate to the many-electron wavefunctions [22]. The M-electron wavefunction Φ^M in the LSDA-SIC may be given by a single Slater determinant. We can define two different types of many-electron wavefunctions; one is constructed of a set of localized orbitals $\{\phi^M_{i\sigma}\}$ and the other is of the canonical orbitals $\{\varphi^M_{k_i\sigma}\}$ as

$$\Phi^M_{\text{local}} = \frac{1}{\sqrt{M!}} \begin{vmatrix} \phi^M_{1\uparrow}(\mathbf{r}_1) & \phi^M_{2\downarrow}(\mathbf{r}_1) & \phi^M_{3\uparrow}(\mathbf{r}_1) & \cdots & \phi^M_{M\downarrow}(\mathbf{r}_1) \\ \phi^M_{1\uparrow}(\mathbf{r}_2) & \phi^M_{2\downarrow}(\mathbf{r}_2) & \phi^M_{3\uparrow}(\mathbf{r}_2) & \cdots & \phi^M_{M\downarrow}(\mathbf{r}_2) \\ \vdots & & & & \vdots \\ \phi^M_{1\uparrow}(\mathbf{r}_M) & \phi^M_{2\downarrow}(\mathbf{r}_M) & \phi^M_{3\uparrow}(\mathbf{r}_M) & \cdots & \phi^M_{M\downarrow}(\mathbf{r}_M) \end{vmatrix}$$

$$\equiv ||\phi^M_{1\uparrow}\phi^M_{2\downarrow}\phi^M_{3\uparrow}\cdots\phi^M_{M\downarrow}|| \tag{27}$$

and

$$\Phi^M_{canonical} = ||\varphi^M_{k_1\uparrow}\varphi^M_{k_2\downarrow}\varphi^M_{k_3\uparrow}\cdots\varphi^M_{k_M\downarrow}|| \,. \tag{28}$$

Here, $1\uparrow,\ 2\downarrow,\ \cdots,\ M\downarrow$ are the indices of localized orbitals and the number of sites where the wavefunctions are mainly localized and the arrows mean the electron spin coordinates. In the second Slater determinant $\Phi^M_{canonical}$, $k_1\uparrow,\ k_2\downarrow,\ \cdots,\ k_M\downarrow$ are the wave numbers of canonical orbitals and spin coordinates. Actually the above two wavefunctions Φ^M_{local} and $\Phi^M_{canonical}$ are identical because the sets $\{\phi^M_{i\sigma}\}$ and $\{\varphi^M_{k_i\sigma}\}$ can be converted by a unitary transformation.

Next, we consider the $(M-1)$-electron wavefunction in a hole-doped system, where one down-spin electron is removed. This state may be described by a single Slater determinant Φ^{M-1,R_j}_{local} constructed of the localized orbitals $\{\phi^{M-1,R_j}_{i\sigma}\}$, where R_j denotes the hole-doped site.

$$\Phi^{M-1,R_j}_{local} = ||\phi^{M-1,R_j}_{1\uparrow}\phi^{M-1,R_j}_{2\downarrow}\phi^{M-1,R_j}_{3\uparrow}\cdots\phi^{M-1,R_j}_{M-1\downarrow}|| \,. \tag{29}$$

It must be noticed that the orthogonality relation between $\{\phi^{M-1,R_j}_{i\sigma}\}$ and $\{\phi^{M-1,R_l}_{i\sigma}\}$ holds for $R_j \neq R_l$. The $(M-1)$-electron wavefunction constructed from canonical orbitals can be written similarly as Eq. (28) with $(M-1)$ canonical orbitals and is identical to Φ^{M-1,R_j}_{local}. However, this wavefunction does not hold the translational symmetry of the whole system because the hole-doped site is specified. We can construct the other $(M-1)$-electron wavefunction by the linear combination of Φ^{M-1,R_j}_{local} as

$$\Psi^{M-1}(p\downarrow) = \frac{1}{A}\sum_j \Phi^{M-1,R_j}_{local} \times \exp(-ipR_j) \,, \tag{30}$$

where $(p\downarrow)$ is the momentum and spin of the doped hole.

We can also define another approximate hole state constructed from canonical orbitals of the M-electron system as

$$\Psi_{canonical}(k_l\downarrow) = ||\varphi^M_{k_1\uparrow}\varphi^M_{k_2\downarrow}\cdots\varphi^M_{k_{l-1}\uparrow}\varphi^M_{k_{l+1}\uparrow}\cdots\varphi^M_{k_M\downarrow}|| \tag{31}$$

where the $(k_l \downarrow)$ orbital $\varphi^M_{k_l\downarrow}$ is removed from $\Phi^M_{canonical}$. Two $(M-1)$-electron wavefunctions $\Psi^{M-1}(p\sigma)$ and $\Psi_{canonical}(q\sigma)$ satisfy the relationship:

$$|\langle \Psi_{canonical}(q\sigma)|\Psi^{M-1}(p\sigma')\rangle| = 0, q\sigma \neq p\sigma' ,$$
$$|\langle \Psi_{canonical}(k\downarrow)|\Psi^{M-1}(k\downarrow)\rangle| \simeq 1 , \tag{32}$$

where the momentum k covers over whole range. Therefore, one can conclude that the Slater determinant constructed of the canonical orbitals of the M-electron (half-filled) system well expresses the $(M-1)$-electron wavefunction and each canonical orbital expresses a one-electron wavefunction of a removed electron.

3 SIC in Solid Hydrogen and Transition Metal Monoxides

To calculate the charge density of localized orbitals, the localized basis set is essential and the Linear Muffin-Tin Orbital (LMTO) method is appropriate [23, 24].

The basis functions of the LMTO are written as

$$|\chi_{RL}\rangle = |\phi_{RL}\rangle + \sum_{R',L'} |\dot{\phi}_{R'L'}\rangle h_{R'L'RL} , \tag{33}$$

where R and L denote atomic sites and angular momenta (l, m), respectively. We omit the spin index σ for simplicity. ϕ_{RL} is a solution of the Schrödinger equation with a fixed energy E_ν and $\dot{\phi}_{RL}$ is its energy derivative. These solutions ϕ_{RL} and $\dot{\phi}_{RL}$ are constructed within the LSDA calculation. The localized orbitals ψ_i are then constructed by a linear combination of the LMTO basis functions as

$$|\psi_i\rangle = \sum_{R,L} |\chi_{RL}\rangle U_{RL,i} . \tag{34}$$

The coefficients $U_{RL,i}$ should be obtained so as to minimize the LSDA-SIC total energy. This has been carried out by using the simulated annealing technique applied to the equation:

$$\frac{d}{dt}U_{RL,i} = -\gamma \left\{ \sum_{R',L'} \left(O^{-1}H^i\right)_{RL,R'L'} U_{R'L',i} + \sum_j U_{RL,j}\varepsilon_{ji} \right\} , \tag{35}$$

where

$$H^i_{RL,R'L'} = \langle \chi_{RL}|H_{LSDA} + V^i_{SIC}|\chi_{R'L'}\rangle$$

and

$$O_{RL,R'L'} = \langle \chi_{RL} | \chi_{R'L'} \rangle \, .$$

The Lagrange multiplier ε_{ij} may be chosen as

$$\varepsilon_{ij} = \frac{1}{2}\Big[\langle \psi_i | \{H_{LSDA} + V^i_{SIC}\} | \psi_j \rangle + \langle \psi_i | \{H_{LSDA} + V^j_{SIC}\} | \psi_j \rangle \Big] \tag{36}$$

and this form of ε_{ij} guarantees the orthogonality condition of ψ_i during the iteration procedure. The final solution can be obtained as a solution of $\frac{d}{dt}U_{RL,i} = 0$ and the parameter γ is arbitrarily chosen so that rapid convergence is achieved.

Once we calculate matrix elements of the SIC potential $\langle \psi_n | V^i_{SIC} | \psi_i \rangle$, all other calculation can be performed in k-space [25]. Accurate estimation of matrix elements of the SIC potential is essential to treat the spatial extent of the localized wavefunctions ψ_i and the calculation of the overlap matrix $O_{RL,R'L'}$ should be carefully done. Another equivalent formalism was given by using Wannier states constructed of the LMTO [26].

3.1 Solid Hydrogen

Here, we should address one comment on the SIC energy and character of the ground states. Usually one may believe that, when one can get localized orbitals for a solution of the LSDA-SIC equation, the localized orbitals could give the lowest energy solution. Actually this is not always the case [25, 27]. The localized orbitals must have negative SIC energies to be the lowest energy solution since the SIC energies for extended states are exactly zero.

The reason why the LSDA can give good results in general is, in case of comparing the LSDA energies of two systems, a large part of the errors in the LSDA total energy cancel with each other. On the contrary, when one compares the LSDA-SIC total energy of a system of extended orbitals, which is just the LSDA solution, with that of localized orbitals, which is an essential LSDA-SIC solution, an error in the exchange-correlation energy comes out to be very crucial. Though the SIC potential is always negative, the correlation energy, which is smaller than the exchange energy, is responsible for the sign of the SIC energy. Therefore, the sign of the SIC energy is often very sensitive to the functional form of the exchange-correlation energy. In other words, at the present stage that there are several functional forms for the exchange-correlation energy and no strict criterion for the choice of the functionals of the LSDA-SIC, the energy criteria of the LSDA-SIC problems for the ground states might lose its physical basis. In several cases, we should choose the ground states from a physical viewpoint, but not from the energy criterion. This point would be essential for discussion of the ground states of solid hydrogen [27] or the transition metal monoxides [25].

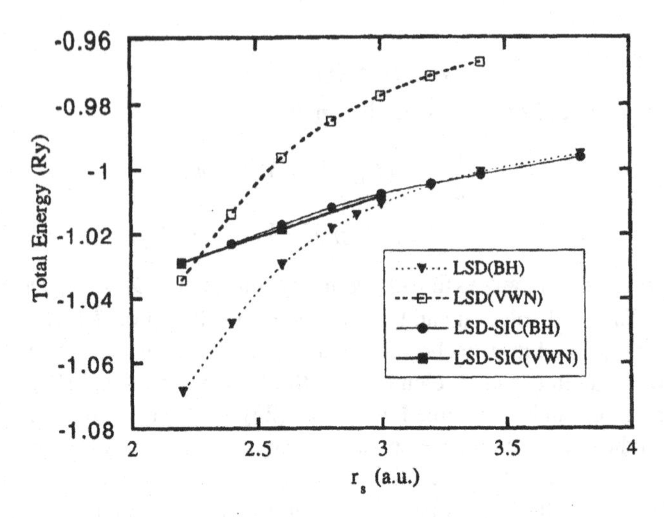

Figure 5: *Total energy of solid hydrogen as a function of the Wigner-Seitz radius.*

The above situation should be mentioned in the discussion of the ground state of solid hydrogen. The LSDA-SIC results for solid hydrogen are shown, using the Barth-Hedin exchange-correlation energy functional, denoted by BH in Fig. 5 [27]. For a larger lattice constant, i.e. larger radius of the Wigner-Seitz sphere r_s, the antiferromagnetic (AF) insulating phase of the localized orbitals gives a lower total energy. As r_s decreases, the lowest energy state changes to the AF insulating phase of the Bloch orbitals at $r_s = 3.2$. The AF insulating phase of the Bloch orbitals closes the band gap at $r_s = 2.8$. Further decreasing r_s, the magnetic moment vanishes at $r_s \approx 2.4$ and the paramagnetic metallic phase becomes the ground state. The transition from localized orbitals to Bloch orbitals within the AF insulating phase is, in itself, unphysical and we should frankly say that we could not discuss the physics near the magnetic and metal-insulator phase transition, by simply following the lowest total energy state.

Different results were obtained by using the exchange-correlation potential by Vosko *et al.* [27, 28], denoted by VWN in Fig. 5. The exchange-correlation energy functional of Vosko *et al.* gives a slightly smaller correlation energy to Bloch orbitals. Thus, it extends the region where the Bloch orbitals have lowest total energy. The transition occurs at $r_s = 2.4$ from the AF insulating phase ($r_s > 2.4$) to the paramagnetic metallic phase ($r_s < 2.4$).

Though the result by using the exchange-correlation functional of Vosko *et al.* describes a physically reasonable metal-insulator transition, this fact would not mean that the functional form of Vosko *et al.* should be always much better than that of Barth-Hedin. Any functional form of the exchange-correlation energy is obtained by simulating a uniform electron-gas system which is free from the SIC and, therefore, there are no reasons to accept some specific functional form for the LSDA-SIC. We presumably say that for the insulating phase one should apply the LSDA-SIC formalism to localized orbitals and that detail discussion near the metal-insulator transition has some ambiguity of energy comparison.

3.2 Transition Metal Compounds

The electronic structure of transition metal monoxides was first calculated in the LSDA [29, 30], where the d-bands of transition metal ions are split due to the exchange-correlation potential and the crystal field, where MnO and NiO would be insulators with band gaps between occupied and unoccupied d-bands. However, the calculated band gaps are very small compared with experimental results and we cannot get bands gaps for FeO and CoO. These failures are also attributed to incomplete treatment of the on-site Coulomb repulsion U as described in Section 2. The LSDA also fails to predict the character of the highest occupied states. Following photoemission experiments [31, 32], NiO is assigned as a charge transfer type insulator where the highest occupied states have a strong character of oxygen p-orbitals. MnO is an insulator intermediate between the charge transfer type and the Mott type. The LSDA-SIC electronic structure in transition metal monoxides were first calculated by Svane and Gunnarsson [33] and by Szotek *et al.* [34] with energy criteria which adopt the state of the lowest LSDA-SIC total energy as the ground state. As already explained in the preceding subsection, these energy criteria would often lead to unphysical results. Actually these authors chose the states constructed of localized transition metal d-orbitals and extended oxygen p-orbitals and concluded that the ground states in MnO, FeO, CoO and NiO would be all of the charge transfer type.

We present here the electronic structure of the transition metal monoxides calculated with the formalism of the LSDA-SIC explained in the previous section [25]. Magnetic moments of the transition metal atoms on a [111] plane are aligned ferromagnetically and they are antiferromagnetic in successive planes. The electronic structure was first calculated in the LSDA, where the oxygen $2s$ bands were also selfconsistently determined. Then the core and semi-core charge densities are fixed as the frozen core approximation. The $3d$ orbitals of transition metal atoms and $2p$ orbitals of oxygen atoms are treated as valence orbitals in the LSDA-SIC calculation. All the following calculations use the Barth-Hedin exchange-correlation functional. There are three types of solutions after the selfconsistent LSDA-SIC calculation; all orbitals are delocalized in the Type I solution, all orbitals are localized in the Type II solution, and transition metal d-orbitals are localized and oxygen p-orbitals are delocalized in the Type III solution [25]. One can summarize these three solution as follows and in Table 3.

Table 3: *Band gaps and magnetic moments.*

	Band gap (eV)				Magnetic moment (μ_B)			
	Type I (LSDA)	Type II	Type III	Expt.[a]	Type I (LSDA)	Type II	Type III	Expt.[a]
VO	0.0	3.0	2.3	0.0	0.0	2.8	2.8	0.0
MnO	1.1	6.5	3.4	3.6–3.8	4.5	4.7	4.7	4.79, 4.58
FeO	0.0	6.1	3.4		3.2	3.7	3.6	3.32
CoO	0.0	5.3	2.7	2.4	2.0	2.7	2.6	3.35, 3.38
NiO	0.0	5.6	2.8	4.3, 4.0	0.7	1.7	1.5	1.77, 1.64, 1.93

M. Arai and T. Fujiwara, *Phys. Rev.* **B51**, 1477 (1995).
a) A. Svane and O. Gunnarsson, *Phys. Rev. Lett.* **65**, 1148 (1990).

Type I solution: This is just the LSDA result where all orbitals are extended. Split narrow d-bands appear in the energy region above the oxygen p-bands for all compounds. The band gap of MnO is about 30% of the experimental value and the highest occupied and lowest unoccupied bands have the same character as the transition metal d-orbitals. There is no band gap for other compounds and then the transition metal monoxides except MnO are metallic. The magnetic moments are also underestimated from the observed ones because the wavefunctions are delocalized in those other than MnO. The energy bands and the partial density of states are shown in Fig. 6.

Type II solution: This is the result of localized orbitals for all valence states. The transition metal d-orbitals are more localized than those of oxygen p-orbitals. Because the SIC potentials are larger in transition metal d-orbitals than oxygen p-orbitals, the energies of transition metal d-orbitals are pulled down significantly and the transition metal d-band and oxygen p-band are strongly hybridized. The energy gaps are overestimated but the magnetic moments are well estimated. The lowest unoccupied band keeps mainly a character of the transition metal d-orbitals and the highest occupied band is a strongly hybridized one of transition metal d-orbitals and oxygen p-orbitals, which is in good agreement with the observed results of orbital character. The energy bands and the partial density of states are shown in Fig. 7.

Type III solution: In this solution, the transition metal d-orbitals are localized and the oxygen p-orbitals are Bloch type functions free from the SIC potentials. Then the oxygen p-orbitals do not shift from the position of the LSDA results and, on the contrary, the transition metal d-orbitals are pulled down below the oxygen p-orbitals. Therefore, the highest occupied band has the character of

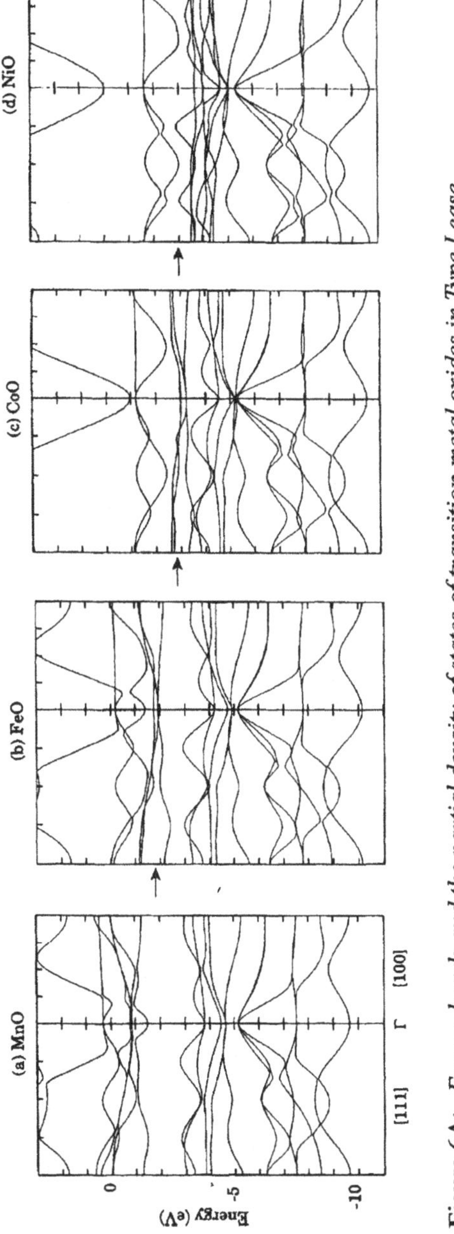

Figure 6A: *Energy bands and the partial density of states of transition metal oxides in Type I case.*

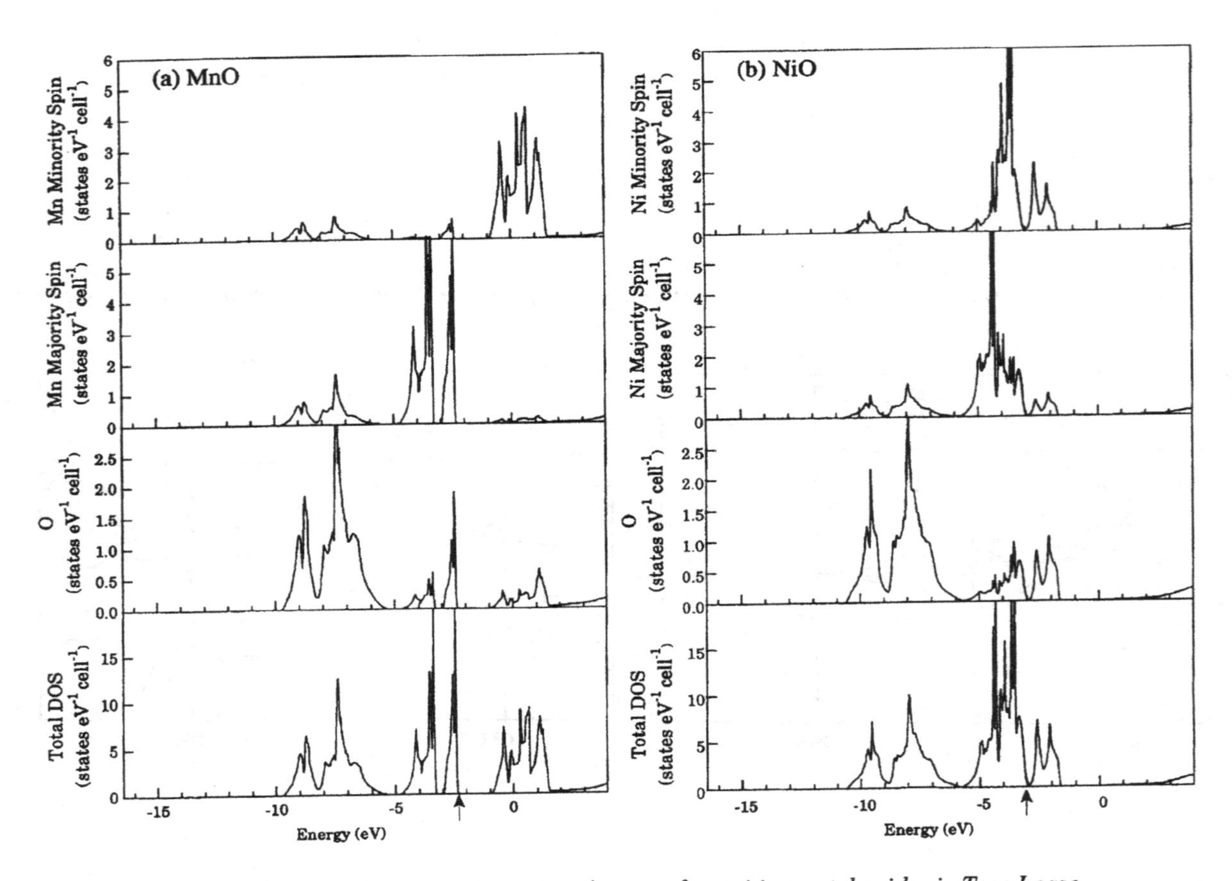

Figure 6B: *Energy bands and the partial density of states of transition metal oxides in Type I case.*

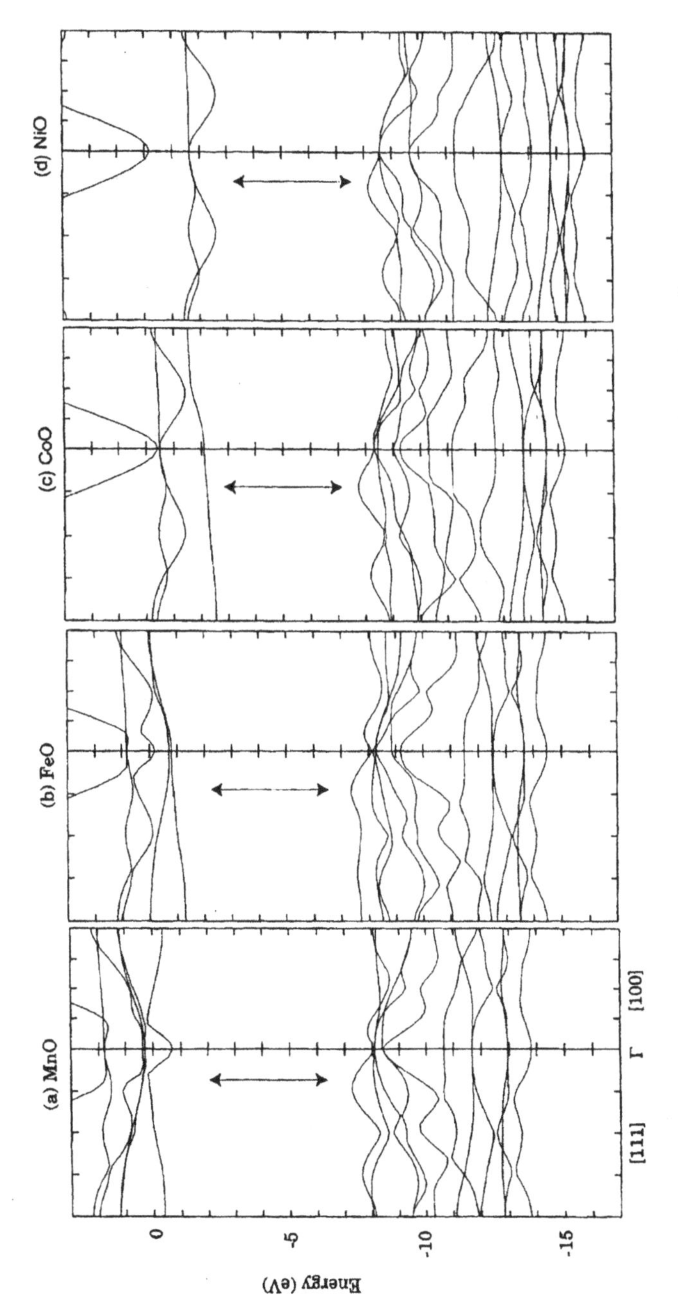

Figure 7A: *Energy bands and the partial density of states of transition metal oxides in Type II case.*

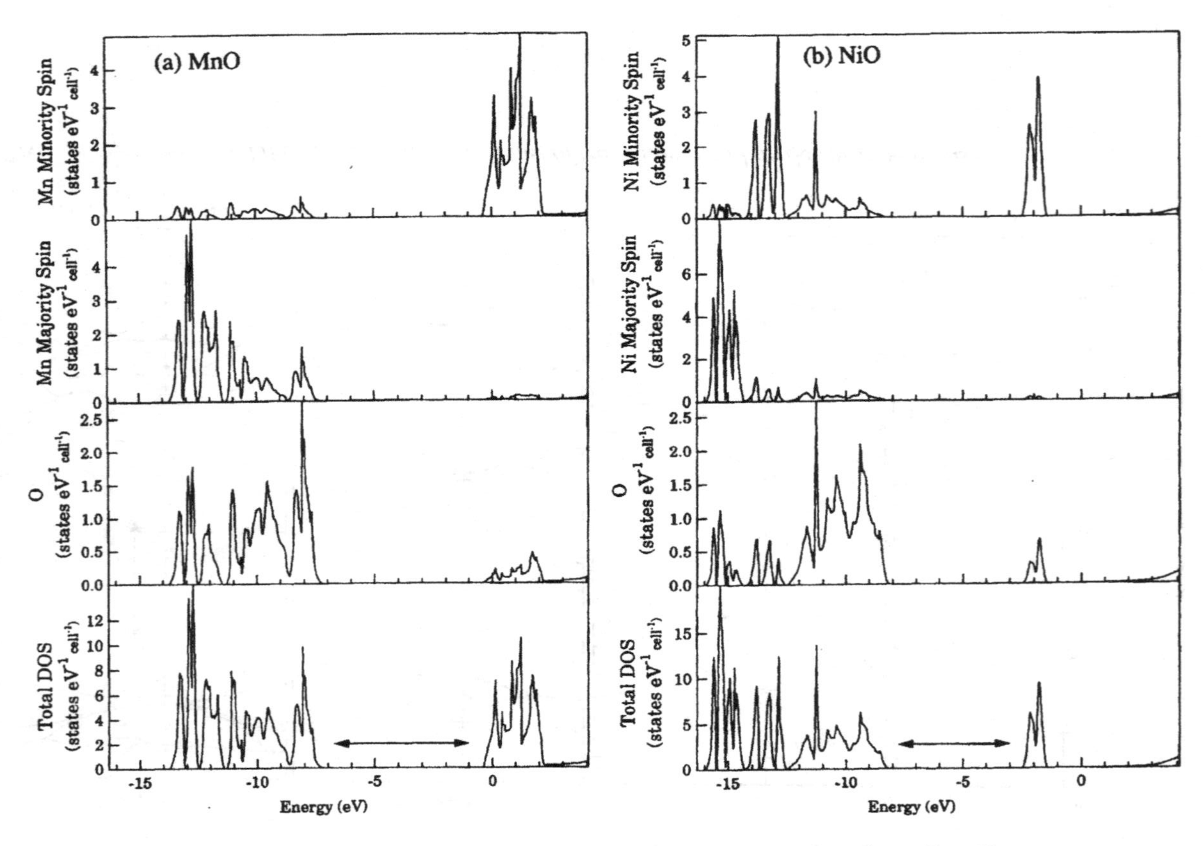

Figure 7B: *Energy bands and the partial density of states of transition metal oxides in Type II case.*

oxygen p-orbitals and the lowest unoccupied band keeps mainly a character of the transition metal d-orbitals. Then the transition metal monoxides are the charge transfer type. The values of the band gaps and the magnetic moments are in rather good agreement with experimental results but the orbital character of the highest occupied band is not consistent with the observed results, especially for MnO. The energy bands and the partial density of states are shown in Fig. 8.

The character of occupied bands is well reproduced by the Type II solution. Let us discuss the origin of the overestimation of band gaps for this solution. The diagonal elements of the Lagrange multipliers $\varepsilon_{ij}^{(\sigma)}$ can be understood to be a one-electron energy neglecting hopping between orbitals. The averaged values of such diagonal elements $\langle\epsilon_d\rangle$ for occupied and unoccupied transition metal d-orbitals are summarized together with the difference $\delta\epsilon_d$, the atomic Coulomb integral and the screened Coulomb repulsion U [43], in Table 4. The atomic Coulomb integral is calculated by using the spin-unpolarized LDA wavefunction in an isolated atom. The atomic Coulomb integral increase gradually from V to Ni according to the spacial extent of the atomic wavefunctions. On the other hand, the Coulomb integrals in solid oxides is slightly smaller and their variation is also rather slow. This may be due to screening in the bulk and the almost identical volume of the atomic spheres. Now we call the atomic Coulomb integral the unscreened one. The difference of the band energies between unoccupied and occupied transition metal d-orbitals $\delta\epsilon_d = \langle\epsilon_d^{unoc}\rangle - \langle\epsilon_d^{occ}\rangle$ does not change whether we use Type II and Type III solutions. The additional energy decrease in the SIC can be estimated as

$$-2E_H[n_{i\sigma}] + \frac{4}{3}E_{xc}^{LSDA}[n_{i\sigma}, 0] \, ,$$

where $E_H[n_{i\sigma}]$ and $E_{xc}^{LSDA}[n_{i\sigma}, 0]$ almost cancel with each other;

$$E_{xc}^{LSDA}[n_{i\sigma}] \simeq -(0.90 \sim 0.95)E_H[n_{i\sigma}] \, .$$

Thus, the SIC splits the occupied and unoccupied d-bands by the Coulomb repulsion U or slightly less than that. The discrepancy for the values of the band gap in the Type II case is due to the overestimate of the Coulomb repulsion in the solid in the LSDA-SIC. This is an essential shortcoming of the LSDA-SIC results where the SIC energy functional is only rigorous in the dilute limit of the electron density. In condensed systems, the electron-electron Coulomb interaction is screened and the Coulomb repulsion should be much reduced. Actually once we shifted the occupied levels by the difference of the unscreened and screened Coulomb repulsion, we would obtain excellent results even for the band gaps in the transition metal monoxides. From these considerations, we should conclude that the LSDA-SIC functional form should be reformulated from the view point of the screening in condensed systems.

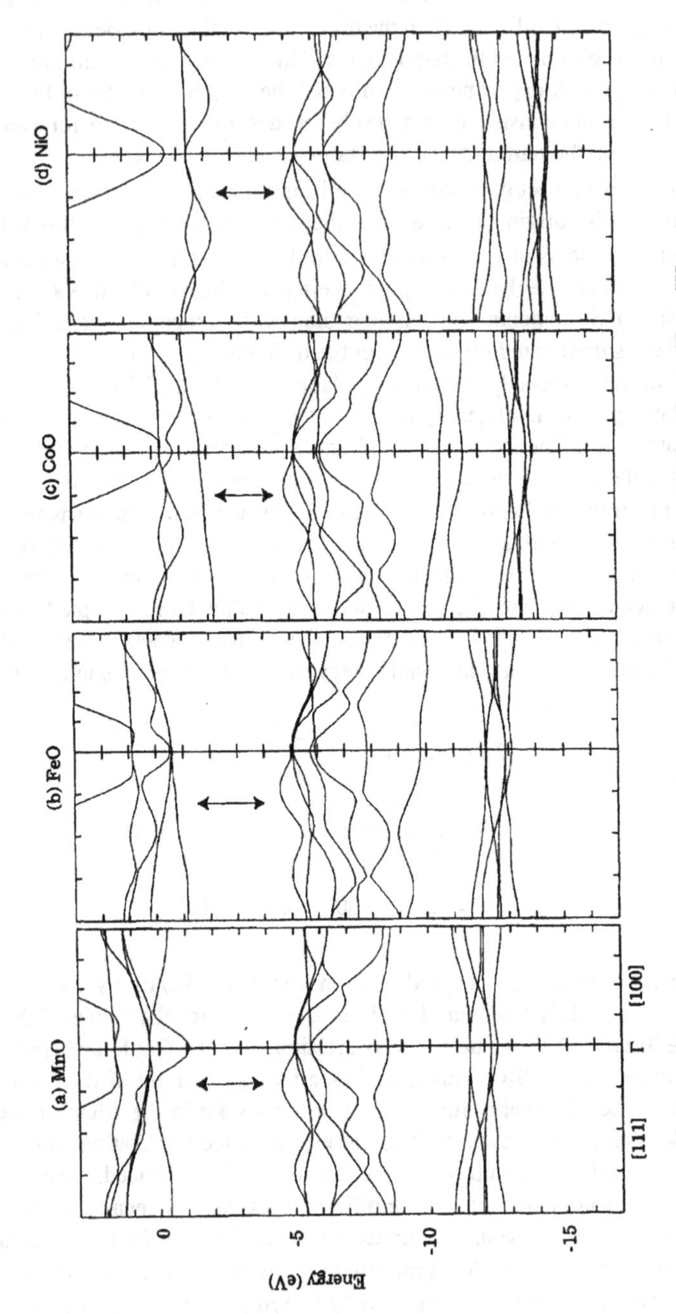

Figure 8A: *Energy bands and the partial density of states of transition metal oxides in Type III case.*

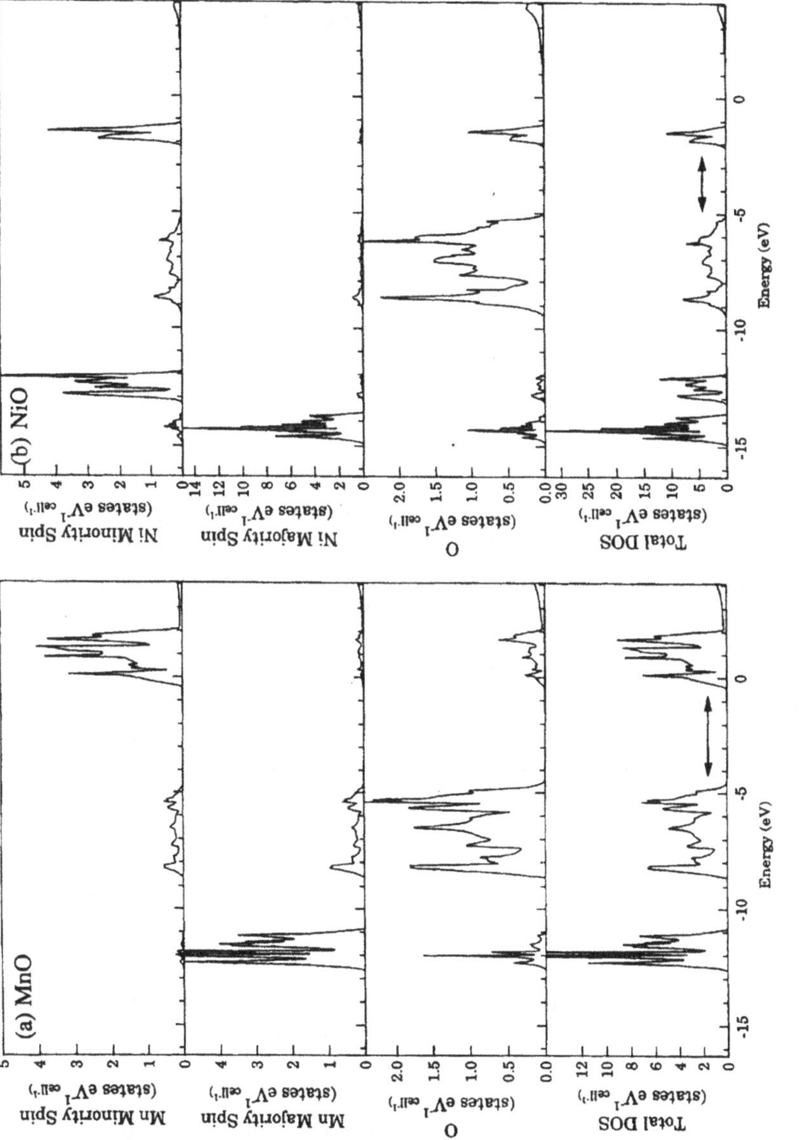

Figure 8B: *Energy bands and the partial density of states of transition metal oxides in Type III case.*

Table 4: *Averaged values of the diagonal elements of the Lagrange multipliers of unoccupied and occupied transition metal d-orbitals,* $\langle\epsilon_d^{unoc}\rangle$ *and* $\langle\epsilon_d^{occ}\rangle$*, respectively, and its difference* $\delta\epsilon_d = \langle\epsilon_d^{unoc}\rangle - \langle\epsilon_d^{occ}\rangle$*. The last two columns show the atomic unscreened and solid screened Couomb repulsion* U*.*

	Type II eV			Type III eV			U eV	
	$\langle\epsilon_d^{unoc}\rangle$	$\langle\epsilon_d^{occ}\rangle$	$\delta\epsilon_d$	$\langle\epsilon_d^{unoc}\rangle$	$\langle\epsilon_d^{occ}\rangle$	$\delta\epsilon_d$	atomic*	oxide**
VO	8.4	−3.7	12.1	8.6	−3.5	12.1	7.9	6.7
MnO	2.3	−11.5	13.8	2.4	−11.3	13.7	9.6	6.9
FeO	1.2	−11.7	13.0	1.5	−11.4	12.9	10.4	6.8
CoO	0.1	−12.9	13.0	0.5	−12.3	12.7	11.2	7.8
NiO	−0.4	−13.9	13.5	0.0	−13.3	13.3	11.9	8.0

*) Coulomb integral calculated by using the spin-unpolarized LDA atomic wavefunction.
) V. I. Anisimov, J. Zaanen and O. K. Andersen, *Phys. Rev.* B44**, 943 (1991).

The electronic structure of tetragonal La_2CuO_4 was also calculated by the LSDA-SIC by adopting localized Cu d-orbitals and delocalized O p-orbitals [35]. The calculated results show, in contrast to the LSDA, the ground state to be AF insulator with a band gap of 1.04 eV and the Cu magnetic moment $0.47\mu_B$ per atom, which is in fair agreement with observed results. The lowest unoccupied states are mainly of Cu d-orbitals. The highest occupied states are constructed of almost equally weighted on-plane and out-of-plane O p-orbitals and the occupied Cu d-orbitals are pulled down about 12 eV below the valence band top. On the contrary, the observation shows the strongly hybridized valence bands of Cu d-orbitals and O p-orbitals [36].

4 SIC in Rare Earth Metals and Rare Earth Metal Compounds

In rare earth metals and their compounds, the correlation among f-electrons and between f-electrons and conduction s, p and d-electrons is very important.

In La $(Xe(6s)^2(5d))$ or Ce $(Xe(6s)^2(4f)(5d))$ compounds, the delocalized electron picture does work with the same standpoint on s, p, d and f-electrons. Here, because the core configuration of rare earth metals is the same Xe atom, we just write 'Xe' for the inner core configuration. In Gd $(Xe(6s)^2(4f)^7(5d))$, because it is a strong ferromagnet with perfectly polarized spins, the delocalized electron picture for f-electrons does work, though the binding energy of f-electrons is not in agreement with experiments. In other rare earth elements and their compounds, all or a part of the f-electrons should be treated as the frozen core, i.e. localized electrons, in the conventional LSDA band structure theory, otherwise the Fermi energy would be pinned at the position of the rare earth f-levels.

4.1 Rare Earth Metals

4.1.1 Praseodymium

The LSDA-SIC formalism, though it might be still unsatisfactory, may be very promising for the unified treatment, from the first principle electronic structure calculations, of the rare earth metals or their compounds. Praseodymium (Pr: Xe$(6s)^2(4f)^4$) is the first rare earth metal studied by the LSDA-SIC [26]. Here s, p, d and f-electrons are treated as delocalized orbitals in the LSDA and, on the contrary, s, p and d-electrons are treated as delocalized electrons and f-electrons as the localized orbitals explicitly in the LSDA-SIC. In Pr metal it is the unique treatment in the LSDA-SIC because the s, p and d-bands cross the Fermi energy and the occupied parts of their orbitals could not form localized orbitals.

The resulting total and partial density of states by the LSDA-SIC are shown in Fig. 9 [26]. The f-bands in the LSDA appear just on the Fermi energy. The f-band in the LSDA-SIC appears at a binding energy of 7.5 eV for occupied orbitals and the unoccupied band appears, crossing the Fermi energy, with the peak at about 1 eV above the Fermi energy. The occupied f-band shows a large gap below the Fermi energy, and then the occupied f-bands can form localized orbitals. This LSDA-SIC result may provide a reasonable unified scenario of both localized and delocalized f-electrons on the same footing. Furthermore, the delocalized f-electrons are hybridized strongly with s, p, and d-electrons and cause the well-known mass enhancement of band electrons.

4.1.2 Cerium

Cerium is a characteristic element which shows an isostructural phase transition (fcc $\rightarrow$ fcc) and was studied in detail by the LSDA-SIC [37].

At absolute zero temperature and zero pressure, the α fcc phase is stable. At room temperatures, the γ phase fcc cerium is more stable and shows the phase transition at 8 kbar with a large volume reduction of 14.8% of volume to the α fcc phase. The f-electrons of a Ce atom are localized in the γ phase with a magnetic moment 5/2 μ_B contributed by spin and orbital polarization, showing Curie-Weiss susceptibility, and are delocalized in the α phase with Pauli paramagnetism.

This phenomenon has been explained by several models. One is the promotion model [38] where localized f-electrons in the γ phase are promoted into non-f bands in the α phase, but this model cannot explain the cohesive properties of Ce solid and the photoemission spectra. Another model is the Kondo singlet formation

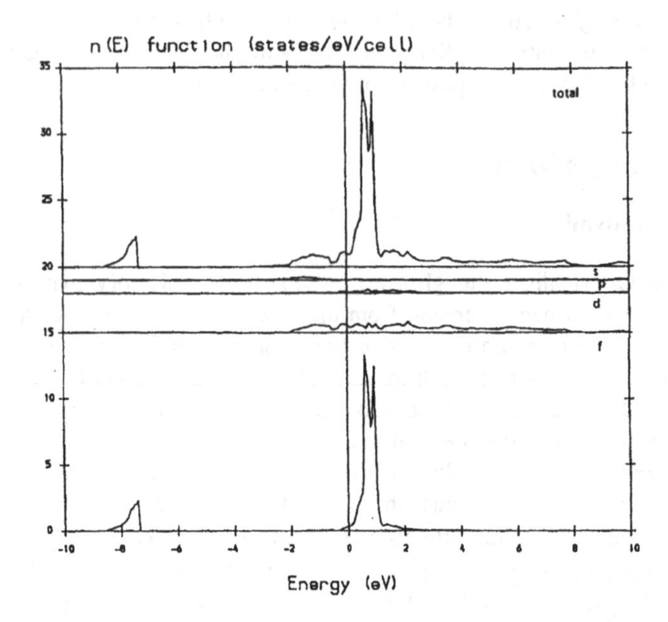

Figure 9: *Total and partial density of states of Pr by the LSDA-SIC.*

model [39], where f-electrons are not screened in the γ phase because of spin fluctuations. They are fully screened in the α phase by neighboring s, p and d-electrons, which form the spin singlet. The Kondo model can explain the overall observed feature.

One other model is the Mott transition model [40]. The f-electrons are localized in the γ phase and delocalized in the α phase. Here the localized f-electrons are favored due to the $f-f$ on-site Coulomb repulsion and, because the configuration of delocalized electrons is the same as La, the equilibrium atomic volume should be equal to the La solid. Then, increasing hopping integrals of f-orbitals due to increasing the hybridization between f and s, p, d-orbitals caused by the volume reduction, the delocalization energy of the f-electrons overcomes the localization energy. The feature of these cohesive properties shares the same standpoint with the Kondo singlet formation model.

Figure 10 shows the equation of state of fcc solid Ce [37], compared with experimental results [41]. The α and the γ phases are represented by the delocalized LSDA and localized LSDA-SIC f-orbitals, respectively. The volume reduction at the $\gamma \rightarrow \alpha$ phase transition is estimated at 24%, contrary to the observed 14.8%. The paramagnetic α phase is calculated to be more stable at $T = 0$ and the volume expansion stabilizes the magnetic γ phase. Except for the underestimate of the equilibrium volume, the overall feature of the equation of state is satisfactory.

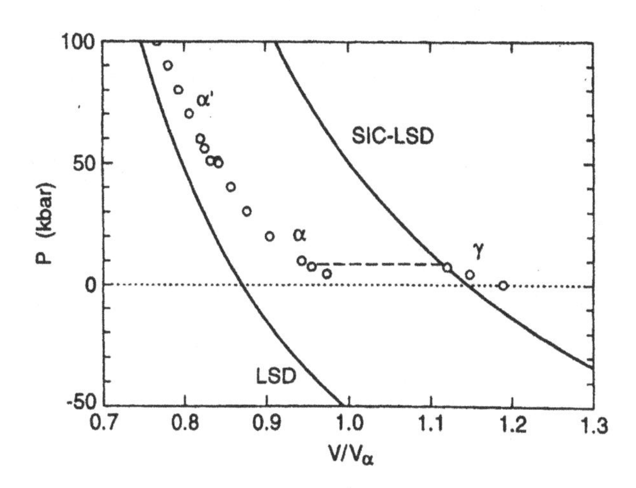

Figure 10: *Equation of state of fcc Ce. The volume is normalized by that of the alpha-phase, both in calculated and experimental. The solid lines are the calculated ones and circles are the experimental ones.*

The spin magnetic moment in γ phase is 1.32 μ_B per Ce atom and, therefore, 0.32 μ_B is contributed by the spin polarization of delocalized conduction electrons. The orbital polarization was not included in the calculations and therefore the total magnetic moment could not be discussed.

This author also discussed the phase equilibrium at finite temperatures by using the random alloy model, where the equilibrium state is assumed to be a random alloy of the γ and α Ce. The total energy $E(x, V)$ and the entropy $S(x, V)$ are estimated by the following equation:

$$\begin{aligned} E(x, V) &= xE_\gamma(V) + (1-x)E_\alpha(V) \\ S(x, V) &= xS_\gamma(V) + (1-x)S_\alpha(V) + S_{mix}(x) \\ S_\gamma(V) &= k_B \ln 6 \\ S_\alpha(V) &= 0 \\ S_{mix}(x)/k_B &= -x\ln x - (1-x)\ln(1-x), \end{aligned} \tag{37}$$

where x is the fraction of γ Ce. The entropy of γ Ce is estimated from the degree of freedom by using localized total moment of $J = 5/2$. This model can give the $\alpha \to \gamma$ phase transition at elevated temperatures, though the critical point is greatly overestimated as (P_c, T_c) =(47 kbar, 1300 K), compared with the experimental one (20 kbar, 600K). This discrepancy may be mainly due to the overestimate of the expansion of equilibrium volume at the $\gamma \to \alpha$ transition.

Table 5: *Calculated phase transition under the pressure. B1 and B2 are the NaCl and CsCl structures. The asterisks * means the distorted structure. 'l' and 'd' denote the states of localized and delocalized f-orbitals, respectively. This table is from Ref. 43.*

	transition	Pressure (kbar)	
		calculated	observed
CeN	B1(d)→ B2(d)	620	
CeP	B1(l)→ B1(d)	71	90, 55
CeP	B1(d)→ B2(d)	113	150(40)
CeAs	B1(l)→ B2(d)	114	140(20)
CeSb	B1(d)→ B2*(l)	70	85(2)
CeSb	B2*(l)→ B2*(d)	252	
CeBi	B1(l)→ B2*(l)	88	90(40)
CeBi	B2*(l)→ B2*(d)	370	

4.2 Rare Earth Compounds - Cerium Pnictides

The cerium pnictides CeN, CeP, CeAs, CeSb and CeBi, are unique rare earth compounds which show a variety of structural phase transition under pressure. They all crystallize in the NaCl structure at zero temperature. The Ce atom in CeN is close to tetravalent, similar to the collapsed α-Ce metal with delocalized f-electrons. Other Ce pnictides are similar to the γ-Ce with localized f-electrons. Under pressure, they all show a structural phase transition, except CeN which is not well known, and the transition is very similar to the $\gamma \rightarrow \alpha$ phase transition of Ce with about 10% volume reduction. Magnetic properties of Ce pnictides are very complicated with several AF ordered phases. The reader should refer the magnetic structure in Ref. [42].

The Ce pnictides were studied very recently by using the LSDA-SIC within the LMTO framework [43]. The Ce f-electrons are treated as localized orbitals in the LSDA or localized orbitals in the LSDA-SIC, and other orbitals are all treated as delocalized orbitals. The magnetic structure is assumed to be in ferromagnetic and in antiferromagnetic order. The main results are summarized in Table 5. These results are obtained from a wide variety of calculations, not only the localized and delocalized character of Ce f-orbitals, but several assumed spin orderings of ferromagnetic or antiferromagnetic spin alignment.

For example, in CeSb, the first transition occurs from the NaCl structure to the CsCl structure, both with localized f-electrons. The second transition cannot be excluded in the calculation at higher pressure from localized to delocalized f-electron states. The magnetic ground state in CeP and CeSb is AF I, antiferromagnetic I, structure which consists of an antiferromagnetic stacking of (100) planes, and within the (100) plane, ferromagnetic ordered. The ground state is not the structure of localized f-electrons with AF II (antiferromagnetic II) alignment which consists of antiferromagnetic stacking of (111) planes and ferromagnetic ordered within the (111) plane. With increasing atomic number of ligand from CeN to CeBi, the specific volume at zero pressure increases and the ionicity of compounds decreases. At the same time, the delocalization of Ce f-electrons is less favorable with decreasing ionicity from CeN to CeBi.

Again, the LSDA-SIC calculation may offer a unified scenario of the structural and magnetic phase transition under pressure in the Ce pnictides.

5 Conclusion

We have reviewed the LSDA-SIC calculations in atoms, molecules and solids. The early stage of the LSDA-SIC, only the SIC term is included in atoms, molecules and even solids but the results suffer from arbitrariness of the basis wavefunctions. The notion of *localized* and *canonical* orbitals is of great help for the unique choice for the basis functions. However still the result is not completely unique, especially in transition metal compounds because there still remains an arbitrariness which orbitals should be localized or delocalized. We propose a principle that all orbitals should be localized in the LSDA-SIC unless there is no arbitrariness for a choice of the exchange-correlation energy functional which could allow a highly accurate energy criteria for the ground state. The rare earth metals and rare earth metal compounds may be free from this arbitrariness because a wide band gap may open between occupied and unoccupied f-electron bands which is always the key scenario for coexistence of localized f-electron spins and itinerant f-electrons.

The Kohn-Sham potential in the LSDA+U can be written down as [43, 44]

$$\begin{aligned} V^{m\sigma} =& V^{LDA} + U \sum_{m'} (n_{m'-\sigma} - n_d) \\ &+ (U - J) \sum_{m \neq m'} (n_{m'\sigma} - n_d) \end{aligned} \tag{38}$$

and this equation contains, for occupied orbitals, the loss of the Coulomb energy and the gain of the exchange energy which is an effective discontinuity of the exchange-correlation potential. The essential difference between the LDA+U and our LSDA-SIC is the estimation of the Coulomb repulsion U. In the present LSDA-SIC, the SIC is only rigorous in the dilute limit and the Coulomb repulsion for the SIC is unscreened. On the contrary, the LDA+U uses an effective U which is much reduced due to the screening effects in the solid. Therefore, though it is not fully satisfactory in the present form, the LSDA-SIC may be more promising once one could establish the SIC energy functional including sufficient screening effects.

Acknowledgements

This work is supported by Grant-in-Aid for COE Research, and Grant-in-Aid from the Japan Ministry of Education, Science and Culture.

References

[1] P. C. Hohenberg and W. Kohn, *Phys. Rev.* **136**, B864 (1964).
[2] W. Kohn and L. J. Sham, *Phys. Rev.* **140**, A1133 (1965).
[3] O. Gunnarsson and R.O. Jones, *Rev. Mod. Phys.* **61**, 689. (1989).
[4] V. L. Morruzi, J. F. Janak and A. R. Williams, *Calculated Electronic Properties of Metals*, (Pergamon, New York, 1978).
[5] J. P. Perdew, R. G. Par, M. Levy and J. L. Balduz, *Phys. Rev. Lett.* **49**, 1691 (1982); J. P. Perdew and M. Levy, *Phys. Rev. Lett.* **51**, 1884 (1982).
[6] L. J. Sham and M. Schlüter, *Phys. Rev. Lett.* **51**, 1884 (1983).
[7] W. Kohn, *Phys. Rev.* B**33**, 4331 (1986).
[8] J.P. Perdew, and A. Zunger, *Phys. Rev.* B**23**, 5048 (1981).
[9] K. Schönhammer and O. Gunnarson, *J. Phys.* C**20**, 3675 (1987).
[10] O. Gunnarsson, O. K. Andersen, O. Jepsen and J. Zaanen, *Phys. Rev.* B**39**, 1789 (1989); A. M. McMahan, R. M. Martin and S. Satpathy, *Phys. Rev.* B**38**, 6650 (1988).
[11] J. F. Janak, *Phys. Rev.* B**18**, 7165 (1978).
[12] Y. Ishii, *Computational Physics as a New Frontier in Condensed Matter Research*, edited by H. Takayama *et al.* (The Physical Society of Japan, Tokyo, 1995), p.57.
[13] J. P. Perdew and Y. Wang, *Phys. Rev.* B**33**, 8800 (1986), A. D. Beck, *Phys. Rev.* A**38**, 3098 (1988); J. P. Perdew, *Phys. Rev.* B**33**, 8822 (1986); *ibid* **B34**, 7406 (1986).
[14] R. A. Heaton, J. G. Harrison and C. C. Lin, *Phys. Rev.* B**28**, 5992 (1983).
[15] N. Hamada and S. Ohnishi, *Phys. Rev.* B**34**, 9042 (1986).
[16] Y. Hatsugai and T. Fujiwara, *Phys. Rev.* B**37**, 1280 (1988).
[17] W. A. Harrison and S. Ciraci, *Phys. Rev.* B**10**, 1516 (1973).
[18] M. Pederson, R. A. Heaton and C.C. Lin, *J. Chem. Phys.* **80**, 1972 (1984); *ibid* **82**, 2688 (1985).
[19] Y. Ishii and K. Terakura, *Phys. Rev.* **B42**, 10924 (1990).
[20] A. Svane and O. Gunnarsson, *Phys. Rev.* B**37**, 9919 (1988); *Europhys. Lett.* **7**, 171 (1988).
[21] T. Miyazaki, K. Terakura and Y. Ishii, *Phys. Rev.* B**48**, 16992 (1993).
[22] S. Kobayashi, M. Arai and T. Fujiwara, *Phys. Rev.* B**52**, 13718 (1995).
[23] O. K. Andersen, *Phys. Rev.* B**12**, 3060 (1975).
[24] O. K. Andersen and O. Jepsen, *Phys. Rev. Lett.* **53**, 2571 (1984).
[25] M. Arai and T. Fujiwara, *Phys. Rev.* B**51**, 1477 (1995).

[26] Z. Szotek, W. M. Temmerman and H. Winter, *Physica* **B172**, 19 (1991).
[27] M. Arai and T. Fujiwara, *Computer Aided Innovation of New Materials II* (North Holland, Tokyo, 1992).
[28] A. Svane and O. Gunnarsson, *Solid State Comm.* **76**, 851 (1990).
[29] L. F. Mattheis, *Phys. Rev.* B**5**, 290 (1972).
[30] K. Terakura, A. R. Williams, T. Oguchi, and J. Kübler, *Phys. Rev. Lett.* **52**, 1830 (1984); T. Oguchi and K. Terakura, *J. Appl. Phys.* **55**, 2318 (1984).
[31] A. Fujimori and F. Minami, *Phys. Rev.* B**30**, 957 (1984).
[32] G. A. Swatzky and J. W. Allen, *Phys. Rev. Lett.* **53**, 2239 (1984).
[33] A. Svane and O. Gunnarsson, *Phys. Rev. Lett.* **65**, 1148 (1990).
[34] Z. Szotek, W. M. Temmerman and H. Winter, *Phys. Rev.* B**47**, 3771 (1992).
[35] E. U. Condon and H. Odabaş i, *Atomic Structure*, (Cambridge Univ. Press, Cambridge, 1980), p. 227.
[36] A. Svane, *Phys. Rev. Lett.* **68**, 1900 (1992).
[37] M. T. Czyzyk and G. A. Sawatzky, *Phys. Rev.* B**49**, 14211 (1994), and references therein.
[38] A. Svane, *Phys. Rev. Lett.* **72**, 1248 (1994); *Phys. Rev.* B**53**, 4275 (1996).
[39] R. Ramirez and L. M. Falicov, *Phys. Rev.* B**3**, 1225 (1971).
[40] J. W. Allen and R. M. Martin, *Phys. Rev. Lett.* **49**, 1106 (1982).
[41] B. Johansson, *Phil. Mag.* **30** 469 (1974).
[42] W. H. Zachariasen and F. H. Ellinger, *Acta Crystallogr.* A**33**, 155 (1977).
[43] A. Svane, Z. Szotek, W. M. Temmerman and H. Winter, to be published in *Phys. Rev. Lett.*
[44] V. I. Anisimov, J. Zaanen and O. K. Andersen, *Phys. Rev.* **B44**, 943 (1991).
[45] V. I. Anisimov, M. A. Korotin, J. Zaanen and O. K. Andersen, *Phys. Rev. Lett.* **68**, 345 (1992).

4. Orbital Functionals in Density Functional Theory: The Optimized Effective Potential Method

T. GRABO, T. KREIBICH, S. KURTH and E.K.U. GROSS

1. Introduction

Density functional theory (DFT) is a powerful quantum mechanical method for calculating the electronic structure of atoms, molecules and solids [1–3]. The success of DFT hinges on the availability of good approximations for the total-energy functional. In this article we shall review a particular approach to the construction of such approximations which involves explicitly orbital-dependent functionals. Before describing the nature of this approach we first briefly review the foundations of DFT.

We are concerned with Coulomb systems described by Hamiltonians of the type

$$\hat{H} = \hat{T} + \hat{W}_{\text{Clb}} + \hat{V} \tag{1}$$

where (atomic units are used throughout this article)

$$\hat{T} = \sum_{i=1}^{N}\left(-\frac{1}{2}\nabla_i^2\right) \tag{2}$$

denotes the kinetic-energy operator,

$$\hat{W}_{\text{Clb}} = \frac{1}{2}\sum_{\substack{i,j=1\\ i\neq j}}^{N} \frac{1}{|\mathbf{r}_i - \mathbf{r}_j|} \tag{3}$$

represents the Coulomb interaction between the particles, and

$$\hat{V} = \sum_{i=1}^{N} v(\mathbf{r}_i) \tag{4}$$

contains all external potentials of the system, typically the Coulomb potentials of the nuclei.

Modern DFT is based on the celebrated theorem of Hohenberg and Kohn (HK) [4] which, for systems with nondegenerate ground states, may be summarized by the following three statements:

1. The ground-state density ρ uniquely determines the ground-state wave function $\Psi[\rho]$ as well as the external potential $v = v[\rho]$. As a consequence, any observable of a static many-particle system is a functional of its ground-state density.
2. The total-energy functional

$$E_{v_0}[\rho] := < \Psi[\rho]|\hat{T} + \hat{W}_{Clb} + \hat{V}_0|\Psi[\rho] > \tag{5}$$

of a particular physical system characterized by the external potential v_0 is equal to the exact ground-state energy E_0 if and only if the exact ground-state density ρ_0 is inserted. For all other densities $\rho \neq \rho_0$ the inequality

$$E_0 < E_{v_0}[\rho] \tag{6}$$

holds. Consequently, the exact ground-state density ρ_0 and the exact ground-state energy E_0 can be determined by solving the Euler-Lagrange equation

$$\frac{\delta}{\delta\rho(\mathbf{r})} E_{v_0}[\rho] = 0. \tag{7}$$

3. The functional

$$F[\rho] := < \Psi[\rho]|\hat{T} + \hat{W}_{\mathrm{Clb}}|\Psi[\rho] > \tag{8}$$

is universal in the sense that it is independent of the external potential v_0 of the particular system considered, i.e. it has the same functional form for all systems with a fixed particle-particle interaction ($\hat{W}_{\mathrm{Clb}}$ in our case).

The proof of the HK theorem does not depend on the particular form of the particle-particle interaction. It is valid for *any* given particle-particle interaction $\hat{W}$, in particular also for $\hat{W} \equiv 0$, i.e. for non-interacting systems described by Hamiltonians of the form

$$\hat{H}_S = \hat{T} + \hat{V}_S. \tag{9}$$

Hence the potential $V_S(\mathbf{r})$ is uniquely determined by the ground-state density:

$$V_S(\mathbf{r}) = V_S[\rho](\mathbf{r}). \tag{10}$$

As a consequence, all single-particle orbitals satisfying the Schrödinger equation

$$\left(-\frac{\nabla^2}{2} + V_S[\rho](\mathbf{r})\right)\varphi_j(\mathbf{r}) = \varepsilon_j\varphi_j(\mathbf{r}) \tag{11}$$

are functionals of the density as well:

$$\varphi_j(\mathbf{r}) = \varphi_j[\rho](\mathbf{r}). \tag{12}$$

The HK total-energy functional of non-interacting particles is given by

$$E_S[\rho] = T_S[\rho] + \int d^3r\, \rho(\mathbf{r}) V_S(\mathbf{r}) \tag{13}$$

where $T_S[\rho]$ is the kinetic-energy functional of non-interacting particles:

$$T_S[\rho] = \sum_{\substack{i=1 \\ \text{lowest } \varepsilon_i}}^{N} \int d^3r\, \varphi_i^*[\rho](\mathbf{r}) \left(-\frac{\nabla^2}{2}\right) \varphi_i[\rho](\mathbf{r}). \tag{14}$$

We emphasize that the quantity (14) really represents a *functional of the density:* Functional means that we can assign a unique number $T_S[\rho]$ to any function $\rho(\mathbf{r})$. This is done by first calculating that very potential $V_S(\mathbf{r})$ which uniquely corresponds to $\rho(\mathbf{r})$. Several numerical schemes have been devised to achieve this task [5–10]. Then we take this potential, solve the Schrödinger equation (11) with it to obtain a set of orbitals $\{\varphi_j(\mathbf{r})\}$ and use those to calculate the number T_S by evaluating the right-hand side of Eq. (14). As a matter of fact, by the same chain of arguments, *any orbital functional is an (implicit) functional of the density,* provided the orbitals come from a local, i.e. multiplicative potential.

Returning to the *interacting* system of interest we now define the so-called exchange-correlation (xc) energy functional by

$$E_{\mathrm{xc}}[\rho] := F[\rho] - \frac{1}{2}\int d^3r \int d^3r'\, \frac{\rho(\mathbf{r})\rho(\mathbf{r}')}{|\mathbf{r}-\mathbf{r}'|} - T_S[\rho]. \tag{15}$$

The HK total-energy functional (5) can then be written as

$$E_{v_0}[\rho] = T_S[\rho] + \int d^3r\, \rho(\mathbf{r}) v_0(\mathbf{r}) + \frac{1}{2}\int d^3r \int d^3r'\, \frac{\rho(\mathbf{r})\rho(\mathbf{r}')}{|\mathbf{r}-\mathbf{r}'|} + E_{\mathrm{xc}}[\rho]. \tag{16}$$

In historical retrospective we may identify three generations of density functional schemes which may be classified according to the level of approximations used for the universal functionals $T_S[\rho]$ and $E_{\mathrm{xc}}[\rho]$.

In what we call the *first generation of DFT, explicitly* density-dependent functionals are used to approximate both $T_S[\rho]$ and $E_{\mathrm{xc}}[\rho]$. The simplest approximation of this kind is the Thomas-Fermi model, where $E_{\mathrm{xc}}[\rho]$ is neglected completely and $T_S[\rho]$ is approximated by

$$T_S^{TF}[\rho] = \frac{3}{10}\left(3\pi^2\right)^{2/3} \int d^3r\, \rho(\mathbf{r})^{5/3} \tag{17}$$

yielding

$$E_{v_0}^{TF}[\rho] = \frac{3}{10}\left(3\pi^2\right)^{2/3} \int d^3r\, \rho(\mathbf{r})^{5/3} + \int d^3r\, v_0(\mathbf{r})\rho(\mathbf{r}) + \frac{1}{2}\int d^3r \int d^3r'\, \frac{\rho(\mathbf{r})\rho(\mathbf{r}')}{|\mathbf{r}-\mathbf{r}'|} \tag{18}$$

as approximate expression for the total-energy functional. For functionals of this type the HK variational principle (7) can be used directly, leading to equations of the Thomas-Fermi type. As these equations only contain one basic variable, namely the density $\rho(\mathbf{r})$ of the system, they are readily solved numerically. The results obtained in this way, however, are generally of only moderate accuracy in T_S, yielding unacceptably large errors in E_0.

The *second generation* of DFT employs the *exact* functional (14) for the non-interacting kinetic energy and an approximate density functional for the xc energy:

$$E_{v_0}^{KS}[\rho] = T_S^{exact}[\rho] + \int d^3r\, v_0(\mathbf{r})\rho(\mathbf{r}) + \frac{1}{2}\int d^3r \int d^3r'\, \frac{\rho(\mathbf{r})\rho(\mathbf{r}')}{|\mathbf{r}-\mathbf{r}'|} + E_{\mathrm{xc}}[\rho]. \tag{19}$$

This total-energy expression leads to the Kohn-Sham (KS) version of DFT [11] as will be shown in the following. Plugging Eq. (19) into the variational principle (7) yields

$$0 = \frac{\delta T_{\mathrm{S}}^{\mathrm{exact}}[\rho]}{\delta\rho(\mathbf{r})} + v_0(\mathbf{r}) + \int d^3r'\, \frac{\rho(\mathbf{r}')}{|\mathbf{r}-\mathbf{r}'|} + \frac{\delta E_{\mathrm{xc}}[\rho]}{\delta\rho(\mathbf{r})}.$$

The variation of the non-interacting kinetic-energy functional is given by

$$\begin{aligned} \delta T_S^{\mathrm{exact}}[\rho] &= \delta \sum_{i=1}^{N} < \varphi_i[\rho]| - \frac{\nabla^2}{2}|\varphi_i[\rho] > \\ &= \delta\left[\sum_{i=1}^{N} \varepsilon_i[\rho] - \int d^3r'\, V_S[\rho](\mathbf{r}')\rho(\mathbf{r}')\right] \end{aligned} \tag{21}$$

where the single-particle equation (11) has been used. Since the HK theorem ensures a one-to-one mapping between the density and the single-particle potential, a variation $\delta\rho$ of the former corresponds to a unique variation δV_S of the latter. Therefore, the variation of the single-particle energies ε_i can be calculated using first-order perturbation theory yielding

$$\delta\varepsilon_i =< \varphi_i[\rho]|\delta V_S[\rho]|\varphi_i[\rho] > . \tag{22}$$

Using this result in (21) gives

$$\delta T_S^{\mathrm{exact}}[\rho] = -\int d^3r'\, V_S[\rho](\mathbf{r}')\delta\rho(\mathbf{r}') \tag{23}$$

which, combined with Eq. (20) leads to

$$V_S[\rho](\mathbf{r}) = v_0(\mathbf{r}) + \int d^3r' \, \frac{\rho(\mathbf{r}')}{|\mathbf{r}-\mathbf{r}'|} + V_{\mathrm{xc}}[\rho](\mathbf{r}) \tag{24}$$

where we have defined the xc potential as

$$V_{\mathrm{xc}}[\rho](\mathbf{r}) := \frac{\delta E_{\mathrm{xc}}[\rho]}{\delta\rho(\mathbf{r})}. \tag{25}$$

Being the HK variational equation of the *interacting* system, Eq. (20) determines the exact ground-state density of the interacting system. Since Eq. (24), on the other hand, is equivalent to Eq. (20), the density

$$\rho(\mathbf{r}) = \sum_{\substack{i=1 \\ N \text{ lowest } \varepsilon_i}}^{N} |\varphi_i(\mathbf{r})|^2 \tag{26}$$

resulting from the solution of the Schrödinger equation (11) with the potential (24) must be identical with the ground-state density of the *interacting* system of interest. Eqs. (11), (24), (25) and (26) are known as Kohn-Sham equations. In practice, these equations have to be solved self-consistently employing approximate but explicitly density-dependent functionals for $E_{\mathrm{xc}}[\rho]$. The resulting scheme is still easy to solve numerically and gives – especially for sophisticated density-gradient-dependent approximations of $E_{\mathrm{xc}}[\rho]$ – excellent results for a wide range of atomic, molecular and solid-state systems.

Finally, in the *third generation of DFT,* one employs, in addition to the *exact* expression for T_S, also the *exact* expression for the exchange energy given by

$$E_{\mathrm{x}}^{\mathrm{exact}}[\rho] = -\frac{1}{2} \sum_{\sigma=\uparrow,\downarrow} \sum_{j,k=1}^{N_\sigma} \int d^3r \int d^3r' \, \frac{\varphi_{j\sigma}^*(\mathbf{r})\varphi_{k\sigma}^*(\mathbf{r}')\varphi_{k\sigma}(\mathbf{r})\varphi_{j\sigma}(\mathbf{r}')}{|\mathbf{r}-\mathbf{r}'|}. \tag{27}$$

Only the correlation part of $E_{\mathrm{xc}}[\rho]$ needs to be approximated in this approach. In contrast to the conventional second-generation KS scheme, the third generation allows for the treatment of explicitly orbital-dependent functionals for E_c as well, giving more flexibility in the construction of such approximations.

The central equation in the third generation of DFT is still the KS equation (11). The difference between the second and third generation lies in the level of approximation to the xc-energy. As a consequence of the orbital dependence of E_{xc} in the third generation of DFT the calculation of $V_{\mathrm{xc}}[\rho](\mathbf{r})$ from Eq. (25) is somewhat more complicated. A detailed derivation will be given in the following section for the spin-dependent version of DFT. The result is an integral equation determining the xc potential. This integral equation, known as the optimized effective potential (OEP) equation [12,13], is difficult to solve. To avoid a full-scale numerical solution, Krieger, Li and Iafrate (KLI) [14-22] have devised a semi-analytical scheme for solving the OEP integral equation approximately. This scheme is described in the subsequent section. After that, some rigorous properties of the OEP and KLI solutions will be deduced and the relation of the OEP method to the Hartree-Fock (HF) scheme will be discussed.

For heavier systems, relativistic effects become more and more important. For example, the nonrelativistic (x-only) ground-state energy of the mercury atom is 18408 a.u. while the relativistic value is 19649 a.u. This demonstrates that a relativistic treatment is indispensable for heavier systems. In section 3 a relativistic generalization of the OEP and KLI methods will be developed. A selection of numerical results for atoms, molecules and solids, including both relativistic and nonrelativistic calculations, will be presented in section 4.

The total-energy functional (16) can also be written as

$$E_{v_0}[\rho] = G[\rho] + \frac{1}{2}\int d^3r \int d^3r' \, \frac{\rho(\mathbf{r})\rho(\mathbf{r}')}{|\mathbf{r}-\mathbf{r}'|} + \int d^3r \, \rho(\mathbf{r}) v_0(\mathbf{r}) \tag{28}$$

where $G[\rho]$ encompasses all the non-trivial parts of the functional $E_{v_0}[\rho]$. The functional $G[\rho]$ can be expanded in powers of the particle-particle interaction $\hat{W}_{\mathrm{Clb}}$:

$$G[\rho] = G^{(0)}[\rho] + G^{(1)}[\rho] + G^{(2)}[\rho] + \ldots \tag{29}$$

where the superscript denotes the order in the coupling constant e^2. (The densities ρ that are inserted in these functionals may or may not depend on e^2 as well. This latter dependence is not of interest here). In the first generation of DFT, the whole functional $G[\rho]$ is approximated by a simple explicit functional of the density. The logical development of DFT towards more and more accurate functionals requires that more and more parts of $G[\rho]$ be treated exactly. This inevitably leads to the use of orbital functionals because all the *exact* functionals of DFT are explicitly orbital-dependent and thereby only implicit functionals of the density. The second generation of DFT, equivalent to the Kohn-Sham method, treats the zero-order term of $G[\rho]$ exactly:

$$G^{(0)}[\rho] = T_S^{\mathrm{exact}}[\rho]. \tag{30}$$

The third generation of DFT, leading to the OEP method, additionally employs the exact first-order term of $G[\rho]$:

$$G^{(1)}[\rho] = E_x^{\mathrm{exact}}[\rho]. \tag{31}$$

Since the expressions (30) and (31) are easily expressed in terms of the KS Green function, the expansion (29) suggests that a systematic orbital representation of the correlation energy

$$E_c[\rho] = G^{(2)}[\rho] + G^{(3)}[\rho] + \ldots \tag{32}$$

can be achieved using the techniques of many-body perturbation theory. Some future directions along these lines will be presented in section 5.

We finally mention that a time-dependent generalization of the OEP has recently been developed [23] to deal with explicitly time-dependent situations such as atoms in strong laser pulses [24]. In the linear-response regime this method has led to a successful procedure [25] to calculate excitation energies from the poles of the frequency-dependent density response. Time-dependent applications of this kind will not be discussed in the present article. The interested reader is referred to recent reviews of time-dependent DFT [26–28].

2 The OEP method, basic formalism

2.1 Derivation of the OEP equations

We are going to derive the OEP equations for the spin-dependent version of DFT [29,30], where the basic variables are the spin-up and spin-down densities $\rho_\uparrow(\mathbf{r})$ and $\rho_\downarrow(\mathbf{r})$, respectively. They are obtained by self-consistently solving the single-particle Schrödinger equations

$$\left(-\frac{\nabla^2}{2} + V_{S\sigma}[\rho_\uparrow, \rho_\downarrow](\mathbf{r})\right)\varphi_{j\sigma}(\mathbf{r}) = \varepsilon_{j\sigma}\varphi_{j\sigma}(\mathbf{r}) \qquad j = 1, \ldots, N_\sigma \qquad \sigma = \uparrow, \downarrow \tag{33}$$

where

$$\rho_\sigma(\mathbf{r}) = \sum_{i=1}^{N_\sigma} |\varphi_{i\sigma}(\mathbf{r})|^2. \tag{34}$$

For convenience we shall assume in the following that infinitesimal symmetry-breaking terms have been added to the external potential to remove any possible degeneracies. The KS orbitals can then be labeled such that

$$\varepsilon_{1\sigma} < \varepsilon_{2\sigma} < \ldots < \varepsilon_{N_\sigma\sigma} < \varepsilon_{(N_\sigma+1)\sigma} < \ldots \tag{35}$$

The Kohn-Sham potentials $V_{S\sigma}(\mathbf{r})$ may be written in the usual way as

$$V_{S\sigma}(\mathbf{r}) = v_0(\mathbf{r}) + \int d^3r' \, \frac{\rho(\mathbf{r}')}{|\mathbf{r} - \mathbf{r}'|} + V_{\mathrm{xc}\sigma}(\mathbf{r}), \tag{36}$$

where

$$\rho(\mathbf{r}) = \sum_{\sigma=\uparrow,\downarrow} \rho_\sigma(\mathbf{r}) \tag{37}$$

and

$$V_{\mathrm{xc}\sigma}(\mathbf{r}) := \frac{\delta E_{\mathrm{xc}}\left[\rho_\uparrow, \rho_\downarrow\right]}{\delta\rho_\sigma(\mathbf{r})}. \tag{38}$$

The starting point of the OEP method is the total-energy functional

$$E_{v_0}^{\mathrm{OEP}}\left[\rho_{\uparrow},\rho_{\downarrow}\right] = \sum_{\sigma=\uparrow,\downarrow}\sum_{i=1}^{N_\sigma}\int d^3r\,\varphi_{i\sigma}^{*}(\mathbf{r})\left(-\frac{1}{2}\nabla^2\right)\varphi_{i\sigma}(\mathbf{r}) + \int d^3r\,v_0(\mathbf{r})\rho(\mathbf{r}) + \frac{1}{2}\int d^3r\int d^3r'\,\frac{\rho(\mathbf{r})\rho(\mathbf{r}')}{|\mathbf{r}-\mathbf{r}'|} + E_{\mathrm{xc}}^{\mathrm{OEP}}\left[\{\varphi_{j\tau}\}\right] \tag{39}$$

where, in contrast to ordinary spin DFT, the xc energy is an *explicit* (approximate) functional of spin orbitals and therefore only an *implicit* functional of the spin densities $\rho_\uparrow$ and $\rho_\downarrow$. In order to calculate the xc potentials defined in Eq. (38) we use the chain rule for functional derivatives to obtain

$$V_{\mathrm{xc}\sigma}^{\mathrm{OEP}}(\mathbf{r}) = \frac{\delta E_{\mathrm{xc}}^{\mathrm{OEP}}\left[\{\varphi_{j\tau}\}\right]}{\delta\rho_\sigma(\mathbf{r})} = \sum_{\alpha=\uparrow,\downarrow}\sum_{i=1}^{N_\alpha}\int d^3r'\,\frac{\delta E_{\mathrm{xc}}^{\mathrm{OEP}}\left[\{\varphi_{j\tau}\}\right]}{\delta\varphi_{i\alpha}(\mathbf{r}')}\frac{\delta\varphi_{i\alpha}(\mathbf{r}')}{\delta\rho_\sigma(\mathbf{r})} + \mathrm{c.c.} \tag{40}$$

and, by applying the functional chain rule once more,

$$V_{\mathrm{xc}\sigma}^{\mathrm{OEP}}(\mathbf{r}) = \sum_{\alpha=\uparrow,\downarrow}\sum_{\beta=\uparrow,\downarrow}\sum_{i=1}^{N_\alpha}\int d^3r'\int d^3r''\times \left(\frac{\delta E_{\mathrm{xc}}^{\mathrm{OEP}}\left[\{\varphi_{j\tau}\}\right]}{\delta\varphi_{i\alpha}(\mathbf{r}')}\frac{\delta\varphi_{i\alpha}(\mathbf{r}')}{\delta V_{S\beta}(\mathbf{r}'')} + c.c.\right)\frac{\delta V_{S\beta}(\mathbf{r}'')}{\delta\rho_\sigma(\mathbf{r})}. \tag{41}$$

The last term on the right-hand side is the inverse $\chi_S^{-1}\left(\mathbf{r},\mathbf{r}'\right)$ of the static density response function of a system of non-interacting particles

$$\chi_{S\alpha,\beta}\left(\mathbf{r},\mathbf{r}'\right) := \frac{\delta\rho_\alpha(\mathbf{r})}{\delta V_{S\beta}(\mathbf{r}')}. \tag{42}$$

This quantity is diagonal with respect to the spin variables so that Eq. (41) reduces to

$$V_{\mathrm{xc}\sigma}^{\mathrm{OEP}}(\mathbf{r}) = \sum_{\alpha=\uparrow,\downarrow}\sum_{i=1}^{N_\alpha}\int d^3r'\int d^3r''\times \left(\frac{\delta E_{\mathrm{xc}}^{\mathrm{OEP}}\left[\{\varphi_{j\tau}\}\right]}{\delta\varphi_{i\alpha}(\mathbf{r}')}\frac{\delta\varphi_{i\alpha}(\mathbf{r}')}{\delta V_{S\sigma}(\mathbf{r}'')} + \mathrm{c.c.}\right)\chi_{S\sigma}^{-1}\left(\mathbf{r}'',\mathbf{r}\right). \tag{43}$$

Acting with the response operator (42) on both sides of Eq. (43) one obtains

$$\int d^3r'\, V_{xc\sigma}^{OEP}(\mathbf{r}')\chi_{S\sigma}\left(\mathbf{r}',\mathbf{r}\right) = \sum_{\alpha=\uparrow,\downarrow}\sum_{i=1}^{N_\alpha}\int d^3r'\, \frac{\delta E_{xc}^{OEP}\left[\{\varphi_{j\tau}\}\right]}{\delta\varphi_{i\alpha}(\mathbf{r}')}\frac{\delta\varphi_{i\alpha}(\mathbf{r}')}{\delta V_{S\sigma}(\mathbf{r})} + \text{c.c.} \tag{44}$$

To further evaluate this equation, we note that the first functional derivative on the right-hand side of Eq. (44) is readily computed once an explicit expression for E_{xc}^{OEP} in terms of single-particle orbitals is given. The remaining functional derivative on the right-hand side of Eq. (44) is calculated using first-order perturbation theory. This yields

$$\frac{\delta\varphi_{i\alpha}(\mathbf{r}')}{\delta V_{S\sigma}(\mathbf{r})} = \delta_{\alpha,\sigma}\sum_{\substack{k=1\\k\neq i}}^{\infty}\frac{\varphi_{k\sigma}(\mathbf{r}')\varphi_{k\sigma}^*(\mathbf{r})}{\varepsilon_{i\sigma}-\varepsilon_{k\sigma}}\varphi_{i\sigma}(\mathbf{r}). \tag{45}$$

Using this equation, the response function

$$\chi_{S\alpha,\beta}\left(\mathbf{r},\mathbf{r}'\right) = \frac{\delta}{\delta V_{S\beta}(\mathbf{r}')}\left(\sum_{i=1}^{N_\alpha}\varphi_{i\alpha}^*(\mathbf{r})\varphi_{i\alpha}(\mathbf{r})\right) \tag{46}$$

is readily expressed in terms of the orbitals as

$$\chi_{S\sigma}\left(\mathbf{r},\mathbf{r}'\right) = \sum_{i=1}^{N_\sigma}\sum_{\substack{k=1\\k\neq i}}^{\infty}\frac{\varphi_{i\sigma}^*(\mathbf{r})\varphi_{k\sigma}(\mathbf{r})\varphi_{k\sigma}^*(\mathbf{r}')\varphi_{i\sigma}(\mathbf{r}')}{\varepsilon_{i\sigma}-\varepsilon_{k\sigma}} + \text{c.c.} \tag{47}$$

Inserting (45) and (47) in Eq. (44), we obtain the standard form of the OEP integral equation:

$$\sum_{i=1}^{N_\sigma}\int d^3r'\varphi_{i\sigma}^*(\mathbf{r}')\left(V_{xc\sigma}^{OEP}(\mathbf{r}') - u_{xci\sigma}(\mathbf{r}')\right)G_{Si\sigma}\left(\mathbf{r}',\mathbf{r}\right)\varphi_{i\sigma}(\mathbf{r}) + \text{c.c.} = 0 \tag{48}$$

where

$$u_{xci\sigma}(\mathbf{r}) := \frac{1}{\varphi_{i\sigma}^*(\mathbf{r})}\frac{\delta E_{xc}^{OEP}\left[\{\varphi_{j\tau}\}\right]}{\delta\varphi_{i\sigma}(\mathbf{r})} \tag{49}$$

and

$$G_{Si\sigma}\left(\mathbf{r},\mathbf{r}'\right) := \sum_{\substack{k=1\\k\neq i}}^{\infty}\frac{\varphi_{k\sigma}(\mathbf{r})\varphi_{k\sigma}^*(\mathbf{r}')}{\varepsilon_{i\sigma}-\varepsilon_{k\sigma}}. \tag{50}$$

The derivation of the OEP integral equation (48) described here was given by Shaginyan [31] and by Görling and Levy [32]. It is important to note that the same expression results [12, 13, 16, 19, 33, 34] if one demands that the local single-particle potential appearing in Eq. (33) be the *optimized* one yielding orbitals minimizing the total-energy functional (39), i.e. that

$$\left.\frac{\delta E_{v_0}^{\mathrm{OEP}}}{\delta V_{S\sigma}(\mathbf{r})}\right|_{V_{S\sigma}=V^{\mathrm{OEP}}} = 0. \tag{51}$$

This equation is the historical origin [12] of the name *optimized effective potential.* As was first pointed out by Perdew and co-workers [35, 36], Eq. (51) is equivalent to the HK variational principle. This is most easily seen by applying the functional chain rule to Eq. (51) yielding

$$0 = \frac{\delta E_{v_0}^{\mathrm{OEP}}}{\delta V_{S\sigma}(\mathbf{r})} = \sum_{\alpha} \int d^3r' \, \frac{\delta E_{v_0}^{\mathrm{OEP}}}{\delta \rho_\alpha(\mathbf{r}')} \frac{\delta \rho_\alpha(\mathbf{r}')}{\delta V_{S\sigma}(\mathbf{r})}. \tag{52}$$

Once again, the last term on the right-hand side of Eq. (52) can be identified with the static KS response function (42). Hence, acting with the inverse response operator on Eq. (52) leads to the HK variational principle

$$0 = \frac{\delta E_{v_0}^{\mathrm{OEP}}}{\delta \rho_\sigma(\mathbf{r})}. \tag{53}$$

2.2 Approximation of Krieger, Li and Iafrate

In order to use the OEP method derived in the previous section, Eq. (48) has to be solved for the xc potential $V_{\mathrm{xc}\sigma}^{\mathrm{OEP}}$. Unfortunately, there is no known analytic solution of $V_{\mathrm{xc}\sigma}^{\mathrm{OEP}}$ depending explicitly on the set of single-particle orbitals $\{\varphi_{j\tau}\}$. Thus, one needs to solve the full integral equation numerically, which is a rather demanding task and has been achieved so far only for systems of high symmetry such as spherical atoms [13, 15, 19, 21, 37, 38] and for solids within the linear muffin tin orbitals atomic sphere approximation [39–41].

However, Krieger, Li and Iafrate [14, 16] recently proposed a transformation of Eq. (48) that leads to an alternative but still exact form of the OEP integral equation which lends itself as a starting point for a highly accurate approximation for $V_{\mathrm{xc}\sigma}^{\mathrm{OEP}}$. Following Krieger, Li and Iafrate [14, 16], we define

$$\psi_{i\sigma}^*(\mathbf{r}) := \int d^3r' \varphi_{i\sigma}^*(\mathbf{r}') \left(V_{\mathrm{xc}\sigma}^{\mathrm{OEP}}(\mathbf{r}') - u_{\mathrm{xc}i\sigma}(\mathbf{r}')\right) G_{Si\sigma}(\mathbf{r}', \mathbf{r}), \tag{54}$$

such that the OEP integral equation (48) can be rewritten as

$$\sum_{i=1}^{N_\sigma} \psi_{i\sigma}^*(\mathbf{r})\varphi_{i\sigma}(\mathbf{r}) + \text{c.c.} = 0. \tag{55}$$

Since the KS orbitals $\{\varphi_{j\tau}\}$ span an orthonormal set, one readily concludes from Eq. (54) that the function $\psi_{i\sigma}(\mathbf{r})$ is orthogonal to $\varphi_{i\sigma}(\mathbf{r})$:

$$\int d^3r\, \psi_{i\sigma}^*(\mathbf{r})\varphi_{i\sigma}(\mathbf{r}) = 0. \tag{56}$$

The quantity $G_{Si\sigma}(\mathbf{r}', \mathbf{r})$ given by Eq. (50) is the Green function of the KS equation projected onto the subspace orthogonal to $\varphi_{i\sigma}(\mathbf{r})$, i.e. it satisfies the equation

$$\left(\hat{h}_{S\sigma}(\mathbf{r}) - \varepsilon_{i\sigma}\right) G_{Si\sigma}(\mathbf{r}', \mathbf{r}) = -\left(\delta(\mathbf{r}' - \mathbf{r}) - \varphi_{i\sigma}(\mathbf{r}')\varphi_{i\sigma}^*(\mathbf{r})\right) \tag{57}$$

where $\hat{h}_{S\sigma}(\mathbf{r})$ is a short-hand notation for the KS Hamiltonian

$$\hat{h}_{S\sigma}(\mathbf{r}) := -\frac{\nabla^2}{2} + V_{S\sigma}[\rho_\uparrow, \rho_\downarrow](\mathbf{r}). \tag{58}$$

Using Eq. (57), we can act with the operator $(\hat{h}_{S\sigma} - \varepsilon_{i\sigma})$ on Eq. (54), leading to

$$\left(\hat{h}_{S\sigma}(\mathbf{r}) - \varepsilon_{i\sigma}\right) \psi_{i\sigma}^*(\mathbf{r}) = -\left(V_{\text{x}c\sigma}^{\text{OEP}}(\mathbf{r}) - u_{\text{x}ci\sigma}(\mathbf{r}) - (\bar{V}_{\text{x}ci\sigma}^{\text{OEP}} - \bar{u}_{\text{x}ci\sigma})\right)\varphi_{i\sigma}^*(\mathbf{r}) \tag{59}$$

where $\bar{V}_{\text{x}ci\sigma}$ denotes the average of $V_{\text{x}c\sigma}(\mathbf{r})$ with respect to the ith orbital, i.e.

$$\bar{V}_{\text{x}ci\sigma}^{\text{OEP}} := \int d^3r\, \varphi_{i\sigma}^*(\mathbf{r}) V_{\text{x}c\sigma}^{\text{OEP}}(\mathbf{r})\varphi_{i\sigma}(\mathbf{r}) \tag{60}$$

and

$$\bar{u}_{\text{x}ci\sigma} := \int d^3r\, \varphi_{i\sigma}^*(\mathbf{r}) u_{\text{x}ci\sigma}(\mathbf{r})\varphi_{i\sigma}(\mathbf{r}). \tag{61}$$

The differential equation (59) has the structure of a KS equation with an additional inhomogeneity term. Eq. (59) plus the boundary condition that $\psi_{i\sigma}^*(\mathbf{r})$ tends to zero as $r \to \infty$ uniquely determines $\psi_{i\sigma}^*(\mathbf{r})$. We can prove this statement by assuming that there are two independent solutions $\psi_{i\sigma,1}^*(\mathbf{r})$ and $\psi_{i\sigma,2}^*(\mathbf{r})$ of Eq. (59). Then the difference between these two solutions, $\Psi_{i\sigma}^*(\mathbf{r}) := \psi_{i\sigma,1}^*(\mathbf{r}) - \psi_{i\sigma,2}^*(\mathbf{r})$, satisfies the homogeneous KS equation

$$(\hat{h}_{S\sigma} - \varepsilon_{i\sigma})\Psi_{i\sigma}^*(\mathbf{r}) = 0, \tag{62}$$

which has a unique solution

$$\Psi^*_{i\sigma}(\mathbf{r}) = \varphi^*_{i\sigma}(\mathbf{r}), \tag{63}$$

if the above boundary condition is fulfilled. However, this solution leads to a contradiction with the orthogonality relation (56) so that $\Psi^*_{i\sigma}(r)$ can only be the trivial solution of Eq. (62),

$$\Psi^*_{i\sigma}(\mathbf{r}) \equiv 0. \tag{64}$$

This completes the proof.

At this point it is useful to attach some physical meaning to the quantity $\psi_{i\sigma}$: From Eq. (54) it is obvious that $\psi_{i\sigma}$ is the usual first-order shift in the wave function caused by the perturbing potential $\delta V_{i\sigma} = V^{\mathrm{OEP}}_{\mathrm{xc}\sigma} - u_{\mathrm{xc}i\sigma}$. This fact also motivates the boundary condition assumed above. In x-only theory, $u_{\mathrm{x}i\sigma}$ is the local, orbital-dependent HF exchange potential so that $\psi_{i\sigma}$ is the first-order shift of the KS wave function towards the HF wave function. One has to realize, however, that the first-order change of the orbital dependent potential $u_{\mathrm{x}i\sigma}[\{\varphi_{j\tau}\}]$ has been neglected. This change can be expected to be small compared to $\delta V_{i\sigma}$ [16].

To further transform the OEP integral equation (55), we solve Eq. (59) for $V_{S\sigma}(\mathbf{r})\psi^*_{i\sigma}(\mathbf{r})$:

$$\begin{aligned} V_{S\sigma}(\mathbf{r})\psi^*_{i\sigma}(\mathbf{r}) = & -\Big(V^{\mathrm{OEP}}_{\mathrm{xc}\sigma}(\mathbf{r}) - u_{\mathrm{xc}i\sigma}(\mathbf{r}) - (\bar{V}^{\mathrm{OEP}}_{\mathrm{xc}i\sigma} - \bar{u}_{\mathrm{xc}i\sigma})\Big)\varphi^*_{i\sigma}(\mathbf{r}) \\ & + \left(\frac{\nabla^2}{2} + \varepsilon_{i\sigma}\right)\psi^*_{i\sigma}(\mathbf{r}). \end{aligned} \tag{65}$$

We then multiply Eq. (55) by the KS potential $V_{S\sigma}(\mathbf{r})$:

$$\sum_{i=1}^{N_\sigma} V_{S\sigma}(\mathbf{r})\psi^*_{i\sigma}(\mathbf{r})\varphi_{i\sigma}(\mathbf{r}) + \text{c.c.} = 0, \tag{66}$$

and employ Eq. (65) to obtain

$$\begin{aligned} 0 = \sum_{i=1}^{N_\sigma} \Big\{ & \Big(V^{\mathrm{OEP}}_{\mathrm{xc}\sigma}(\mathbf{r}) - u_{\mathrm{xc}i\sigma}(\mathbf{r}) - (\bar{V}^{\mathrm{OEP}}_{\mathrm{xc}i\sigma} - \bar{u}_{\mathrm{xc}i\sigma})\Big)\varphi^*_{i\sigma}(\mathbf{r}) \\ & - \left(\frac{\nabla^2}{2} + \varepsilon_{i\sigma}\right)\psi^*_{i\sigma}(\mathbf{r})\Big\}\varphi_{i\sigma}(\mathbf{r}) + \text{c.c.} \end{aligned} \tag{67}$$

Solving this equation for $V^{\mathrm{OEP}}_{\mathrm{xc}\sigma}$ yields

$$\begin{aligned} V^{\mathrm{OEP}}_{\mathrm{xc}\sigma}(\mathbf{r}) = & \frac{1}{2\rho_\sigma(\mathbf{r})}\sum_{i=1}^{N_\sigma}\Big\{|\varphi_{i\sigma}(\mathbf{r})|^2\Big(u_{\mathrm{xc}i\sigma}(\mathbf{r}) + (\bar{V}^{\mathrm{OEP}}_{\mathrm{xc}i\sigma} - \bar{u}_{\mathrm{xc}i\sigma})\Big) \\ & + \left(\frac{\nabla^2}{2}\psi^*_{i\sigma}(\mathbf{r}) + \varepsilon_{i\sigma}\psi^*_{i\sigma}(\mathbf{r})\right)\varphi_{i\sigma}(\mathbf{r})\Big\} + \text{c.c.} \end{aligned} \tag{68}$$

The second term in the curled brackets may be rewritten by using the KS and the OEP equation again, leading to

$$\sum_{i=1}^{N_\sigma}\left(\frac{\nabla^2}{2}\psi_{i\sigma}^*(\mathbf{r})+\varepsilon_{i\sigma}\psi_{i\sigma}^*(\mathbf{r})\right)\varphi_{i\sigma}(\mathbf{r})+\text{c.c.}$$

$$=\sum_{i=1}^{N_\sigma}\left[\left(\frac{\nabla^2}{2}\psi_{i\sigma}^*(\mathbf{r})\right)\varphi_{i\sigma}(\mathbf{r})-\psi_{i\sigma}^*(\mathbf{r})\left(\frac{\nabla^2}{2}\varphi_{i\sigma}(\mathbf{r})\right)\right]+\text{c.c.}$$

$$=-\sum_{i=1}^{N_\sigma}\nabla\cdot\left(\psi_{i\sigma}^*(\mathbf{r})\nabla\varphi_{i\sigma}(\mathbf{r})\right)+\text{c.c.} \tag{69}$$

In this way Eq. (68) may be written as

$$V_{\text{xc}\sigma}^{\text{OEP}}(\mathbf{r})=\frac{1}{2\rho_\sigma(\mathbf{r})}\sum_{i=1}^{N_\sigma}|\varphi_{i\sigma}(\mathbf{r})|^2\left(v_{\text{xc}i\sigma}(\mathbf{r})+(\bar{V}_{\text{xc}i\sigma}^{\text{OEP}}-\bar{u}_{\text{xc}i\sigma})\right)+\text{c.c.} \tag{70}$$

with

$$v_{\text{xc}i\sigma}(\mathbf{r})=u_{\text{xc}i\sigma}(\mathbf{r})-\frac{1}{|\varphi_{i\sigma}(\mathbf{r})|^2}\nabla\cdot\left(\psi_{i\sigma}^*(\mathbf{r})\nabla\varphi_{i\sigma}(\mathbf{r})\right). \tag{71}$$

Eq. (70) is an exact transformation of the original OEP integral equation (48). The advantage of Eq. (70), although still being an integral equation, lies in the fact that it may serve as a starting point for constructing approximations of $V_{\text{xc}\sigma}^{\text{OEP}}$: We only need to approximate $\psi_{i\sigma}^*$ in Eq. (71) by a suitable functional of the orbitals.

The simplest possible approximation is obtained by completely neglecting the terms involving $\psi_{i\sigma}^*$, i.e. by replacing $v_{\text{xc}i\sigma}$ by $u_{\text{xc}i\sigma}$. At first sight, this approximation might appear rather crude. It can be interpreted [14, 16], however, as a mean-field approximation in the sense that the neglected terms averaged over the ground-state spin density $\rho_\sigma(\mathbf{r})$ vanish. To demonstrate this, we investigate the quantity

$$I=\int d^3r\,\nabla\cdot\frac{1}{2}\sum_{i=1}^{N_\sigma}\left(\psi_{i\sigma}^*(\mathbf{r})\nabla\varphi_{i\sigma}(\mathbf{r})\right)+\text{c.c.} \tag{72}$$

which amounts to the difference between the exact $V_{\text{xc}\sigma}^{\text{OEP}}(\mathbf{r})$ and the approximated xc potential averaged over $\rho_\sigma(\mathbf{r})$. From Eq. (54) one easily derives

$$\begin{aligned} \frac{\nabla^2}{2}\psi^*_{i\sigma}(\mathbf{r}) &= \sum_{\substack{k=1\\k\neq i}}^{\infty} A_{ik\sigma}\left(\frac{\nabla^2}{2}\right)\varphi^*_{k\sigma}(\mathbf{r}) \\ &= \sum_{\substack{k=1\\k\neq i}}^{\infty} A_{ik\sigma}\left(V_{S\sigma}(\mathbf{r})\varphi^*_{k\sigma}(\mathbf{r}) - \varepsilon_{k\sigma}\varphi^*_{k\sigma}(\mathbf{r})\right) \\ &= V_{S\sigma}(\mathbf{r})\psi^*_{i\sigma}(\mathbf{r}) - \sum_{\substack{k=1\\k\neq i}}^{\infty} A_{ik\sigma}\varepsilon_{k\sigma}\varphi^*_{k\sigma}(\mathbf{r}) \end{aligned} \tag{73}$$

where we have used the abbreviation

$$A_{ik\sigma} = \frac{1}{\varepsilon_{i\sigma} - \varepsilon_{k\sigma}} \int d^3r' \varphi^*_{i\sigma}(\mathbf{r}')\left(V^{\mathrm{OEP}}_{\mathrm{xc}\sigma}(\mathbf{r}') - u_{\mathrm{xc}i\sigma}(\mathbf{r}')\right)\varphi_{k\sigma}(\mathbf{r}') \ . \tag{74}$$

Insertion of Eq. (73) in Eq. (69) leads to

$$\begin{aligned} I = -\frac{1}{2}\int d^3r \sum_{i=1}^{N_\sigma}\Big(& V_{S\sigma}(\mathbf{r})\psi^*_{i\sigma}(\mathbf{r})\varphi_{i\sigma}(\mathbf{r}) \\ & - \sum_{\substack{k=1\\k\neq i}}^{\infty} A_{ik\sigma}\varepsilon_{k\sigma}\varphi^*_{k\sigma}(\mathbf{r})\varphi_{i\sigma}(\mathbf{r}) + \varepsilon_{i\sigma}\psi^*_{i\sigma}(\mathbf{r})\varphi_{i\sigma}(\mathbf{r}) + \mathrm{c.c.}\Big) = 0 \end{aligned} \tag{75}$$

where the last step follows from the orthogonality of the KS orbitals together with Eqs. (55) and (56). Hence, the neglected terms have zero average value.

The resulting equation, known as the KLI approximation, is given by [14–22, 42–44]

$$V^{\mathrm{KLI}}_{\mathrm{xc}\sigma}(\mathbf{r}) = \frac{1}{2\rho_\sigma(\mathbf{r})}\sum_{i=1}^{N_\sigma}|\varphi_{i\sigma}(\mathbf{r})|^2\left(u_{\mathrm{xc}i\sigma}(\mathbf{r}) + (\bar{V}^{\mathrm{KLI}}_{\mathrm{xc}i\sigma} - \bar{u}_{\mathrm{xc}i\sigma})\right) + \mathrm{c.c.} \tag{76}$$

which has proven to be an excellent approximation to the full xc potential $V^{\mathrm{OEP}}_{\mathrm{xc}\sigma}(\mathbf{r})$, as will be shown in section 4. We immediately recognize that this form is very similar to the Slater potential. It should be noted that – in contrast to the work of Krieger [16] – we did *not* use any asymptotic properties of $\psi_{i\sigma}$ or $\varphi_{i\sigma}$ in the derivation of Eq. (75). This implies that the KLI approximation is also justified for solid state systems.

In contrast to the full OEP equation (48), the KLI equation, still being an integral equation, can be solved explicitly in terms of the orbitals $\{\varphi_{j\tau}\}$: Multiplying Eq. (76) by $|\varphi_{j\sigma}(\mathbf{r})|^2$ and integrating over space yields

$$\bar{V}^{\mathrm{KLI}}_{\mathrm{xc}j\sigma} = \bar{V}^{S}_{\mathrm{xc}j\sigma} + \sum_{i=1}^{N_\sigma-1} M_{ji\sigma}\left(\bar{V}^{\mathrm{KLI}}_{\mathrm{xc}i\sigma} - \frac{1}{2}\left(\bar{u}_{\mathrm{xc}i\sigma} + \bar{u}^*_{\mathrm{xc}i\sigma}\right)\right), \tag{77}$$

where

$$\bar{V}^{S}_{\text{xc}j\sigma} := \int d^3r \, \frac{|\varphi_{j\sigma}(\mathbf{r})|^2}{\rho_\sigma(\mathbf{r})} \sum_{i=1}^{N_\sigma} |\varphi_{i\sigma}(\mathbf{r})|^2 \frac{1}{2}\Big(u_{\text{xc}i\sigma}(\mathbf{r}) + u^*_{\text{xc}i\sigma}(\mathbf{r})\Big) \tag{78}$$

and

$$M_{ji\sigma} := \int d^3r \, \frac{|\varphi_{j\sigma}(\mathbf{r})|^2 |\varphi_{i\sigma}(\mathbf{r})|^2}{\rho_\sigma(\mathbf{r})}. \tag{79}$$

The term corresponding to the highest occupied orbital $\varphi_{N_\sigma\sigma}$ has been excluded from the sum in Eq. (77) because $\bar{V}^{\text{KLI}}_{\text{xc}N_\sigma\sigma} = \bar{u}_{\text{xc}N_\sigma\sigma}$, which will be proven in the next section. The remaining unknown constants $(\bar{V}^{\text{KLI}}_{\text{xc}i\sigma} - \bar{u}_{\text{xc}i\sigma})$ are determined by the linear equation

$$\sum_{i=1}^{N_\sigma - 1} (\delta_{ji} - M_{ji\sigma}) \left(\bar{V}^{\text{KLI}}_{\text{xc}i\sigma} - \frac{1}{2}\left(\bar{u}_{\text{xc}i\sigma} + \bar{u}^*_{\text{xc}i\sigma}\right)\right) = \left(\bar{V}^{S}_{\text{xc}j\sigma} - \frac{1}{2}\left(\bar{u}_{\text{xc}j\sigma} + \bar{u}^*_{\text{xc}j\sigma}\right)\right), \tag{80}$$

with $j = 1, \ldots N_\sigma - 1$. Solving Eq. (80) and substituting the result into Eq. (76), we obtain an explicitly orbital-dependent functional.

We conclude this section in remarking that the derivation given here differs slightly from the one given by Krieger, Li and Iafrate [14, 16]. The main difference is that we choose to work with the quantity $\psi_{i\sigma}$, which is related to $p_{i\sigma}$ introduced in Ref. [14, 16] by $\psi_{i\sigma} = p_{i\sigma}\varphi_{i\sigma}$. Since both $\psi_{i\sigma}$ and $\varphi_{i\sigma}$ are well-behaved functions, $p_{i\sigma}$ has poles where $\varphi_{i\sigma}$ becomes zero. It is therefore more convenient to work with the well-behaved function $\psi_{i\sigma}$, especially for the considerations of the next section.

We finally note that the KLI equation (76) can also be obtained by a less rigorous derivation, namely by approximating the energy dominator in the Green function (57) by a single constant as was first suggested by Sharp and Horton [12] and further elaborated by Krieger, Li, and Iafrate [15, 17, 19–22].

2.3 Rigorous properties of the OEP and KLI potentials

2.3.1 An important lemma

In this section a number of rigorous statements on the optimized effective potential of finite systems will be derived [45]. For this purpose, the exchange-only potential and the correlation potential have to be treated separately. The exact exchange potential of DFT is defined as

$$V_{\text{x}\sigma}[\rho_\uparrow, \rho_\downarrow](\mathbf{r}) = \frac{\delta E^{\text{exact}}_{\text{x}}}{\delta\rho_\sigma(\mathbf{r})}, \qquad \sigma = \uparrow, \downarrow \tag{81}$$

where the exact exchange-energy functional is given by Eq. (27). In an ordinary OEP calculation, one only determines the potential $V_{x\sigma}[\rho_{\uparrow 0}, \rho_{\downarrow 0}](\mathbf{r})$ corresponding to the self-consistent ground-state spin densities $(\rho_{\uparrow 0}, \rho_{\downarrow 0})$ of the system considered. If one were to calculate $V_{x\sigma}[\rho_{\uparrow}, \rho_{\downarrow}]$ *for an arbitrary given set* $(\rho_{\uparrow}, \rho_{\downarrow})$ *of spin densities* one would have to perform the following three steps:

1. Determine the unique potentials $V_{S\sigma}[\rho_{\uparrow}, \rho_{\downarrow}](\mathbf{r})$, $\sigma = \uparrow, \downarrow$, corresponding to the given spin densities $(\rho_{\uparrow}, \rho_{\downarrow})$.
2. Solve the Schrödinger equation (33) for the spin-up and spin-down orbitals with the potentials of step (1).
3. Plug the orbitals obtained in step (2) into the OEP integral equation

$$\sum_{i=1}^{N_\sigma} \int d^3r' \, \varphi_{i\sigma}^*(\mathbf{r}') \Big(V_{x\sigma}(\mathbf{r}') - u_{xi\sigma}(\mathbf{r}') \Big) G_{Si\sigma}(\mathbf{r}', \mathbf{r}) \varphi_{i\sigma}(\mathbf{r}) + \text{c.c.} = 0 \quad (82)$$

and solve this equation for $V_{x\sigma}$ keeping the orbitals of step (2) fixed.

In this way Filippi, Umrigar and Gonze [46] have recently calculated the exchange potentials corresponding to the *exact* (not the x-only) densities of some atoms where the exact densities were determined in a quantum Monte-Carlo calculation. Likewise, for any given approximate functional $E_c[\{\varphi_{i\sigma}\}]$, the corresponding correlation potential

$$V_{c\sigma}[\rho_{\uparrow}, \rho_{\downarrow}](\mathbf{r}) = \frac{\delta E_c}{\delta \rho_\sigma(\mathbf{r})} \quad (83)$$

is obtained by the above steps (1) and (2), and step (3) replaced by the solution of

$$\sum_{i=1}^{N_\sigma} \int d^3r' \, \varphi_{i\sigma}^*(\mathbf{r}') \Big(V_{c\sigma}(\mathbf{r}') - u_{ci\sigma}(\mathbf{r}') \Big) G_{Si\sigma}(\mathbf{r}', \mathbf{r}) \varphi_{i\sigma}(\mathbf{r}) + \text{c.c.} = 0. \quad (84)$$

Whenever, in the following derivations, the OEP equations (82) and (84) are used or transformed it is understood that the orbitals $\{\varphi_{i\sigma}\}$ are kept fixed so that they always correspond to a unique fixed set $(\rho_{\uparrow}, \rho_{\downarrow})$ of spin densities.

We first prove an important *lemma* concerning the constants defined by Eqs. (60) and (61). The lemma states that

(i)

$$\bar{u}_{xN_\sigma\sigma} = \bar{V}_{xN_\sigma\sigma}$$

is satisfied for

$$u_{xi\sigma}(\mathbf{r}) = \frac{1}{\varphi_{i\sigma}^*(\mathbf{r})} \frac{\delta E_x^{\text{exact}}}{\delta \varphi_{i\sigma}(\mathbf{r})} \quad (85)$$

with the exact exchange-energy functional;

(ii)

$$\bar{u}_{cN_\sigma\sigma} = \bar{V}_{cN_\sigma\sigma}$$

is satisfied for any approximate correlation-energy functional $E_c[\{\varphi_{i\sigma}\}]$ *having the property*

$$u_{ci\sigma}(\mathbf{r}) = \frac{1}{\varphi_{i\sigma}^*(\mathbf{r})} \frac{\delta E_c}{\delta\varphi_{i\sigma}(\mathbf{r})} \overset{r\to\infty}{\longrightarrow} const\ ,\ i = 1...N_\sigma. \tag{86}$$

We begin with the proof of statement (ii). To this end we use Eq. (59) for the correlation part only:

$$\left(-\frac{\nabla^2}{2} + V_{S\sigma}(\mathbf{r}) - \varepsilon_{i\sigma}\right)\psi_{i\sigma}^*(\mathbf{r}) = \left(V_{c\sigma}(\mathbf{r}) - u_{ci\sigma}(\mathbf{r}) - C_{i\sigma}\right)\varphi_{i\sigma}^*(\mathbf{r}) \tag{87}$$

where we have introduced the abbreviation

$$C_{i\sigma} = \bar{V}_{ci\sigma} - \bar{u}_{ci\sigma} \tag{88}$$

(dropping the superscript OEP for notational simplicity). If Eq. (87) is satisfied with potentials $V_{S\sigma}(\mathbf{r})$, $V_{c\sigma}(\mathbf{r})$ and $u_{ci\sigma}(\mathbf{r})$ it will also be satisfied with the constantly shifted potentials

$$\tilde{V}_{S\sigma}(\mathbf{r}) := V_{S\sigma}(\mathbf{r}) + B_{S\sigma} \tag{89}$$

$$\tilde{V}_{c\sigma}(\mathbf{r}) := V_{c\sigma}(\mathbf{r}) + B_{c\sigma} \tag{90}$$

$$\tilde{u}_{ci\sigma}(\mathbf{r}) := u_{ci\sigma}(\mathbf{r}) + B_{i\sigma} \tag{91}$$

and the corresponding eigenvalues $\tilde{\varepsilon}_{i\sigma}$ and the constants $\bar{\tilde{V}}_{ci\sigma}$, $\bar{\tilde{u}}_{ci\sigma}$. The constants $B_{S\sigma}$, $B_{c\sigma}$, $B_{i\sigma}$ cancel out in Eq. (87) because the eigenvalues $\tilde{\varepsilon}_{i\sigma}$ resulting from solving the Schrödinger equation (33) with the potential (89) are given by

$$\tilde{\varepsilon}_{i\sigma} = \varepsilon_{i\sigma} + B_{S\sigma} \tag{92}$$

and the constants $\bar{\tilde{V}}_{c\sigma}$, $\bar{\tilde{u}}_{ci\sigma}$ obtained from the correlation parts of Eqs. (60), (61) with the potentials (90), (91) are

$$\bar{\tilde{V}}_{ci\sigma} = \bar{V}_{ci\sigma} + B_{c\sigma} \tag{93}$$

$$\bar{\tilde{u}}_{ci\sigma} = \bar{u}_{ci\sigma} + B_{i\sigma}. \tag{94}$$

Hence we can assume *without restriction* that

$$V_{S\sigma}(\mathbf{r}) \overset{r\to\infty}{\longrightarrow} 0 \tag{95}$$

$$V_{c\sigma}(\mathbf{r}) \overset{r\to\infty}{\longrightarrow} 0 \tag{96}$$

$$u_{ci\sigma}(\mathbf{r}) \overset{r\to\infty}{\longrightarrow} 0. \tag{97}$$

In the following we shall investigate the asymptotic behavior of the KS orbitals $\varphi_{i\sigma}(\mathbf{r})$ and of the quantities $\psi_{i\sigma}(\mathbf{r})$ determined by Eq. (87). As a shorthand we write

$$\varphi_{i\sigma}(\mathbf{r}) \stackrel{r\to\infty}{\longrightarrow} \Phi_{i\sigma}(r) f_{i\sigma}(\Omega) \tag{98}$$

$$\psi_{i\sigma}(\mathbf{r}) \stackrel{r\to\infty}{\longrightarrow} \Psi_{i\sigma}(r) g_{i\sigma}(\Omega). \tag{99}$$

The aim is to determine the asymptotically dominant functions $\Phi_{i\sigma}(r)$ and $\Psi_{i\sigma}(r)$. The angular parts $f_{i\sigma}(\Omega)$ and $g_{i\sigma}(\Omega)$ are not of interest in the present context. Using the fact that the KS potential of finite neutral systems behaves asymptotically as [47]

$$V_{S\sigma}(\mathbf{r}) \stackrel{r\to\infty}{\longrightarrow} -\frac{1}{r} \tag{100}$$

the KS equation (33) leads to the following asymptotic equation

$$\left(-\frac{1}{2}\frac{1}{r}\frac{d^2}{dr^2}r - \frac{1}{r} - \varepsilon_{i\sigma}\right)\Phi_{i\sigma}(r) = 0. \tag{101}$$

The asymptotic form of $\Phi_{i\sigma}(r)$ is easily found to be

$$\Phi_{i\sigma}(r) \stackrel{r\to\infty}{\longrightarrow} r^{1/\beta_{i\sigma}} \frac{\mathrm{e}^{-\beta_{i\sigma} r}}{r} \tag{102}$$

with

$$\beta_{i\sigma} := \sqrt{-2\varepsilon_{i\sigma}}. \tag{103}$$

By virtue of Eqs. (87) and (102), $\Psi_{i\sigma}(r)$ must satisfy the asymptotic equation

$$\left(-\frac{1}{2}\frac{1}{r}\frac{d^2}{dr^2}r - \frac{1}{r} - \varepsilon_{i\sigma}\right)\Psi_{i\sigma}(r) = (W_{i\sigma}(r) - C_{i\sigma})\, r^{1/\beta_{i\sigma}} \frac{\mathrm{e}^{-\beta_{i\sigma} r}}{r} \tag{104}$$

where we have introduced the quantity $W_{i\sigma}(r)$ defined by

$$\left(V_{c\sigma}(\mathbf{r}) - u_{ci\sigma}(\mathbf{r})\right) \stackrel{r\to\infty}{\longrightarrow} W_{i\sigma}(r) w_{i\sigma}(\Omega). \tag{105}$$

From Eqs. (96) and (97) we know that

$$W_{i\sigma}(r) \stackrel{r\to\infty}{\longrightarrow} 0. \tag{106}$$

Inserting the ansatz

$$\Psi_{i\sigma}(r) = p_{i\sigma}(r)\frac{\mathrm{e}^{-\beta_{i\sigma} r}}{r} \tag{107}$$

in Eq. (104) we find that the function $p_{i\sigma}(r)$ must satisfy the equation

$$\frac{1}{2}p''_{i\sigma} - \beta_{i\sigma}p'_{i\sigma} + \frac{p_{i\sigma}}{r} = C_{i\sigma}r^{1/\beta_{i\sigma}} \qquad \text{if} \quad C_{i\sigma} \neq 0 \tag{108}$$

and

$$\frac{1}{2}p''_{i\sigma} - \beta_{i\sigma}p'_{i\sigma} + \frac{p_{i\sigma}}{r} = -W_{i\sigma}(r)r^{1/\beta_{i\sigma}} \qquad \text{if} \quad C_{i\sigma} = 0. \tag{109}$$

The asymptotic solution of Eq. (108) is immediately recognized as

$$p_{i\sigma}(r) \overset{r\to\infty}{\longrightarrow} -\frac{C_{i\sigma}}{\beta_{i\sigma}}r^{(1/\beta_{i\sigma}+1)} \tag{110}$$

so that

$$\Psi_{i\sigma}(r) = -\frac{C_{i\sigma}}{\beta_{i\sigma}}r^{1/\beta_{i\sigma}}\mathrm{e}^{-\beta_{i\sigma}r} \qquad \text{if} \quad C_{i\sigma} \neq 0. \tag{111}$$

Writing

$$p_{i\sigma}(r) = F_{i\sigma}(r)r^{1/\beta_{i\sigma}+1} \qquad \text{if} \quad C_{i\sigma} = 0 \tag{112}$$

one readily verifies by insertion in Eq. (109) that

$$F_{i\sigma}(r) \overset{r\to\infty}{\longrightarrow} 0 \tag{113}$$

as a consequence of (106).

We now prove statement (ii) of the lemma by reductio ad absurdum: Assume that $C_{N_\sigma\sigma} \neq 0$. Then the asymptotic form of $\Psi_{N_\sigma\sigma}(r)$ is given by (111) and we conclude that

$$\psi^*_{N_\sigma\sigma}(\mathbf{r})\varphi_{N_\sigma\sigma}(\mathbf{r}) \overset{r\to\infty}{\longrightarrow} -\frac{C_{N_\sigma\sigma}}{\beta_{N_\sigma\sigma}}r^{\left(\frac{2}{\beta_{N_\sigma\sigma}}-1\right)}\mathrm{e}^{-2\beta_{N_\sigma\sigma}r} \cdot g^*_{N_\sigma\sigma}(\Omega)f_{N_\sigma\sigma}(\Omega). \tag{114}$$

For $i \neq N_\sigma$, on the other hand, we obtain

$$\psi^*_{i\sigma}(\mathbf{r})\varphi_{i\sigma}(\mathbf{r}) \overset{r\to\infty}{\longrightarrow} G_{i\sigma}(r)r^{\left(\frac{2}{\beta_{i\sigma}}-1\right)}\mathrm{e}^{-2\beta_{i\sigma}r} \cdot g^*_{i\sigma}(\Omega)f_{i\sigma}(\Omega) \tag{115}$$

where

$$G_{i\sigma}(r) = \begin{cases} -C_{i\sigma}/\beta_{i\sigma} & \text{if} \quad C_{i\sigma} \neq 0 \\ F_{i\sigma}(r) \overset{r\to\infty}{\longrightarrow} 0 & \text{if} \quad C_{i\sigma} = 0. \end{cases} \tag{116}$$

From this we conclude that the OEP integral equation

$$\psi^*_{N_\sigma\sigma}(\mathbf{r})\varphi_{N_\sigma\sigma}(\mathbf{r}) + \sum_{i=1}^{N_\sigma-1}\psi^*_{i\sigma}(\mathbf{r})\varphi_{i\sigma}(\mathbf{r}) + \text{c.c.} \equiv 0 \tag{117}$$

is not satisfied for $r \to \infty$ because the dominant term given by (114) cannot be canceled by any of the other contributions (115) which all fall off more rapidly (cf. Eq. (35)). Consequently the $\psi_{j\sigma}$ cannot be solutions of the OEP equation which is the desired contradiction. This implies that $C_{N_\sigma\sigma} = 0$ which completes the proof of statement (ii).

In order to prove statement (i) of the lemma we first investigate the asymptotic form of the quantities $u_{\mathrm{x}i\sigma}(\mathbf{r})$. Employing the exact exchange-energy functional (27) we find

$$u_{\mathrm{x}i\sigma}(\mathbf{r}) = -\sum_{j=1}^{N_\sigma} \frac{\varphi_{j\sigma}^*(\mathbf{r})}{\varphi_{i\sigma}^*(\mathbf{r})} K_{ji\sigma}(\mathbf{r}) \tag{118}$$

with

$$K_{ji\sigma}(\mathbf{r}) := \int d^3r' \, \frac{\varphi_{j\sigma}(\mathbf{r}')\varphi_{i\sigma}^*(\mathbf{r}')}{|\mathbf{r}-\mathbf{r}'|}. \tag{119}$$

Performing a multipole expansion of $K_{ji\sigma}(\mathbf{r})$ and using the orthonormality of the KS orbitals we find

$$K_{ii\sigma}(\mathbf{r}) \stackrel{r\to\infty}{\longrightarrow} \frac{1}{r} \tag{120}$$

$$K_{ji\sigma}(\mathbf{r}) \stackrel{r\to\infty}{\longrightarrow} \frac{1}{r^m} k_{ji\sigma}(\Omega) \qquad i \neq j \tag{121}$$

with some integer $m \geq 2$ that depends on i and j. Hence the sum in Eq. (118) must be dominated asymptotically by the $j = N_\sigma$ term:

$$u_{\mathrm{x}i\sigma}(\mathbf{r}) \stackrel{r\to\infty}{\longrightarrow} -\frac{\varphi_{N_\sigma\sigma}^*(\mathbf{r})}{\varphi_{i\sigma}^*(\mathbf{r})} K_{N_\sigma i\sigma}(\mathbf{r}). \tag{122}$$

Using Eqs. (120), (121) and the asymptotic behavior (102) of the KS orbitals we obtain

$$u_{\mathrm{x}N_\sigma\sigma}(\mathbf{r}) \stackrel{r\to\infty}{\longrightarrow} -\frac{1}{r} \tag{123}$$

and for $i \neq N_\sigma$

$$u_{\mathrm{x}i\sigma}(\mathbf{r}) \stackrel{r\to\infty}{\longrightarrow} -r^{\left(\frac{1}{\beta_{N_\sigma\sigma}}-\frac{1}{\beta_{i\sigma}}-m\right)} e^{(\beta_{i\sigma}-\beta_{N_\sigma\sigma})r}\, \omega_{i\sigma}(\Omega). \tag{124}$$

We recognize that $u_{\mathrm{x}i\sigma}(\mathbf{r})$ diverges exponentionally to $-\infty$ for $i < N_\sigma$. In the x-only case, the quantities $\psi_{i\sigma}(\mathbf{r})$ satisfy the equation

$$\left(-\frac{\nabla^2}{2} + V_{S\sigma}(\mathbf{r}) - \varepsilon_{i\sigma}\right)\psi_{i\sigma}^*(\mathbf{r}) = \left(V_{\mathrm{x}\sigma}(\mathbf{r}) - u_{\mathrm{x}i\sigma}(\mathbf{r}) - C_{i\sigma}\right)\varphi_{i\sigma}^*(\mathbf{r}) \tag{125}$$

where

$$C_{i\sigma} = \bar{V}_{\mathrm{x}i\sigma} - \bar{u}_{\mathrm{x}i\sigma}. \tag{126}$$

In the following we prove statement (i) of the lemma by reductio ad absurdum: Assume that $C_{N_\sigma\sigma} \neq 0$. Then, by Eq. (123), the right-hand side of Eq. (125) for $i = N_\sigma$ is asymptotically dominated by $-C_{N_\sigma\sigma}\varphi^*_{N_\sigma\sigma}(\mathbf{r})$ and we obtain, in complete analogy to the correlation-only case:

$$\Psi_{N_\sigma\sigma}(r) = -\frac{C_{N_\sigma\sigma}}{\beta_{N_\sigma\sigma}} r^{1/\beta_{N_\sigma\sigma}} \mathrm{e}^{-\beta_{N_\sigma\sigma} r} \qquad \text{for} \quad C_{N_\sigma\sigma} \neq 0. \tag{127}$$

For $i < N_\sigma$, the right-hand side of Eq. (125) is dominated by $-u_{\mathrm{x}i\sigma}(\mathbf{r})\varphi^*_{i\sigma}(\mathbf{r})$. Using Eqs. (102) and (124) $\Psi_{i\sigma}(r)$ satisfies the asymptotic differential equation

$$\left(-\frac{1}{2}\frac{1}{r}\frac{d^2}{dr^2}r - \frac{1}{r} - \varepsilon_{i\sigma}\right)\Psi_{i\sigma}(r) = r^{\left(\frac{1}{\beta_{N_\sigma\sigma}}-1-m\right)}\mathrm{e}^{-\beta_{N_\sigma\sigma} r}. \tag{128}$$

From this equation one readily concludes that

$$\Psi_{i\sigma}(r) \overset{r\to\infty}{\longrightarrow} \frac{1}{\varepsilon_{N_\sigma\sigma} - \varepsilon_{i\sigma}} r^{\left(\frac{1}{\beta_{N_\sigma\sigma}}-1-m\right)}\mathrm{e}^{-\beta_{N_\sigma\sigma} r}, \qquad i < N_\sigma. \tag{129}$$

We note in passing that all the functions $\psi_{i\sigma}, i = 1 \ldots N_\sigma$, have the *same* exponential decay, $\mathrm{e}^{-\beta_{N_\sigma\sigma} r}$, determined by the highest occupied orbital energy $\beta_{N_\sigma\sigma} = \sqrt{-2\varepsilon_{N_\sigma\sigma}}$. This fact further supports the interpretation of the quantities $\psi_{i\sigma}$ (in the x-only case) as a shift from the KS orbitals towards the HF orbitals: The HF orbitals $\varphi^{\mathrm{HF}}_{i\sigma}$ are known [48] to be asymptotically dominated by the exponential decay $\mathrm{e}^{-\beta_{N_\sigma\sigma} r}$ of the highest occupied orbital. The same holds true for the shifted KS orbitals $(\varphi_{i\sigma} + \psi_{i\sigma})$.

From Eqs. (102), (127) and (129) we obtain

$$\psi^*_{N_\sigma\sigma}(\mathbf{r})\varphi_{N_\sigma\sigma}(\mathbf{r}) \overset{r\to\infty}{\longrightarrow} -\frac{C_{N_\sigma\sigma}}{\beta_{N_\sigma\sigma}} r^{\left(\frac{2}{\beta_{N_\sigma\sigma}}-1\right)}\mathrm{e}^{-2\beta_{N_\sigma\sigma} r} \cdot g^*_{N_\sigma\sigma}(\Omega) f_{N_\sigma\sigma}(\Omega) \tag{130}$$

and

$$\psi^*_{i\sigma}(\mathbf{r})\varphi_{i\sigma}(\mathbf{r}) \overset{r\to\infty}{\longrightarrow} \frac{1}{\varepsilon_{N_\sigma\sigma} - \varepsilon_{i\sigma}} r^{\left(\frac{1}{\beta_{N_\sigma\sigma}}+\frac{1}{\beta_{i\sigma}}-2-m\right)}\mathrm{e}^{-(\beta_{N_\sigma\sigma}+\beta_{i\sigma})r} \cdot g^*_{i\sigma}(\Omega) f_{i\sigma}(\Omega), \; i < N_\sigma. \tag{131}$$

Once again we conclude that in the OEP equation (117) the asymptotically dominant term (130) cannot be canceled by any of the other terms (131), leading to the contradiction that the $\psi_{j\sigma}(\mathbf{r})$ are not solutions of the OEP integral equation. Hence we conclude that $C_{N_\sigma\sigma} = 0$ which completes the proof of the lemma.

The original proof [13] of statement (i) was based on the asymptotic form of the Green function which is easily accessible only in 1D. Considering the 3D Green function, Krieger, Li and Iafrate [19] made it plausible that the statement holds true in the 3D case as well. An alternative proof was recently given [49] for the x-only case. This proof is based on the scaling properties of the exchange-energy functional and can therefore not be generalized to the case of correlation. The proof presented above for the correlation part of the OEP (statement (ii) of the lemma) is valid for all correlation-energy functionals leading to asymptotically bounded functions $u_{ci\sigma}(\mathbf{r})$. For asymptotically diverging $u_{ci\sigma}(\mathbf{r})$ the lemma might still be valid. In particular, if the divergence is the same as the one (Eq. (124)) found in the exchange case, the proof of statement (i) carries over. The lemma has a number of important consequences which will be discussed in the following two subsections.

2.3.2 Asymptotic form

In this section we shall investigate the asymptotic form of the exchange and correlation potentials. It will be shown that $V_{x\sigma}(\mathbf{r})$ and $u_{xN_\sigma\sigma}(\mathbf{r})$ approach each other exponentially fast for $r \to \infty$, and that the difference between $V_{c\sigma}(\mathbf{r})$ and $u_{cN_\sigma\sigma}(\mathbf{r})$ decays exponentially as well. Using the notation of the last section the detailed statements read as follows:

THEOREM 1

$$V_{x\sigma}(\mathbf{r}) - u_{xN_\sigma\sigma}(\mathbf{r}) \stackrel{r\to\infty}{\longrightarrow} r^{\left(\frac{1}{\beta_{(N_\sigma-1)\sigma}} - \frac{1}{\beta_{N_\sigma\sigma}} - m\right)} e^{-(\beta_{(N_\sigma-1)\sigma} - \beta_{N_\sigma\sigma})r} \tag{132}$$

where m is an integer satisfying $m \geq 2$.

THEOREM 2 *If the constant $C_{(N_\sigma-1)\sigma}$ defined by Eq.* (88) *does not vanish then*

$$V_{c\sigma}(\mathbf{r}) - u_{cN_\sigma\sigma}(\mathbf{r}) \stackrel{r\to\infty}{\longrightarrow} r^{\left(\frac{2}{\beta_{(N_\sigma-1)\sigma}} - \frac{2}{\beta_{N_\sigma\sigma}} + 1\right)} e^{-2(\beta_{(N_\sigma-1)\sigma} - \beta_{N_\sigma\sigma})r} . \tag{133}$$

If $C_{(N_\sigma-1)\sigma} = 0$ the right-hand side of (133) *is an upper bound of $|V_{c\sigma}(\mathbf{r}) - u_{cN_\sigma\sigma}(\mathbf{r})|$ for $r \to \infty$, i.e. for $C_{(N_\sigma-1)\sigma} = 0$, $V_{c\sigma}(\mathbf{r})$ and $u_{cN_\sigma\sigma}(\mathbf{r})$ approach each other even faster than given by the right-hand side of Eq.* (133).

To prove theorem 1 we write

$$\Psi_{N_\sigma\sigma}(r) = q(r)\frac{e^{-\beta_{N_\sigma\sigma}r}}{r} . \tag{134}$$

Using the lemma of the last section ensuring that $C_{N_\sigma\sigma} = 0$, $q(r)$ must satisfy the following asymptotic differential equation:

$$-\frac{1}{2}r^{-\frac{1}{\beta_{N_\sigma\sigma}}}q''(r) + \beta_{N_\sigma\sigma}r^{-\frac{1}{\beta_{N_\sigma\sigma}}}q'(r) - r^{-\left(\frac{1}{\beta_{N_\sigma\sigma}}+1\right)}q(r) = V_{x\sigma}(\mathbf{r}) - u_{xN_\sigma\sigma}(\mathbf{r}) . \tag{135}$$

This is readily verified by inserting (100), (102) and (134) in Eq. (125). By virtue of Eqs. (35) and (131) the sum

$$\sum_{i=1}^{N_\sigma-1} \psi_{i\sigma}^*(\mathbf{r})\varphi_{i\sigma}(\mathbf{r}) \tag{136}$$

must be asymptotically dominated by the $i = (N_\sigma - 1)$ term which decays as

$$\begin{aligned}&\psi_{(N_\sigma-1)\sigma}^*(\mathbf{r})\varphi_{(N_\sigma-1)\sigma}(\mathbf{r})\\ &\quad\overset{r\to\infty}{\longrightarrow} \frac{1}{\varepsilon_{N_\sigma\sigma} - \varepsilon_{(N_\sigma-1)\sigma}} r^{\left(\frac{1}{\beta_{N_\sigma\sigma}}+\frac{1}{\beta_{(N_\sigma-1)\sigma}}-2-m\right)} \mathrm{e}^{-(\beta_{N_\sigma\sigma}+\beta_{(N_\sigma-1)\sigma})r} .\end{aligned} \tag{137}$$

This term cannot be canceled by any other term of the sum (136). Hence, for the OEP equation (55) to be asymptotically satisfied, the expression (137) must be canceled by the $i = N_\sigma$ term which behaves as

$$\psi_{N_\sigma\sigma}^*(\mathbf{r})\varphi_{N_\sigma\sigma}(\mathbf{r}) \overset{r\to\infty}{\longrightarrow} q(r) r^{\left(\frac{1}{\beta_{N_\sigma\sigma}}-2\right)} \mathrm{e}^{-2\beta_{N_\sigma\sigma} r} . \tag{138}$$

Equating the right-hand side of Eqs. (137) and (138), the function $q(r)$ is readily determined to be

$$q(r) = \frac{1}{\varepsilon_{N_\sigma\sigma} - \varepsilon_{(N_\sigma-1)\sigma}} r^{\left(\frac{1}{\beta_{(N_\sigma-1)\sigma}}-m\right)} \mathrm{e}^{-(\beta_{(N_\sigma-1)\sigma}-\beta_{N_\sigma\sigma})r} . \tag{139}$$

Finally, by inserting this result in the left-hand side of Eq. (135), we confirm that the right-hand side of this equation decays asymptotically as stated in theorem 1.

To prove theorem 2 we write for the correlation-only case

$$\Psi_{N_\sigma\sigma}(r) = p(r)\frac{\mathrm{e}^{-\beta_{N_\sigma\sigma} r}}{r} . \tag{140}$$

Since $C_{N_\sigma\sigma} = 0$, $p(r)$ must satisfy the following asymptotic differential equation (cf. Eq. (109)):

$$-\frac{1}{2} r^{-\frac{1}{\beta_{N_\sigma\sigma}}} p''(r) + \beta_{N_\sigma\sigma} r^{-\frac{1}{\beta_{N_\sigma\sigma}}} p'(r) - r^{-\left(\frac{1}{\beta_{N_\sigma\sigma}}+1\right)} p(r) = V_{c\sigma}(\mathbf{r}) - u_{cN_\sigma\sigma}(\mathbf{r}) . \tag{141}$$

If $C_{(N_\sigma-1)\sigma} \neq 0$, the sum

$$\sum_{i=1}^{N_\sigma-1} \psi_{i\sigma}^*(\mathbf{r})\varphi_{i\sigma}(\mathbf{r}) \tag{142}$$

is asymptotically dominated by the $i = (N_\sigma - 1)$ term which, according to Eqs. (115) and (116), decays as

$$\psi^*_{(N_\sigma-1)\sigma}(\mathbf{r})\varphi_{(N_\sigma-1)\sigma}(\mathbf{r}) \stackrel{r\to\infty}{\longrightarrow} r^{\left(\frac{2}{\beta_{(N_\sigma-1)\sigma}}-1\right)} \mathrm{e}^{-2\beta_{(N_\sigma-1)\sigma}r}. \tag{143}$$

Once again this term cannot be canceled by any other term of the sum (142). Hence it must be canceled asymptotically by the $i = N_\sigma$ term which behaves as

$$\psi^*_{N_\sigma\sigma}(\mathbf{r})\varphi_{N_\sigma\sigma}(\mathbf{r}) \stackrel{r\to\infty}{\longrightarrow} p(r) r^{\left(\frac{1}{\beta_{N_\sigma\sigma}}-2\right)} \mathrm{e}^{-2\beta_{N_\sigma\sigma}r} . \tag{144}$$

Equating the right-hand sides of Eqs. (143) and (144) we can identify the asymptotic form of $p(r)$:

$$p(r) \propto r^{\left(\frac{2}{\beta_{(N_\sigma-1)\sigma}}-\frac{1}{\beta_{N_\sigma\sigma}}+1\right)} \mathrm{e}^{-2(\beta_{(N_\sigma-1)\sigma}-\beta_{N_\sigma\sigma})r}. \tag{145}$$

Insertion of this expression in the left-hand side of Eq. (141) proves Eq. (133) for the case $C_{(N_\sigma-1)\sigma} \neq 0$. If $C_{(N_\sigma-1)\sigma} = 0$ the asymptotic form of $V_{c\sigma}(\mathbf{r}) - u_{cN_\sigma\sigma}(\mathbf{r})$ cannot be stated explicitly. It is clear, however, that the $i = N_\sigma$ term (144) must be canceled asymptotically by some contribution to the sum (142). Since, by Eqs. (115) and (116), *all* contributions to the sum (142) fall off more rapidly than the right-hand side of (143), $p(r)$ must decay more rapidly than the right-hand side of (145). Hence, by Eq. (141), the right-hand side of (133) provides an upper bound of $|V_{c\sigma}(\mathbf{r}) - u_{cN_\sigma\sigma}(\mathbf{r})|$ for $r \to \infty$ if $C_{(N_\sigma-1)\sigma} = 0$. This completes the proof.

Since the asymptotic form of $u_{xN_\sigma\sigma}(\mathbf{r})$, as derived in Eq. (123), is $-\frac{1}{r}$, theorem 1 immediately implies that

$$V_{x\sigma}(\mathbf{r}) \stackrel{r\to\infty}{\longrightarrow} -\frac{1}{r} . \tag{146}$$

This is a well-known result that has been obtained in several different ways [13, 14, 16, 19, 47, 50, 51, 52]. The exact correlation potential of DFT is known [47] to fall off as $-\alpha/(2r^4)$ for atoms with spherical N and $(N-1)$-electron ground states, with α being the static polarizability of the $(N-1)$-electron ground state. Theorem 2 provides a simple way of checking how the OEP correlation-only potential $V_{c\sigma}(\mathbf{r})$ falls off for a given approximate orbital functional $E_c^{\mathrm{approx}}[\{\varphi_{i\sigma}\}]$: One only needs to determine the asymptotic decay of $u_{cN_\sigma\sigma}(\mathbf{r})$.

We now turn to the discussion of the KLI potential. We shall demonstrate that the above rigorous properties of the full OEP are preserved by the KLI approximation. To this end we write the KLI approximation (76) separately for the exchange and correlation potentials:

$$\sum_{i=1}^{N_\sigma} |\varphi_{i\sigma}(\mathbf{r})|^2 \left(V_{x\sigma}^{\mathrm{KLI}}(\mathbf{r}) - U_{xi\sigma}(\mathbf{r}) - \left(\bar{V}_{xi\sigma}^{\mathrm{KLI}} - \bar{U}_{xi\sigma}\right)\right) = 0 \tag{147}$$

$$\sum_{i=1}^{N_\sigma} |\varphi_{i\sigma}(\mathbf{r})|^2 \left(V_{c\sigma}^{\mathrm{KLI}}(\mathbf{r}) - U_{ci\sigma}(\mathbf{r}) - \left(\bar{V}_{ci\sigma}^{\mathrm{KLI}} - \bar{U}_{ci\sigma}\right)\right) = 0 \tag{148}$$

where, for convenience, we have introduced

$$U_{xi\sigma}(\mathbf{r}) = \frac{1}{2}\left(u_{xi\sigma}(\mathbf{r}) + u_{xi\sigma}^*(\mathbf{r})\right) \tag{149}$$

and

$$U_{ci\sigma}(\mathbf{r}) = \frac{1}{2}\left(u_{ci\sigma}(\mathbf{r}) + u_{ci\sigma}^*(\mathbf{r})\right) \tag{150}$$

in order to deal with real-valued quantities only. Following the argument given in the beginning of section 2.3.1 (Eqs. (89)–(97)) we can assume *without restriction* that

$$V_{x\sigma}^{\mathrm{KLI}}(\mathbf{r}) \stackrel{r\to\infty}{\longrightarrow} 0 \tag{151}$$

$$V_{c\sigma}^{\mathrm{KLI}}(\mathbf{r}) \stackrel{r\to\infty}{\longrightarrow} 0 \tag{152}$$

$$U_{ci\sigma}(\mathbf{r}) \stackrel{r\to\infty}{\longrightarrow} 0\,. \tag{153}$$

This is because the structure of Eqs. (147) and (148) is again such that an additive constant in the potentials (151)–(153) cancels out. Of course, Eq. (153) is valid only for those approximate orbital functionals $E_c[\{\varphi_{i\sigma}\}]$ leading to bounded functions $u_{ci\sigma}(\mathbf{r})$ for $r \to \infty$ (cf. condition (86)).

In order to determine the asymptotic form of the KLI-x-only potential $V_{x\sigma}^{\mathrm{KLI}}(\mathbf{r})$, we first investigate the asymptotic behavior of the term $\sum_{i=1}^{N_\sigma} |\varphi_{i\sigma}(\mathbf{r})|^2 u_{xi\sigma}(\mathbf{r})$ appearing in the KLI equation (147): By Eqs. (118) and (119) the expression

$$|\varphi_{N_\sigma\sigma}(\mathbf{r})|^2 u_{xN_\sigma\sigma}(\mathbf{r}) + \sum_{i=1}^{N_\sigma-1} |\varphi_{i\sigma}(\mathbf{r})|^2 u_{xi\sigma}(\mathbf{r})$$

can be written as

$$= |\varphi_{N_\sigma\sigma}(\mathbf{r})|^2 u_{xN_\sigma\sigma}(\mathbf{r}) + \sum_{i=1}^{N_\sigma-1}\sum_{j=1}^{N_\sigma} \varphi_{i\sigma}(\mathbf{r})\varphi_{j\sigma}^*(\mathbf{r})K_{ji\sigma}(\mathbf{r})$$

$$= |\varphi_{N_\sigma\sigma}(\mathbf{r})|^2 \left(u_{\mathrm{x}N_\sigma\sigma}(\mathbf{r}) + \sum_{i=1}^{N_\sigma-1} \sum_{j=1}^{N_\sigma} \left(\frac{\varphi_{i\sigma}(\mathbf{r})}{\varphi_{N_\sigma\sigma}(\mathbf{r})} \right) \left(\frac{\varphi^*_{j\sigma}(\mathbf{r})}{\varphi^*_{N_\sigma\sigma}(\mathbf{r})} \right) K_{ji\sigma}(\mathbf{r}) \right) .$$

Since $K_{ji\sigma}(\mathbf{r})$ decays as an inverse power the double sum over i and j must be asymptotically dominated by the term with $i = N_\sigma - 1$, $j = N_\sigma$ so that

$$\overset{r\to\infty}{\longrightarrow} |\varphi_{N_\sigma\sigma}(\mathbf{r})|^2 \left(u_{\mathrm{x}N_\sigma\sigma}(\mathbf{r}) + \frac{\varphi_{(N_\sigma-1)\sigma}(\mathbf{r})}{\varphi_{N_\sigma\sigma}(\mathbf{r})} K_{N_\sigma(N_\sigma-1)\sigma}(\mathbf{r}) \right) .$$

The KLI equation (147) then yields

$$\begin{aligned}
&\sum_{i=1}^{N_\sigma} |\varphi_{i\sigma}(\mathbf{r})|^2 \left[V^{\mathrm{KLI}}_{\mathrm{x}\sigma}(\mathbf{r}) - U_{\mathrm{x}i\sigma}(\mathbf{r}) - \left(\bar{V}^{\mathrm{KLI}}_{\mathrm{x}i\sigma} - \bar{U}_{\mathrm{x}i\sigma} \right) \right] \\
&\quad \overset{r\to\infty}{\longrightarrow} |\varphi_{N_\sigma\sigma}(\mathbf{r})|^2 \left[V^{\mathrm{KLI}}_{\mathrm{x}\sigma}(\mathbf{r}) - U_{\mathrm{x}N_\sigma\sigma}(\mathbf{r}) - \left(\bar{V}^{\mathrm{KLI}}_{\mathrm{x}N_\sigma\sigma} - \bar{U}_{\mathrm{x}N_\sigma\sigma} \right) \right. \\
&\quad + \left(\frac{\varphi_{(N_\sigma-1)\sigma}(\mathbf{r})}{\varphi_{N_\sigma\sigma}(\mathbf{r})} K_{N_\sigma(N_\sigma-1)\sigma}(\mathbf{r}) + \text{c.c.} \right) \\
&\quad \left. + \sum_{i=1}^{N_\sigma-1} \frac{|\varphi_{i\sigma}(\mathbf{r})|^2}{|\varphi_{N_\sigma\sigma}(\mathbf{r})|^2} \left(V^{\mathrm{KLI}}_{\mathrm{x}\sigma}(\mathbf{r}) - U_{\mathrm{x}i\sigma}(\mathbf{r}) - \left(\bar{V}^{\mathrm{KLI}}_{\mathrm{x}i\sigma} - \bar{U}_{\mathrm{x}i\sigma} \right) \right) \right] \equiv 0 .
\end{aligned} \tag{154}$$

Since the KLI equation must be satisfied in the asymptotic region, the expression in square brackets on the right-hand side of Eq. (154) must vanish identically for $r \to \infty$. The term involving $\varphi_{(N_\sigma-1)\sigma}(\mathbf{r})/\varphi_{N_\sigma\sigma}(\mathbf{r})$ cannot be canceled by any of the terms involving $|\varphi_{i\sigma}(\mathbf{r})|^2/|\varphi_{N_\sigma\sigma}(\mathbf{r})|^2$ because the latter decay more rapidly. From this we conclude that

$$\begin{aligned}
&V^{\mathrm{KLI}}_{\mathrm{x}\sigma}(\mathbf{r}) - U_{\mathrm{x}N_\sigma\sigma}(\mathbf{r}) - \left(\bar{V}^{\mathrm{KLI}}_{\mathrm{x}N_\sigma\sigma} - \bar{U}_{\mathrm{x}N_\sigma\sigma} \right) \\
&\quad \overset{r\to\infty}{\longrightarrow} -\frac{\varphi_{(N_\sigma-1)\sigma}(\mathbf{r})}{\varphi_{N_\sigma\sigma}(\mathbf{r})} K_{N_\sigma(N_\sigma-1)\sigma}(\mathbf{r}) + \text{c.c.} \\
&\quad \overset{r\to\infty}{\longrightarrow} -r^{\left(\frac{1}{\beta_{(N_\sigma-1)\sigma}} - \frac{1}{\beta_{N_\sigma\sigma}} - m\right)} \mathrm{e}^{-(\beta_{(N_\sigma-1)\sigma} - \beta_{N_\sigma\sigma})r}
\end{aligned} \tag{155}$$

where, in the second step, we have used Eqs. (102) and (121). $U_{\mathrm{x}N_\sigma\sigma}(\mathbf{r})$ goes to zero asymptotically (cf. Eq. (123)) and the arbitrary additive constant in $V^{\mathrm{KLI}}_{\mathrm{x}\sigma}(\mathbf{r})$ had been fixed in such a way that $V^{\mathrm{KLI}}_{\mathrm{x}\sigma}(\mathbf{r})$ vanishes asymptotically (cf. Eq. (151)). Hence Eq. (155) immediately implies that [14, 16, 19]

$$\bar{V}^{\mathrm{KLI}}_{\mathrm{x}N_\sigma\sigma} = \bar{U}_{\mathrm{x}N_\sigma\sigma} \tag{156}$$

and thereby

$$V^{\mathrm{KLI}}_{\mathrm{x}\sigma}(\mathbf{r}) - U_{\mathrm{x}N_\sigma\sigma}(\mathbf{r}) \overset{r\to\infty}{\longrightarrow} -r^{\left(\frac{1}{\beta_{(N_\sigma-1)\sigma}} - \frac{1}{\beta_{N_\sigma\sigma}} - m\right)} \mathrm{e}^{-(\beta_{(N_\sigma-1)\sigma} - \beta_{N_\sigma\sigma})r} . \tag{157}$$

We thus conclude that both the lemma of section 2.3.1 and the theorem 1 are preserved in the KLI approximation. Once again, Eqs. (123) and (157) immediately imply that [14, 16, 19]

$$V_{x\sigma}^{\mathrm{KLI}} \overset{r\to\infty}{\longrightarrow} -\frac{1}{r}\,. \tag{158}$$

For the correlation potential $V_{c\sigma}^{\mathrm{KLI}}(\mathbf{r})$ the considerations are even simpler. Dividing the KLI equation by $|\varphi_{N_\sigma\sigma}(\mathbf{r})|^2$ we find:

$$0 \equiv V_{c\sigma}^{\mathrm{KLI}}(\mathbf{r}) - U_{cN_\sigma\sigma}(\mathbf{r}) - C_{N_\sigma\sigma} + \sum_{i=1}^{N_\sigma-1} \frac{|\varphi_{i\sigma}(\mathbf{r})|^2}{|\varphi_{N_\sigma\sigma}(\mathbf{r})|^2}\left(V_{c\sigma}^{\mathrm{KLI}}(\mathbf{r}) - U_{ci\sigma}(\mathbf{r}) - C_{i\sigma}\right) \tag{159}$$

where

$$C_{i\sigma} := \bar{V}_{ci\sigma}^{\mathrm{KLI}} - \bar{U}_{ci\sigma}\,. \tag{160}$$

By Eqs. (102), (152) and (153) all the r-dependent functions in (159) vanish asymptotically. Since the KLI equation (159) must be satisfied for $r \to \infty$ as well we readily conclude that

$$C_{N_\sigma\sigma} = 0 \tag{161}$$

so that

$$V_{c\sigma}^{\mathrm{KLI}}(\mathbf{r}) - U_{cN_\sigma\sigma}(\mathbf{r}) = \sum_{i=1}^{N_\sigma-1} \frac{|\varphi_{i\sigma}(\mathbf{r})|^2}{|\varphi_{N_\sigma\sigma}(\mathbf{r})|^2}\left(C_{i\sigma} - V_{c\sigma}(\mathbf{r}) + U_{ci\sigma}^{\mathrm{KLI}}(\mathbf{r})\right)\,. \tag{162}$$

If $C_{(N_\sigma-1)\sigma} \neq 0$, the right-hand side of (162) is asymptotically dominated by the $i = (N_\sigma - 1)$ term and we obtain

$$V_{c\sigma}^{\mathrm{KLI}}(\mathbf{r}) - U_{cN_\sigma\sigma}(\mathbf{r}) \overset{r\to\infty}{\longrightarrow} r^{\left(\frac{2}{\beta_{(N_\sigma-1)\sigma}} - \frac{2}{\beta_{N_\sigma\sigma}}\right)} e^{-2(\beta_{(N_\sigma-1)\sigma} - \beta_{N_\sigma\sigma})r} \tag{163}$$

If $C_{(N_\sigma-1)\sigma} = 0$, the right-hand side of (163) is an upper bound of $|V_{c\sigma}^{\mathrm{KLI}}(\mathbf{r}) - U_{cN_\sigma\sigma}(\mathbf{r})|$ for $r \to \infty$. We note that $V_{c\sigma}^{\mathrm{KLI}}(\mathbf{r})$ and $U_{cN_\sigma\sigma}(\mathbf{r})$ approach each other exponentially fast for $r \to \infty$ with the same exponential function as theorem 2 predicts for the full OEP. However, the power of r multiplying the exponential function in (163) differs by 1 from the power in theorem 2.

2.3.3 Derivative discontinuities

In the early eighties an unexpected property of the exact xc potential was discovered [53, 54, 55]: Writing the density $\rho(\mathbf{r}) = M\,\xi(\mathbf{r})$ with a shape function $\xi(\mathbf{r})$ integrating to 1 and allowing for arbitrary (fractional) particle numbers M, the exact xc potential $V_{xc}[M\xi](\mathbf{r})$ is a discontinuous function of M. None of the standard density functionals such as the LDA, gradient expansions or GGAs show these discontinuities. Once again KLI were the first to point out [16, 17, 19] that the OEP correctly reproduces the required discontinuities, as will be discussed below.

DFT is readily extended to systems where the density ρ integrates to a non-integer particle number

$$M = N + \omega = \int d^3r\, \rho(\mathbf{r}) \quad \text{with } N \in \mathbf{N},\ \ 0 \leq \omega \leq 1 \tag{164}$$

(see [2] for a detailed description). In this generalization, a system with fractional particle number $N + \omega$ is described by an ensemble consisting of the N and $N + 1$ particle systems. Specifically, the ensemble density $\rho_{N+\omega}$ of the system with non-integer particle number $N + \omega$ is given by

$$\rho_{N+\omega}(\mathbf{r}) = (1 - \omega)\, \rho_N(\mathbf{r}) + \omega \rho_{N+1}(\mathbf{r}) \tag{165}$$

where ρ_N and ρ_{N+1} are the ground-state densities of the N and $N + 1$ particle systems, respectively. The ensemble energy is given by

$$E_{N+\omega} = (1 - \omega)\, E_N + \omega E_{N+1}, \tag{166}$$

i.e., the energy for fractional particle number is obtained by connecting the ground-state energies of integer particle numbers by straight lines. As a consequence,

$$\mu(M) = \frac{\partial E_M}{\partial M},\ M \in \mathbf{R}, \tag{167}$$

jumps discontinuously if M passes through an integer. From (166) one obtains

$$\mu(M) = \begin{cases} -I(Z): & Z - 1 < M < Z \\ -A(Z): & Z < M < Z + 1 \end{cases} \tag{168}$$

where $I(Z)$ is the first ionization potential

$$I(Z) = E_{Z-1}(Z) - E_Z(Z) \tag{169}$$

and $A(Z)$ the electron affinity

$$A(Z) = E_Z(Z) - E_{Z+1}(Z) \tag{170}$$

of a system with nuclear charge Z, electron number M and energy $E_M(Z)$. Using the HK variational equation

$$\left.\frac{\delta E[\rho]}{\delta \rho(\mathbf{r})}\right|_M = \mu(M) \tag{171}$$

we can express the derivative discontinuity of the total-energy functional

$$\Delta := \lim_{\omega \to 0} \left\{ \left.\frac{\delta E[\rho]}{\delta \rho(\mathbf{r})}\right|_{N+\omega} - \left.\frac{\delta E[\rho]}{\delta \rho(\mathbf{r})}\right|_{N-\omega} \right\}_{\rho_N} \tag{172}$$

in terms of the chemical potential

$$= \lim_{\omega \to 0} \{\mu(N+\omega) - \mu(N-\omega)\}. \tag{173}$$

By virtue of Eq. (168) we obtain

$$\Delta = I(N) - A(N). \tag{174}$$

This fundamental equation shows that the derivative discontinuity of the total-energy functional is identical with the *band gap* of an infinite insulator and, in the case of finite species, with twice the chemical *hardness*. A glance at the expression for the total energy (16) shows that only two terms may contribute to the derivative discontinuity, namely T_S and E_{xc}. In order to further evaluate the derivative discontinuity of the interacting system, we first observe that for a system of N non-interacting particles the derivative discontinuity Δ_{nonint} of the total energy

$$\begin{aligned}\Delta_{\text{nonint}} &= \lim_{\omega \to 0} \left\{ \left.\frac{\delta E[\rho]}{\delta\rho(\mathbf{r})}\right|_{N+\omega} - \left.\frac{\delta E[\rho]}{\delta\rho(\mathbf{r})}\right|_{N-\omega} \right\} \\ &= \lim_{\omega \to 0} \left\{ \left.\frac{\delta T_S[\rho]}{\delta\rho(\mathbf{r})}\right|_{N+\omega} - \left.\frac{\delta T_S[\rho]}{\delta\rho(\mathbf{r})}\right|_{N-\omega} \right\}\end{aligned} \tag{175}$$

is given by

$$\Delta_{\text{nonint}} = I_{\text{nonint}} - A_{\text{nonint}} = \varepsilon_{N+1}(N) - \varepsilon_N(N) \tag{176}$$

where $\varepsilon_m(M)$ denotes the m-th single-particle orbital of the M-particle system. If the right-hand side of Eq. (175) is evaluated at the ground-state density ρ_N of the interacting N-particle system, the non-interacting system leading to this density is the KS system and the resulting discontinuity is

$$\Delta^{KS}_{\text{nonint}} = \varepsilon^{KS}_{N+1}(N) - \varepsilon^{KS}_N(N). \tag{177}$$

Hence the total discontinuity of the interacting system

$$\begin{aligned}\Delta &= \lim_{\omega \to 0} \left\{ \left.\frac{\delta E[\rho]}{\delta\rho(\mathbf{r})}\right|_{N+\omega} - \left.\frac{\delta E[\rho]}{\delta\rho(\mathbf{r})}\right|_{N-\omega} \right\}_{\rho_N} \\ &= \lim_{\omega \to 0} \left\{ \left.\frac{\delta T_S[\rho]}{\delta\rho(\mathbf{r})}\right|_{N+\omega} - \left.\frac{\delta T_S[\rho]}{\delta\rho(\mathbf{r})}\right|_{N-\omega} \right\}_{\rho_N} \\ &\quad + \lim_{\omega \to 0} \left\{ \left.\frac{\delta E_{xc}[\rho]}{\delta\rho(\mathbf{r})}\right|_{N+\omega} - \left.\frac{\delta E_{xc}[\rho]}{\delta\rho(\mathbf{r})}\right|_{N-\omega} \right\}_{\rho_N}\end{aligned} \tag{178}$$

can be written as

$$\Delta = \Delta_{\text{nonint}}^{\text{KS}} + \Delta_{\text{xc}} \tag{179}$$

where

$$\Delta_{\text{xc}} = \lim_{\omega \to 0} \left\{ \left. \frac{\delta E_{\text{xc}}[\rho]}{\delta \rho(\mathbf{r})} \right|_{N+\omega} - \left. \frac{\delta E_{\text{xc}}[\rho]}{\delta \rho(\mathbf{r})} \right|_{N-\omega} \right\}_{\rho_N} . \tag{180}$$

As it is known [47] that

$$I = -\varepsilon_N^{\text{KS}}(N) \tag{181}$$

and

$$A = -\varepsilon_{N+1}^{\text{KS}}(N+1) \tag{182}$$

it follows from Eqs. (174), (177) and (179) that

$$\Delta_{\text{xc}} = \varepsilon_{N+1}^{\text{KS}}(N+1) - \varepsilon_{N+1}^{\text{KS}}(N). \tag{183}$$

We will now investigate the xc-potential. First we observe that the exact xc-energy functional may be written in terms of the (coupling-constant averaged) pair-correlation function $g_{\sigma\sigma'}[\rho](\mathbf{r}, \mathbf{r}')$ as

$$E_{\text{xc}}[\rho] = \frac{1}{2} \sum_{\sigma\sigma'} \int \int d^3r \, d^3r' \, \frac{\rho_\sigma(\mathbf{r})\rho_{\sigma'}(\mathbf{r}')}{|\mathbf{r} - \mathbf{r}'|} \left(g_{\sigma\sigma'}[\rho](\mathbf{r}, \mathbf{r}') - 1 \right). \tag{184}$$

Following van Leeuwen, Gritsenko and Baerends [56], the xc-potential may be split up in the following manner:

$$V_{\text{xc}\sigma}(\mathbf{r}) = V_{\text{xc}\sigma}^{\text{scr}}(\mathbf{r}) + V_{\text{xc}\sigma}^{\text{resp}}(\mathbf{r}) \tag{185}$$

where the screening potential is defined as

$$V_{\text{xc}\sigma}^{\text{scr}}(\mathbf{r}) = \sum_{\sigma'} \int d^3r' \, \frac{\rho_{\sigma'}(\mathbf{r}')}{|\mathbf{r} - \mathbf{r}'|} \left(g_{\sigma\sigma'}[\rho](\mathbf{r}, \mathbf{r}') - 1 \right) \tag{186}$$

and the screening response potential as

$$V_{\text{xc}\sigma}^{\text{resp}}(\mathbf{r}) = \frac{1}{2} \sum_{\sigma'\sigma''} \int \int d^3r' \, d^3r'' \, \frac{\rho_{\sigma'}(\mathbf{r}')\rho_{\sigma'}(\mathbf{r}'')}{|\mathbf{r}' - \mathbf{r}''|} \frac{\delta g_{\sigma'\sigma''}[\rho](\mathbf{r}', \mathbf{r}'')}{\delta \rho_\sigma(\mathbf{r})}. \tag{187}$$

As the xc-energy may be written as

$$E_{\mathrm{xc}}[\rho] = \frac{1}{2} \sum_{\sigma} \int d^3r \; \rho_\sigma(\mathbf{r}) V_{\mathrm{xc}\sigma}^{\mathrm{scr}}(\mathbf{r}) \tag{188}$$

it is obvious that the discontinuity of the xc-potential will only show up in the screening response potential and not in the screening potential, as the xc-energy itself must be continuous as a function of particle number.

In the x-only limit, the pair-correlation function is given by

$$g_{\mathrm{x}\,\sigma\sigma'}[\rho](\mathbf{r}, \mathbf{r}') = \left(1 - \frac{\sum_{i,j=1}^{N_\sigma} \omega_{i\sigma}\omega_{j\sigma}\varphi_{i\sigma}(\mathbf{r})\varphi_{i\sigma}^*(\mathbf{r}')\varphi_{j\sigma}(\mathbf{r}')\varphi_{j\sigma}^*(\mathbf{r})}{\rho_\sigma(\mathbf{r})\rho_\sigma(\mathbf{r}')}\right)\delta_{\sigma\sigma'} \tag{189}$$

where $\omega_{i\sigma}$ denotes the possibly fractional occupation number of the orbital $i\sigma$. Substitution into (186) gives for the exchange part of the screening potential

$$V_{\mathrm{x}\sigma}^{\mathrm{scr}}(\mathbf{r}) = \frac{1}{\rho_\sigma(\mathbf{r})} \sum_{i=1}^{N_\sigma} u_{\mathrm{x}i\sigma}(\mathbf{r}) |\varphi_{i\sigma}(\mathbf{r})|^2 \tag{190}$$

which is responsible for the $1/r$ behavior of $V_{\mathrm{x}\sigma}$ for large r. By comparison with the exact potential in the form of Eq. (70) the exchange part of the screening response potential is identified as

$$V_{\mathrm{x}\sigma}^{\mathrm{resp}}(\mathbf{r}) = \frac{1}{\rho_\sigma(\mathbf{r})} \sum_{i=1}^{N_\sigma} \left[\left(\bar{V}_{\mathrm{x}i\sigma} - \bar{u}_{\mathrm{x}i\sigma}\right) |\varphi_{i\sigma}(\mathbf{r})|^2 + \nabla\left(\psi_{i\sigma}(\mathbf{r})\nabla\varphi_{i\sigma}(\mathbf{r})\right)\right]. \tag{191}$$

The last term in the above expression is the one omitted in KLI approximation and therefore known to be small. The first term, on the other hand, shows a clear step structure. It is almost constant within the atomic shells where the orbitals vary little and changes rapidly at the atomic shell boundaries, where the orbitals decay. This is clearly visible in Figure 1 where the screening and screening-response potentials from the OEP and KLI method are plotted for the Calcium atom. The step structure, first discovered by van Leeuwen et. al. [56], is responsible for the discontinuity of the total exchange potential as a function of particle number. As the particle number changes through an integer N to $N + \omega$ with $\omega << 1$ and a new outer shell is started to be filled, a new step is added to $V_{\mathrm{x}\sigma}^{\mathrm{resp}}(\mathbf{r})$, shifting the total potential by a constant, i.e. by Δ_{x}. In Figure 2 this discontinuity is plotted for the Calcium ion. Clearly, the OEP and KLI methods give very similar results. This constant shift has no effect on the KS orbitals in the limit of vanishing ω as the KS equation (33) is invariant under additive constants to the KS potential. This is consistent with the fact that the density, being just the sum of the absolute squares of the occupied orbitals, is a continuous function of the particle number.

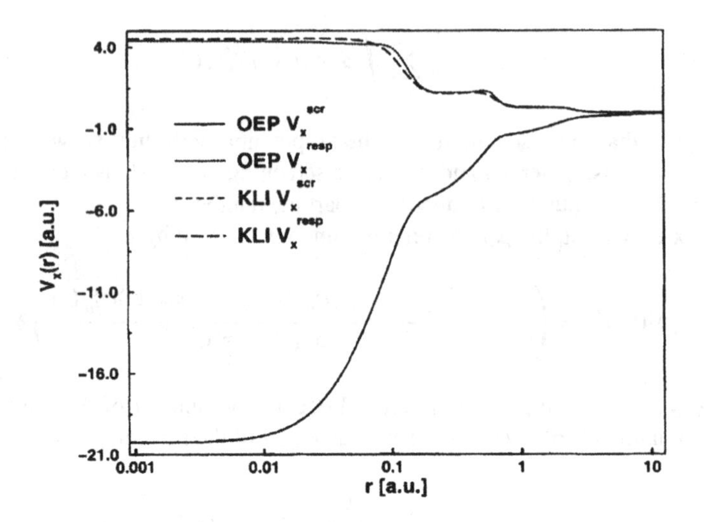

Figure 1: *Screening and screening response potentials* $V_{\mathrm{x}}^{\mathrm{scr}}(\mathbf{r})$ *and* $V_{\mathrm{x}}^{\mathrm{resp}}(\mathbf{r})$ *for Ca from the OEP and KLI method.*

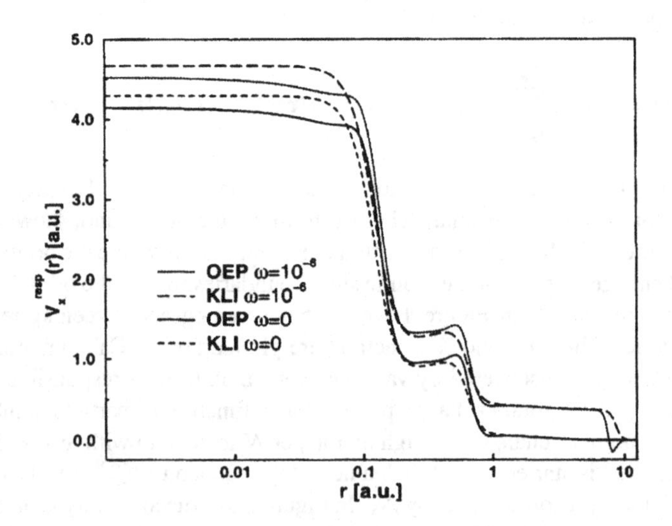

Figure 2: *Derivative discontinuities in the screening response potential* $V_{\mathrm{x}\uparrow}^{\mathrm{resp}}(\mathbf{r})$ *for* Ca^+ *for* $N_\uparrow = 9 + \omega$ *electrons from the x-only OEP and KLI method.*

2.4 Hartree-Fock versus x-only OEP, a comparison

In this section we compare the x-only OEP scheme with the Hartree-Fock (HF) method. Both approaches are based on the same total-energy functional:

$$\begin{aligned} E_{\text{tot}}\left[\{\varphi_{j\tau}\}\right] = & \sum_{\sigma=\uparrow,\downarrow}\sum_{i=1}^{N_\sigma}\int d^3r\ \varphi_{i\sigma}^*(\mathbf{r})\left(-\frac{1}{2}\nabla^2\right)\varphi_{i\sigma}(\mathbf{r}) \\ & + \int d^3r\ v_{ext}(\mathbf{r})\rho(\mathbf{r}) + \frac{1}{2}\int d^3r\int d^3r'\ \frac{\rho(\mathbf{r})\rho(\mathbf{r}')}{|\mathbf{r}-r'|} \\ & - \frac{1}{2}\sum_{\sigma=\uparrow,\downarrow}\int d^3r\int d^3r'\ \frac{\rho_\sigma(\mathbf{r},\mathbf{r}')\rho_\sigma(\mathbf{r}',\mathbf{r})}{|\mathbf{r}-r'|} \end{aligned} \tag{192}$$

where

$$\rho_\sigma(\mathbf{r},\mathbf{r}') = \sum_{i=1}^{N_\sigma}\varphi_{i\sigma}^*(\mathbf{r}')\varphi_{i\sigma}(\mathbf{r}). \tag{193}$$

In HF, this total-energy functional is minimized without restriction (except for orthonormality of the orbitals), leading to the variational equation

$$\begin{aligned} & \left(-\frac{\nabla^2}{2} + v_{ext}(\mathbf{r}) + \int d^3r\ \frac{\rho(\mathbf{r}')}{|\mathbf{r}-\mathbf{r}'|}\right)\varphi_{i\sigma}^{\text{HF}}(\mathbf{r}) \\ & - \int d^3r'\ \frac{\rho_\sigma^{\text{HF}}(\mathbf{r},\mathbf{r}')}{|\mathbf{r}-\mathbf{r}'|}\varphi_{i\sigma}^{\text{HF}}(\mathbf{r}') = \varepsilon_{i\sigma}^{\text{HF}}\varphi_{i\sigma}^{\text{HF}}(\mathbf{r}). \end{aligned} \tag{194}$$

This single-particle Schrödinger equation features a non-local effective potential. By contrast, in the OEP method the total-energy functional (192) is minimized under the subsidiary condition that the orbitals come from a *local* potential, i.e.,

$$\left(-\frac{\nabla^2}{2} + v_{ext}(\mathbf{r}) + \int d^3r\ \frac{\rho(\mathbf{r}')}{|\mathbf{r}-\mathbf{r}'|} + V_{x\sigma}^{\text{OEP}}(\mathbf{r})\right)\varphi_{i\sigma}^{\text{KS}}(\mathbf{r}) = \varepsilon_{i\sigma}^{\text{KS}}\varphi_{i\sigma}^{\text{KS}}(\mathbf{r}), \tag{195}$$

where $V_{x\sigma}^{\text{OEP}}(\mathbf{r})$ is determined by the integral equation (82). Since the self-consistent HF solutions of Eq. (194) yield the lowest possible value of (192) the energy obtained from the self-consistent OEP scheme is necessarily higher:

$$E_{\text{tot}}^{\text{HF}} \leq E_{\text{tot}}^{\text{OEP}}. \tag{196}$$

The difference between the two, however, turns out to be very small as we shall see in section 4.

A well-known consequence [48] of the non-locality of the HF potential is the fact that the *occupied* HF orbitals all have the *same* exponential decay:

$$\varphi_{i\sigma}^{\text{HF}}(r) \xrightarrow{r\to\infty} \mathrm{e}^{-\beta_{N_\sigma\sigma}r} \qquad \text{for all } i \leq N_\sigma. \tag{197}$$

The orbitals resulting form the OEP, on the other hand, fall off each with its own orbital energy

$$\varphi_{i\sigma}^{\mathrm{KS}}(r) \stackrel{r\to\infty}{\longrightarrow} \mathrm{e}^{-\beta_{i\sigma}r} \quad \text{for all } i. \tag{198}$$

In HF, the effective potential acting on an unoccupied orbital $\varphi_{u\sigma}^{\mathrm{HF}}(\mathbf{r})$ falls off exponentially. This is most easily seen by rewriting the exchange term as

$$\begin{aligned}\left(\hat{v}_{\mathrm{x}\sigma}^{\mathrm{HF}}\varphi_{u\sigma}^{\mathrm{HF}}\right)(\mathbf{r}) &= -\int d^3r'\,\frac{\rho_\sigma^{\mathrm{HF}}(\mathbf{r},\mathbf{r}')}{|\mathbf{r}-\mathbf{r}'|}\varphi_{u\sigma}^{\mathrm{HF}}(\mathbf{r}') \\ &= \left\{-\sum_{i=1}^{N_\sigma}\frac{\varphi_{i\sigma}^{\mathrm{HF}}(\mathbf{r})}{\varphi_{u\sigma}^{\mathrm{HF}}(\mathbf{r})}K_{iu\sigma}^{\mathrm{HF}}(\mathbf{r})\right\}\varphi_{u\sigma}^{\mathrm{HF}}(\mathbf{r})\end{aligned} \tag{199}$$

where

$$K_{iu\sigma}^{\mathrm{HF}}(\mathbf{r}) = \int d^3r'\,\frac{\varphi_{i\sigma}^{*\mathrm{HF}}(\mathbf{r}')\varphi_{u\sigma}^{\mathrm{HF}}(\mathbf{r}')}{|\mathbf{r}-\mathbf{r}'|}. \tag{200}$$

The term in curly brackets can be interpreted as a local exchange potential (which depends on the orbital it acts on). Since $K_{iu\sigma}^{\mathrm{HF}}$ falls off with a power law (cf. Eq. (121)) the local exchange potential in curly brackets falls off exponentionally if $\varphi_{u\sigma}^{\mathrm{HF}}(\mathbf{r})$ is a virtual orbital, and hence the total HF potential falls off exponentially as well. As a consequence the HF potential can support very few (if any) unoccupied bound states which are a very poor starting point if used as a lowest-order approximation for excited states. By contrast, the x-only OEP falls off as $-1/r$ (cf. Eq. (146)) for *all* orbitals, including the unoccupied ones. In recent calculations of excitation energies, the unoccupied OEP eigenvalues were found to be an excellent starting point [25–27, 57, 58]. Likewise, the band gap of semiconductors and insulators is much too large in HF while the x-only OEP yields band gaps rather close to experiment (see also section 4.4).

3 Relativistic generalization of the OEP and KLI methods

In the previous section the use of orbital-dependent xc-functionals was discussed within the context of nonrelativistic DFT. Relativistic effects were completely neglected. However, if heavier elements come into play, relativistic contributions become more and more important: For example, the ground-state energies of high-Z atoms or the bond lengths of molecules are changed considerably [59]. Also, the importance of relativistic effects in solids was recognized long ago [60].

In this section, we present the generalization of the OEP and KLI methods to the realm of relativistic systems. Before doing so, we first briefly outline the fundamental ideas of relativistic DFT (RDFT).

Similar to nonrelativistic DFT, a HK theorem can be proven which can be summarized as follows [61]: The renormalized ground-state four current $j^{\nu}(\mathbf{r})$ of an interacting system of Dirac particles uniquely determines – up to within a gauge transformation – the external static four potential $A^{\mu}_{ext}[j^{\nu}]$ as well as the ground-state wave function $\Psi[j^{\nu}]$. As a consequence, any observable of the relativistic many-body system under consideration is a unique functional of the ground-state four current. (For a more detailed discussion of RDFT, also including questions of renormalization, the reader is referred to recent reviews [62, 63].) As usual, the exact four current of the interacting system can in principle be obtained from an auxiliary non-interacting system – the relativistic Kohn-Sham (RKS) system [62, 64, 65, 66]:

$$j^{\nu}(\mathbf{r}) = \sum_{-c^2<\varepsilon_k\leq\varepsilon_N} \bar{\varphi}_k(\mathbf{r})\gamma^{\nu}\varphi_k(\mathbf{r}) \tag{201}$$

Here and in the following, we assume that vacuum contributions can be neglected. This means that we restrict ourselves to the calculation of relativistic effects and ignore radiative corrections. Since we aim at electronic structure calculations for atoms, molecules and solids, the neglected terms are expected to be small.

The four-component spinors $\varphi_k(\mathbf{r})$ are obtained from the single-particle Dirac equation

$$\gamma_0\left(-ic\boldsymbol{\gamma}\cdot\nabla + c^2 + \gamma_{\mu}A^{\mu}_{S}[j^{\nu}](\mathbf{r})\right)\varphi_k(\mathbf{r}) = \varepsilon_k\varphi_k(\mathbf{r}) \tag{202}$$

(for notational and metric conventions cf. [67]). The local effective potential $A^{\mu}_{S}[j^{\nu}](\mathbf{r})$ is given by

$$A^{\mu}_{S}[j^{\nu}](\mathbf{r}) = A^{\mu}_{0}(\mathbf{r}) + \int d^3r'\,\frac{j^{\mu}(\mathbf{r}')}{|\mathbf{r}-\mathbf{r}'|} + A^{\mu}_{\rm xc}[j^{\nu}](\mathbf{r}), \tag{203}$$

where the first term is the static external potential, for example the potential of the nuclei assumed at rest. The second term represents the Hartree potential, whereas the last term denotes the xc four potential defined by

$$A^{\mu}_{\rm xc}[j^{\nu}](\mathbf{r}) := \frac{\delta E_{\rm xc}[j^{\nu}]}{\delta j_{\mu}(\mathbf{r})}, \tag{204}$$

with the relativistic xc energy functional $E_{\rm xc}[j^{\nu}]$ now being a functional of the four current j^{ν}. Eqs. (201)–(204) represent the relativistic KS (RKS) scheme which has to be solved self-consistently.

3.1 Relativistic optimized effective potential method

Similar to the nonrelativistic case, the HK theorem, applied to the non-interacting system, guarantees that the RKS spinors are unique functionals of the ground-state four current. Therefore, every relativistic xc functional

$$E_{\mathrm{xc}} = E_{\mathrm{xc}}[\{\varphi_j\}] \tag{205}$$

depending *explicitly* on the set of single-particle spinors $\{\varphi_j\}$ is an *implicit* functional of j^ν. To calculate the corresponding xc four potential $A^\mu_{\mathrm{xc}}(\mathbf{r})$, one has to resort to the optimized potential method, now generalized to the realm of relativistic systems subject to static but otherwise arbitrary external four potentials [68].

The relativistic OEP (ROEP) integral equation can be derived in close analogy to its nonrelativistic counterpart: We again start out from the very definition of the xc four potential, Eq. (204). By applying the chain rule for functional derivatives, one obtains

$$\begin{aligned} A^{\mathrm{ROEP}}_{\mathrm{xc}\mu}(\mathbf{r}) &= \frac{\delta E^{\mathrm{ROEP}}_{\mathrm{xc}}[\{\varphi_j\}]}{\delta j^\mu(\mathbf{r})} \\ &= \sum_{-c^2<\varepsilon_k\le\varepsilon_N} \int d^3r' \int d^3r'' \left(\frac{\delta E^{\mathrm{ROEP}}_{\mathrm{xc}}[\{\varphi_j\}]}{\delta\varphi_k(\mathbf{r}')} \frac{\delta\varphi_k(\mathbf{r}')}{\delta A_{S\nu}(\mathbf{r}'')} + \mathrm{c.c.}\right) \frac{\delta A_{S\nu}(\mathbf{r}'')}{\delta j^\mu(\mathbf{r})}. \end{aligned} \tag{206}$$

As in the nonrelativistic expression (41), the first functional derivative is readily calculated once an approximation for $E_{\mathrm{xc}}[\{\varphi_j\}]$ is given. The last functional derivative in Eq. (206) is identified with the inverse of the static response function of a system of non-interacting Dirac particles, defined as

$$\chi^{\mu\nu}_S(\mathbf{r},\mathbf{r}') := \frac{\delta j^\mu(\mathbf{r})}{\delta A_{S\nu}(\mathbf{r}')} \tag{207}$$

such that Eq. (206) can be rewritten as

$$A^{\mathrm{ROEP}}_{\mathrm{xc}\mu}(\mathbf{r}) = \sum_{-c^2<\varepsilon_k\le\varepsilon_N} \int d^3r' \int d^3r'' \left(\frac{\delta E^{\mathrm{ROEP}}_{\mathrm{xc}}[\{\varphi_j\}]}{\delta\varphi_k(\mathbf{r}')} \frac{\delta\varphi_k(\mathbf{r}')}{\delta A_{S\nu}(\mathbf{r}'')} + \mathrm{c.c.}\right) \chi^{-1}_{S\nu\mu}(\mathbf{r}'',\mathbf{r}). \tag{208}$$

Acting with the response operator (207) on both sides of Eq. (208) and using the identity

$$\int d^3r\, \chi^{-1}_{S\nu\mu}(\mathbf{r}'',\mathbf{r})\, \chi^{\mu\sigma}_S(\mathbf{r},\mathbf{r}') = \delta^\sigma_\nu\, \delta(\mathbf{r}''-\mathbf{r}') \tag{209}$$

leads (after rearranging the indices) to

$$\int d^3r'\, A^{\mathrm{ROEP}}_{\mathrm{xc}\nu}(\mathbf{r}')\, \chi_S^{\nu\mu}(\mathbf{r}',\mathbf{r}) = \sum_{-c^2<\varepsilon_k\le\varepsilon_N} \int d^3r'\, \frac{\delta E^{\mathrm{ROEP}}_{\mathrm{xc}}[\{\varphi_j\}]}{\delta\varphi_k(\mathbf{r}')} \frac{\delta\varphi_k(\mathbf{r}')}{\delta A_{S\mu}(\mathbf{r})} + \mathrm{c.c.} \tag{210}$$

The remaining functional derivative $\delta\varphi_k/\delta A_{S\nu}$ can be calculated by using first-order perturbation theory, yielding

$$\frac{\delta\varphi_k(\mathbf{r}')}{\delta A_{S\mu}(\mathbf{r})} = \sum_{\substack{\varepsilon_l\\ l\neq k}} \frac{\varphi_l(\mathbf{r}')}{\varepsilon_k-\varepsilon_l} \bar{\varphi}_l(\mathbf{r})\gamma^\mu\varphi_k(\mathbf{r}). \tag{211}$$

Once again, this expression can be used to express the response function

$$\chi_S^{\mu\nu}(\mathbf{r},\mathbf{r}') := \frac{\delta}{\delta A_{S\nu}(\mathbf{r}')} \left(\sum_{-c^2<\varepsilon_k\le\varepsilon_N} \bar{\varphi}_k(\mathbf{r})\gamma^\mu\varphi_k(\mathbf{r}) \right) \tag{212}$$

in terms of the RKS spinors:

$$\chi_S^{\mu\nu}(\mathbf{r}',\mathbf{r}) = \sum_{-c^2<\varepsilon_k\le\varepsilon_N} \sum_{\substack{\varepsilon_l\\ l\neq k}} \frac{\bar{\varphi}_k(\mathbf{r}')\gamma^\mu\varphi_l(\mathbf{r}')\, \bar{\varphi}_l(\mathbf{r})\gamma^\nu\varphi_k(\mathbf{r})}{\varepsilon_k-\varepsilon_l} + \mathrm{c.c.} \tag{213}$$

Finally, putting Eqs. (210), (211) and (213) together leads to the relativistic generalization of the OEP integral equation:

$$\sum_{-c^2<\varepsilon_k\le\varepsilon_N} \int d^3r' \left(\bar{\varphi}_k(\mathbf{r}')\gamma^\nu A^{\mathrm{ROEP}}_{\mathrm{xc}\nu}(\mathbf{r}') - \frac{\delta E^{\mathrm{ROEP}}_{\mathrm{xc}}[\{\varphi_j\}]}{\delta\varphi_k(\mathbf{r}')} \right) \times$$
$$G_{Sk}(\mathbf{r}',\mathbf{r})\,\gamma^0\gamma^\mu\varphi_k(\mathbf{r}) + \mathrm{c.c.} = 0$$
$$\mu = 0,1,2,3 \tag{214}$$

where

$$G_{Sk}(\mathbf{r}',\mathbf{r}) := \sum_{\substack{\varepsilon_l\\ l\neq k}} \frac{\varphi_l(\mathbf{r}')\varphi_l^\dagger(\mathbf{r})}{\varepsilon_k-\varepsilon_l}. \tag{215}$$

These four integral equations determine the local xc four potential $A^\mu_{\mathrm{xc}}(\mathbf{r})$ – up to an arbitrary constant which can be specified by requiring $A^{\mathrm{ROEP}}_{\mathrm{xc}\mu}(\mathbf{r})$ to vanish asymptotically (for finite systems) – and have to be solved self-consistently with the RKS equation (202).

3.2 Relativistic KLI approximation

In close analogy to the nonrelativistic situation, one has to deal with the ROEP integral equations numerically. Owing to the four-component structure of Eq. (214) and to the fact that four integral equations have to be solved, considerably more effort is needed to determine the xc four potential $A^{\mu}_{\mathrm{xc}}(\mathbf{r})$ as compared to the nonrelativistic case. Therefore, a simplified scheme for the calculation of $A^{\mu}_{\mathrm{xc}}(\mathbf{r})$, leading to the relativistic generalization of the KLI approximation discussed in section 2.2, is presented in the following [68].

By defining

$$\psi_k^{\dagger}(\mathbf{r}) := \int d^3r' \left(\bar{\varphi}_k(\mathbf{r}')\gamma^{\nu} A^{\mathrm{ROEP}}_{\mathrm{xc}\nu}(\mathbf{r}') - \frac{\delta E^{\mathrm{ROEP}}_{\mathrm{xc}}}{\delta \varphi_k(\mathbf{r}')}\right) G_{Sk}(\mathbf{r}', \mathbf{r}) \tag{216}$$

one can rewrite the ROEP integral equation as

$$\sum_{-c^2 < \varepsilon_k \leq \varepsilon_N} \bar{\psi}_k(\mathbf{r})\gamma^{\mu}\varphi_k(\mathbf{r}) + \mathrm{c.c.} = 0, \tag{217}$$

where the adjoint spinor $\bar{\psi}_k(\mathbf{r})$ is defined in the usual way, i.e.

$$\bar{\psi}_k(\mathbf{r}) := \psi_k^{\dagger}(\mathbf{r})\gamma^0. \tag{218}$$

The quantity $\psi_k^{\dagger}(\mathbf{r})$, although being a four-component object, closely resembles its namesake introduced in Eq. (54): One readily proves the orthogonality relation

$$\int d^3r\, \psi_k^{\dagger}(\mathbf{r})\varphi_k(\mathbf{r}) = 0. \tag{219}$$

Furthermore, its physical interpretation discussed in section 2.2 remains valid in the relativistic domain. Again, a differential equation that uniquely determines $\psi_k^{\dagger}(\mathbf{r})$ can be derived. To demonstrate this, we use the defining property of $G_{Sk}(\mathbf{r}', \mathbf{r})$

$$G_{Sk}(\mathbf{r}', \mathbf{r})\left(\hat{h}_D^{\dagger} - \varepsilon_k\right) = -\left(\delta(\mathbf{r}' - \mathbf{r}) - \varphi_k(\mathbf{r}')\varphi_k^{\dagger}(\mathbf{r})\right) \tag{220}$$

where the operator $\hat{h}_D^{\dagger}$ denotes the hermitian conjugate of the RKS Hamiltonian, i.e.

$$\hat{h}_D^{\dagger} := \gamma^0\left(ic\boldsymbol{\gamma}\cdot\overleftarrow{\nabla} + c^2 + \gamma^{\nu}A_{S\nu}(\mathbf{r})\right), \tag{221}$$

acting from the right on the unprimed variable of $G_{Sk}(\mathbf{r}', \mathbf{r})$ (the arrow on top of the gradient indicates the direction in which the derivative has to be taken). Using Eq. (220), we can act with the operator $(\hat{h}_D^\dagger - \varepsilon_k)$ on Eq. (216), leading to the differential equation determining $\psi_k^\dagger(\mathbf{r})$:

$$\begin{aligned}\psi_k^\dagger(\mathbf{r})\left(\hat{h}_D^\dagger - \varepsilon_k\right) &= -\int d^3r' \left(\bar{\varphi}_k(\mathbf{r}')\gamma_\nu A_{\mathrm{xc}}^\nu(\mathbf{r}') - \frac{\delta E_{\mathrm{xc}}^{\mathrm{ROEP}}}{\delta\varphi_k(\mathbf{r}')}\right) G_{Sk}(\mathbf{r}', \mathbf{r})\left(\hat{h}_D^\dagger - \varepsilon_k\right) \\ &= -\left(\bar{\varphi}_k(\mathbf{r})\gamma_\nu A_{\mathrm{xc}}^\nu(\mathbf{r}) - \frac{\delta E_{\mathrm{xc}}^{\mathrm{ROEP}}}{\delta\varphi_k(\mathbf{r})}\right) + \left(\bar{A}_{\mathrm{xc}k}^{\mathrm{ROEP}} - \bar{u}_{\mathrm{xc}k}\right)\varphi_k^\dagger(\mathbf{r}),\end{aligned} \tag{222}$$

with the constants $\bar{A}_{\mathrm{xc}k}^{\mathrm{ROEP}}$ and $\bar{u}_{\mathrm{xc}k}$ introduced in accordance to Eq. (60), i.e.

$$\bar{A}_{\mathrm{xc}k}^{\mathrm{ROEP}} := \int d^3r\, \bar{\varphi}_k(\mathbf{r})\gamma^\nu A_{\mathrm{xc}\nu}^{\mathrm{ROEP}}(\mathbf{r})\varphi_k(\mathbf{r}) \tag{223}$$

and

$$\bar{u}_{\mathrm{xc}k} := \int d^3r\, \frac{\delta E_{\mathrm{xc}}^{\mathrm{ROEP}}}{\delta\varphi_k(\mathbf{r}')}\varphi_k(\mathbf{r}). \tag{224}$$

Eq. (222) can now be used to further transform the ROEP integral equation (217). We therefore multiply Eq. (217) by $A_S^0(\mathbf{r})$:

$$\sum_{-c^2<\varepsilon_k\le\varepsilon_N} A_S^0(\mathbf{r})\bar{\psi}_k(\mathbf{r})\gamma^\mu\varphi_k(\mathbf{r}) + \text{c.c.} = 0 \tag{225}$$

and employ Eq. (222), solved for $A_S^0(\mathbf{r})\psi_k^\dagger(\mathbf{r})$, to obtain

$$\begin{aligned}\sum_{-c^2<\varepsilon_k\le\varepsilon_N} &\left(\bar{\varphi}_k(\mathbf{r})\gamma^\nu A_{\mathrm{xc}\nu}^{\mathrm{ROEP}}(\mathbf{r}) - \frac{\delta E_{\mathrm{xc}}^{\mathrm{ROEP}}}{\delta\varphi_k(\mathbf{r})} - \left(\bar{A}_{\mathrm{xc}k}^{\mathrm{ROEP}} - \bar{u}_{\mathrm{xc}k}\right)\varphi_k^\dagger(\mathbf{r})\right. \\ &+ \left.\bar{\psi}_k(\mathbf{r})\left(ic\boldsymbol{\gamma}\cdot\overleftarrow{\nabla} + c^2 - \boldsymbol{\gamma}\cdot\mathbf{A}_S(\mathbf{r}) - \gamma^0\varepsilon_k\right)\right)\gamma^0\gamma^\mu\varphi_k(\mathbf{r}) + \text{c.c.} = 0.\end{aligned} \tag{226}$$

Defining the 4×4-matrix

$$\mathcal{J}^{\mu\nu}(\mathbf{r}) := \frac{1}{2}\sum_{-c^2<\varepsilon_k\le\varepsilon_N}\left(\bar{\varphi}_k(\mathbf{r})\gamma^\nu\gamma^0\gamma^\mu\varphi_k(\mathbf{r}) + \text{c.c.}\right) \tag{227}$$

we rewrite Eq. (226) as

$$\mathcal{J}^{\mu\nu}(\mathbf{r})A_{\mathrm{xc}\nu}^{\mathrm{ROEP}}(\mathbf{r}) = \frac{1}{2}\sum_{-c^2<\varepsilon_k\le\varepsilon_N}\left(a_{\mathrm{xc}k}^\mu(\mathbf{r}) + \bar{\varphi}_k(\mathbf{r})\gamma^\mu\varphi_k(\mathbf{r})\left(\bar{A}_{\mathrm{xc}k}^{\mathrm{ROEP}} - \bar{u}_{\mathrm{xc}k}\right)\right) + \text{c.c.}, \tag{228}$$

where $a^{\mu}_{\mathrm{x}ck}(\mathbf{r})$ is a shorthand notation for

$$a^{\mu}_{\mathrm{x}ck}(\mathbf{r}) := \frac{\delta E^{\mathrm{ROEP}}_{\mathrm{xc}}}{\delta\varphi_k(\mathbf{r})}\gamma^0\gamma^{\mu}\varphi_k(\mathbf{r}) \\ - \bar{\psi}_k(\mathbf{r})\left(ic\gamma\cdot\overleftarrow{\nabla} + c^2 - \gamma\cdot\mathbf{A}_S(\mathbf{r}) - \gamma^0\varepsilon_k\right)\gamma^0\gamma^{\mu}\varphi_k(\mathbf{r}). \tag{229}$$

In order to solve Eq. (228) for $A^{\mathrm{ROEP}}_{\mathrm{xc}\mu}(\mathbf{r})$, we first have to demonstrate that the 4×4-matrix $\mathcal{J}(\mathbf{r})$ defined by Eq. (227) is nonsingular, i.e. that the inverse $\mathcal{J}^{-1}(\mathbf{r})$ exists. Using the commutator algebra of the γ-matrices

$$\{\gamma^{\nu},\gamma^{\mu}\} = 2\,g^{\nu\mu}, \tag{230}$$

where $g^{\nu\mu}$ denotes the metric tensor and the fact that the hermitian conjugate of γ^{ν} is given by

$$\gamma^{\nu^{\dagger}} = \gamma^0\gamma^{\nu}\gamma^0, \tag{231}$$

$\mathcal{J}^{\mu\nu}(\mathbf{r})$ can be rewritten as

$$\mathcal{J}^{\mu\nu}(\mathbf{r}) = j^{\mu}(\mathbf{r})\,g^{\nu0} - j^0(\mathbf{r})\,g^{\nu\mu} + j^{\nu}(\mathbf{r})\,g^{\mu0}. \tag{232}$$

From this equation the determinant of $\mathcal{J}(\mathbf{r})$ is easily calculated:

$$\det(\mathcal{J}(\mathbf{r})) = \begin{vmatrix} j^0(\mathbf{r}) & j^1(\mathbf{r}) & j^2(\mathbf{r}) & j^3(\mathbf{r}) \\ j^1(\mathbf{r}) & j^0(\mathbf{r}) & & \\ j^2(\mathbf{r}) & & j^0(\mathbf{r}) & \\ j^3(\mathbf{r}) & & & j^0(\mathbf{r}) \end{vmatrix} = \rho^4(\mathbf{r})\left(1 - \frac{\mathbf{j}^2(\mathbf{r})}{c^2\rho^2(\mathbf{r})}\right) \tag{233}$$

where, in the last step, the four current was decomposed into the density and the spatial component of the current according to

$$j^{\mu}(\mathbf{r}) = \left(\rho(\mathbf{r}), \frac{1}{c}\mathbf{j}(\mathbf{r})\right). \tag{234}$$

Since $\rho(\mathbf{r}) > 0$ and $\mathbf{j}(\mathbf{r})/c\rho(\mathbf{r})$ represents the velocity field $v(\mathbf{r}) < 1$ of the system, one obtains

$$\det(\mathcal{J}(\mathbf{r})) > 0 \tag{235}$$

proving that the inverse $\mathcal{J}^{-1}(\mathbf{r})$ exists.

Therefore, Eq. (228) can be solved for the xc four potential:

$$A^{\mathrm{ROEP}}_{\mathrm{xc}\mu}(\mathbf{r}) = \frac{1}{2}\mathcal{J}^{-1}_{\mu\nu}(\mathbf{r}) \sum_{-c^2<\varepsilon_k\leq\varepsilon_N} \left(a^{\nu}_{\mathrm{x}ck}(\mathbf{r}) + \bar{\varphi}_k(\mathbf{r})\gamma^{\nu}\varphi_k(\mathbf{r})\left(\bar{A}^{\mathrm{ROEP}}_{\mathrm{x}ck} - \bar{u}_{\mathrm{x}ck}\right)\right) + \mathrm{c.c.} \tag{236}$$

This equation represents an exact transformation of the ROEP integral equation (214). In particular, due to the appearance of the quantity $\psi_k^\dagger(\mathbf{r})$ in $a_{\mathrm{x}ck}^\nu(\mathbf{r})$, Eq. (236) is still an integral equation. Similar to its nonrelativistic counterpart, Eq. (70), a simple approximation is obtained by completely neglecting all terms involving $\psi_k^\dagger(\mathbf{r})$ in Eq. (229). The resulting equation representing the relativistic generalization of the KLI approximation is then given by

$$A_{\mathrm{xc}\mu}^{\mathrm{RKLI}}(\mathbf{r}) = \frac{1}{2}\,\mathcal{J}_{\mu\nu}^{-1}(\mathbf{r}) \sum_{-c^2<\varepsilon_k\le\varepsilon_N} \left(\frac{\delta E_{\mathrm{xc}}^{\mathrm{ROEP}}}{\delta\varphi_k(\mathbf{r})} \gamma^0\gamma^\nu\varphi_k(\mathbf{r}) + j_k^\nu(\mathbf{r}) \left(\bar{A}_{\mathrm{xc}k}^{\mathrm{RKLI}} - \bar{u}_{\mathrm{xc}k}\right)\right) + \mathrm{c.c.} \tag{237}$$

where the orbital current is defined as

$$j_k^\nu(\mathbf{r}) := \bar{\varphi}_k(\mathbf{r})\,\gamma^\nu\,\varphi_k(\mathbf{r}). \tag{238}$$

Although still being an integral equation, this RKLI equation can be solved explicitly in terms of the RKS spinors. This can be seen by multiplying Eq. (237) by $j_l^\mu(\mathbf{r})$, summing over all μ and integrating over space, yielding

$$\bar{A}_{\mathrm{xc}l}^{\mathrm{RKLI}} = \bar{A}_{\mathrm{xc}l}^{S} + \sum_{-c^2<\varepsilon_k\le\varepsilon_N} M_{lk}\left(\bar{A}_{\mathrm{xc}k}^{\mathrm{RKLI}} - \frac{1}{2}\left(\bar{u}_{\mathrm{xc}k} + \bar{u}_{\mathrm{xc}k}^*\right)\right) \tag{239}$$

where

$$\bar{A}_{\mathrm{xc}l}^{S} := \frac{1}{2}\int d^3r\, j_l^\mu(\mathbf{r})\mathcal{J}_{\mu\nu}^{-1}(\mathbf{r}) \sum_{-c^2<\varepsilon_k\le\varepsilon_N} \left(\frac{\delta E_{\mathrm{xc}}^{\mathrm{ROEP}}}{\delta\varphi_k(\mathbf{r})}\gamma^0\gamma^\nu\varphi_k(\mathbf{r}) + \mathrm{c.c.}\right) \tag{240}$$

and

$$M_{lk} := \int d^3r\, j_l^\mu(\mathbf{r})\mathcal{J}_{\mu\nu}^{-1}(\mathbf{r})\, j_k^\nu(\mathbf{r}). \tag{241}$$

The unknown coefficients $(\bar{A}_{\mathrm{xc}k}^{\mathrm{RKLI}} - \frac{1}{2}\left(\bar{u}_{\mathrm{xc}k} + \bar{u}_{\mathrm{xc}k}^*\right))$ are then determined by the set of linear equations

$$\sum_{-c^2<\varepsilon_k\le\varepsilon_N} \left(\frac{1}{2}\delta_{lk} - M_{lk}\right)\left(\bar{A}_{\mathrm{xc}k}^{\mathrm{RKLI}} + \frac{1}{2}\left(\bar{u}_{\mathrm{xc}k} + \bar{u}_{\mathrm{xc}k}^*\right)\right) = \left(\bar{A}_{\mathrm{xc}l}^{S} - \frac{1}{2}\left(\bar{u}_{\mathrm{xc}l} + \bar{u}_{\mathrm{xc}l}^*\right)\right). \tag{242}$$

Inserting the result in Eq. (237) we finally obtain an expression for the xc four potential $A_{\mathrm{xc}\mu}^{\mathrm{RKLI}}(\mathbf{r})$ that depends explicitly on the set of single-particle spinors $\{\varphi_j\}$.

3.3 Relativistic OEP in the electrostatic case

In the previous section, we discussed the extension of the OEP method to relativistic systems subject to static but otherwise arbitrary external fields. In particular, the ROEP method allows one to deal with external magnetic fields of arbitrary strength. In electronic structure calculations for atoms, molecules and solids, however, we most commonly encounter situations where no magnetic fields are present (in a suitable Lorentz frame, typically the rest frame of the nuclei). In this so-called "electrostatic case", the general approach presented above can be considerably simplified.

We consider the situation where the spatial components of the external four potential vanish, i.e. $\mathbf{A}_{ext}(\mathbf{r}) = 0$. (This also includes a partial fixing of the gauge.) Then, one can derive a simplified Hohenberg-Kohn-Sham scheme [63, 64, 69] stating that the zeroth component $\rho(\mathbf{r}) = j^0(\mathbf{r})$ of the ground-state current density alone uniquely determines the scalar external potential $V = V[\rho]$ as well as the ground-state wave function $\Psi[\rho]$. Consequently, only the scalar effective potential $V_S(\mathbf{r})$, given by

$$V_S[\rho](\mathbf{r}) := V_{ext}(\mathbf{r}) + \int d^3r \, \frac{\rho(\mathbf{r})}{|\mathbf{r} - \mathbf{r}'|} + V_{\text{xc}}[\rho](\mathbf{r}), \tag{243}$$

with

$$V_{\text{xc}}[\rho](\mathbf{r}) := \frac{\delta E_{\text{xc}}[\rho]}{\delta \rho(\mathbf{r})}, \tag{244}$$

is present in the RKS equation (202). By applying the definition for $V_{\text{xc}}(\mathbf{r})$ to the case of explicitly orbital-dependent xc functionals $E_{\text{xc}}[\{\varphi_j[\rho]\}]$, we obtain the ROEP integral equation for the "electrostatic case":

$$\sum_{-c^2 < \varepsilon_k \le \varepsilon_N} \int d^3r' \left(\varphi_k^\dagger(\mathbf{r}') V_{\text{xc}}^{\text{ROEP}}(\mathbf{r}') - \frac{\delta E_{\text{xc}}^{\text{ROEP}}}{\delta \varphi_k(\mathbf{r}')} \right) G_{Sk}(\mathbf{r}', \mathbf{r})\, \varphi_k(\mathbf{r}) + \text{c.c.} = 0. \tag{245}$$

Although this equation, first derived in the x-only limit by Shadwick, Talman and Norman [69], is considerably simpler than the ROEP integral equation (214), its numerical solution is still a very demanding task and has been achieved so far only for spherical atoms [63, 64, 69, 70].

In order to avoid a full numerical treatment of Eq. (245), one can again derive an RKLI approximation. Following the arguments in section 3.2, the RKLI equation for the "electrostatic case" is:

$$V_{\text{xc}}^{\text{RKLI}}(\mathbf{r}) = \frac{1}{2\rho(\mathbf{r})} \sum_{-c^2 < \varepsilon_k \le \varepsilon_N} \left(\frac{\delta E_{\text{xc}}^{\text{ROEP}}}{\delta \varphi_k(\mathbf{r})} \varphi_k(\mathbf{r}) + \rho_k(\mathbf{r}) \left(\bar{V}_{\text{xc}k}^{\text{RKLI}} - \bar{u}_{\text{xc}k} \right) \right) + \text{c.c.}, \tag{246}$$

with $\bar{V}_{xck}^{\rm RKLI}$ defined similar to Eq. (223). As in the nonrelativistic case, this equation can be understood as a mean-field-type approximation to the full ROEP integral equation (245) [68]. Furthermore, the RKLI equation (246) closely resembles its nonrelativistic counterpart, Eq. (76). In section 4.2, we will show that the RKLI approximation (246) is highly accurate when applied to the calculation of relativistic effects in high-Z atoms.

To conclude this section, we emphasize that the results of this section can *not* be obtained as a well-defined limit of the results of the sections 3.1 and 3.2 but must be derived separately within the framework of RDFT in the "electrostatic case" [63, 64]. This is because the basic variables of the underlying HK theorems are different: In the "electrostatic case", the density $\rho(\mathbf{r})$ alone is sufficient to determine all ground-state properties of the relativistic interacting many-body system, whereas the four current $j^{\mu}(\mathbf{r})$ must be used when arbitrary external four potentials are considered. This does not mean that the spatial components $\mathbf{j}(\mathbf{r})$ of the four current vanish in this case; they are simply functionals $\mathbf{j}[\rho](\mathbf{r})$ of the density.

4 Numerical results

4.1 Exchange-only calculations for nonrelativistic systems

4.1.1 Atomic systems

The x-only limit of the xc energy functional is given by the exact Fock term, Eq. (27). As explained in the preceding section, the OEP method then provides the corresponding exchange potential $V_{\rm x}^{\rm exact}(\mathbf{r})$ and therefore represents the exact implementation of x-only DFT. Consequently it provides a benchmark for testing approximate exchange-energy functionals employed within the Kohn-Sham scheme. In this section, we will review selected results of fully self-consistent x-only calculations performed with the OEP method and the KLI approximation to it [13, 14, 16–22, 33, 36–38, 42, 44, 52, 71–74] as described in section 2. For comparison, we list the results from traditional KS calculations using the x-only LDA (xLDA), where the exchange energy is given by

$$E_{\rm x}^{\rm LDA}[\rho] = -\frac{3}{2}\left(\frac{3}{4\pi}\right)^{\frac{1}{3}} \sum_{\sigma=\uparrow,\downarrow} \int d^3r \rho_\sigma^{\frac{4}{3}}(\mathbf{r}), \tag{247}$$

the generalized gradient approximation (GGA) due to Becke (B88) [75],

$$E_{\rm x}^{\rm B88}[\rho] = E_{\rm x}^{LDA}[\rho] - \beta \sum_{\sigma=\uparrow,\downarrow} \int d^3r \rho_\sigma^{\frac{4}{3}}(\mathbf{r}) \frac{x_\sigma^2(\mathbf{r})}{\left(1 + 6\beta x_\sigma(\mathbf{r}) \sinh^{-1} x_\sigma(\mathbf{r})\right)} \tag{248}$$

where

$$x_\sigma(\mathbf{r}) := \frac{|\nabla\rho_\sigma(\mathbf{r})|}{\rho_\sigma^{\frac{4}{3}}(\mathbf{r})} \tag{249}$$

and $\beta = 0.0042$, and the GGA due to Perdew and Wang (xPW91) [76–78], which may be written as

$$E_{\mathrm{x}}^{\mathrm{PW91}}[\rho] = -\frac{3}{2}\left(\frac{3}{4\pi}\right)^{\frac{1}{3}} \sum_{\sigma=\uparrow,\downarrow} \int d^3r \rho_\sigma^{\frac{4}{3}}(\mathbf{r}) F\left(s_\sigma(\mathbf{r})\right) \tag{250}$$

with

$$F(s) = \frac{1 + 0.19645 s \sinh^{-1}(7.7956 s) + \left(0.2743 - 0.1508 \exp(-100 s^2)\right) s^2}{\left(1 + 0.19645 s \sinh^{-1}(7.7956 s) + 0.004 s^4\right)} \tag{251}$$

and

$$s_\sigma(\mathbf{r}) = \frac{|\nabla\rho_\sigma(\mathbf{r})|}{\left(6\pi^2\right)^{\frac{1}{3}} \rho_\sigma^{\frac{4}{3}}(\mathbf{r})}. \tag{252}$$

We will also give results obtained with the xLDA-SIC functional

$$E_{\mathrm{x}}^{\mathrm{LDA\text{-}SIC}}[\{\varphi_{j\sigma}\}] = E_{\mathrm{x}}^{LDA}[\rho_\uparrow, \rho_\downarrow] - \sum_{\sigma=\uparrow,\downarrow} \sum_{i=1}^{N_\sigma} E_{\mathrm{x}}^{LDA}[|\varphi_{i\sigma}|^2, 0] \tag{253}$$

$$-\frac{1}{2} \sum_{\sigma=\uparrow,\downarrow} \sum_{i=1}^{N_\sigma} \int d^3r \int d^3r' \, \frac{|\varphi_{i\sigma}(\mathbf{r})|^2 |\varphi_{i\sigma}(\mathbf{r}')|^2}{|\mathbf{r} - \mathbf{r}'|} \tag{254}$$

employed within the KLI scheme for comparison. We will refer to this method as xLDA-SICKLI, respectively. As it has been found [43] that the exact OEP for the xLDA-SIC functional yields results which differ only marginally from the ones obtained with the KLI approximation, we will not list them here. In addition, we include the results of spin-unrestricted HF (SUHF) calculations, taken from [15].

For the OEP, KLI, B88, PW91 and xLDA calculations we have used a numerical code which solves the radial part of the Schrödinger equation (33) on a logarithmic mesh by the Numerov method as described in [79]. For non-spherical open-shell systems angular averaging was used in order to calculate the density and the exchange energy according to the Fock expression (27), for the latter in the form suggested by Slater [80]. Therefore, an analytical treatment of the angular parts was possible for all systems. The SIC-LDA energy functional was evaluated using spherically averaged total and orbital densities, as it was found [43] that the use of the exact nonspherical densities results only in minor changes of a few tenths of a percent. The OEP integral equation (48) for the radial variable r,

$$\int_0^\infty dr' K_\sigma(r, r') V_{\mathrm{x}\sigma}(r') = I_\sigma(r) \tag{255}$$

with

$$K_\sigma(r, r') = \sum_{i=1}^{N_\sigma} \int \frac{r^2 d\Omega}{4\pi} \int r'^2 d\Omega' \, \varphi_{i\sigma}^*(\mathbf{r}') G_{si\sigma}(\mathbf{r}, \mathbf{r}') \varphi_{i\sigma}(\mathbf{r}) \tag{256}$$

and

$$I_\sigma(r) = \sum_{i=1}^{N_\sigma} \int \frac{r^2 d\Omega}{4\pi} \int r'^2 dr' d\Omega' \, \varphi_{i\sigma}^*(\mathbf{r}') u_{xi\sigma}(\mathbf{r}') G_{si\sigma}(\mathbf{r}, \mathbf{r}') \varphi_{i\sigma}(\mathbf{r}) \tag{257}$$

was solved on the same mesh as the radial Schrödinger equation by numerical quadrature. The radial parts of the functions $G_{si\sigma}(\mathbf{r}, \mathbf{r}')$ were obtained from the radial parts of the orbitals $\varphi_{i\sigma}(\mathbf{r})$ and the corresponding complementary solutions of the radial part of the Schrödinger equation (33). A more detailed description may be found in [13, 37, 72].

Table 1: *Absolute total ground-state energies for H through Ar calculated self-consistently employing various x-only approximations. SUHF results have been taken from [15]. All numbers in atomic units.*

	SUHF	OEP	KLI	B88	PW91	xLDA	xLDA-SICKLI
H	0.5000	0.5000	0.5000	0.4979	0.4953	0.4571	0.5000
He	2.8617	2.8617	2.8617	2.8634	2.8552	2.7236	2.8617
Li	7.4328	7.4325	7.4324	7.4288	7.4172	7.1934	7.4342
Be	14.5730	14.5724	14.5723	14.5664	14.5543	14.2233	14.5784
B	24.5293	24.5283	24.5281	24.5173	24.5035	24.0636	24.5490
C	37.6900	37.6889	37.6887	37.6819	37.6658	37.1119	37.7450
N	54.4046	54.4034	54.4030	54.4009	54.3824	53.7093	54.5064
O	74.8136	74.8121	74.8117	74.8148	74.7964	73.9919	74.9624
F	99.4108	99.4092	99.4087	99.4326	99.4130	98.4740	99.6351
Ne	128.5471	128.5454	128.5448	128.5901	128.5689	127.4907	128.8586
Na	161.8590	161.8566	161.8559	161.8834	161.8613	160.6443	162.2170
Mg	199.6146	199.6116	199.6107	199.6320	199.6120	198.2488	200.0273
Al	241.8768	241.8733	241.8723	241.8829	241.8617	240.3561	242.3409
Si	288.8545	288.8507	288.8495	288.8551	288.8320	287.1820	289.3775
P	340.7193	340.7150	340.7137	340.7107	340.6857	338.8885	341.2989
S	397.5063	397.5016	397.5002	397.4921	397.4665	395.5190	398.1483
Cl	459.4826	459.4776	459.4760	459.4697	459.4426	457.3435	460.1966
Ar	526.8175	526.8122	526.8105	526.7998	526.7710	524.5174	527.5994

Table 2: *Absolute total ground-state energies for K through Kr calculated self-consistently employing various x-only approximations. For elements denoted with an asterisk the SUHF results have been obtained by using wave functions consisting of more than one Slater determinant, whereas the DFT results are obtained with a single-determinant wave function. SUHF results have been taken from [15]. All numbers in atomic units.*

	SUHF	OEP	KLI	B88	xPW91	xLDA
K	599.1649	599.1591	599.1571	599.1483	599.1194	596.7115
Ca	676.7582	676.7519	676.7497	676.7529	676.7262	674.1601
Sc	759.7359	759.7277	759.7249	759.7567	759.7294	757.0083
Ti*	848.4066	848.3802	848.3772	848.4360	848.4078	845.5307
V*	942.8856	942.8569	942.8539	942.9369	942.9078	939.8733
Cr	1043.3568	1043.3457	1043.3422	1043.4917	1043.4564	1040.2732
Mn	1149.8698	1149.8600	1149.8569	1149.9671	1149.9360	1146.5831
Fe	1262.4500	1262.4380	1262.4344	1262.5851	1262.5543	1259.0385
Co*	1381.4186	1381.3818	1381.3781	1381.5710	1381.5400	1377.8606
Ni*	1506.8303	1506.8340	1506.8303	1507.0634	1507.0321	1503.1881
Cu	1638.9642	1638.9523	1638.9481	1639.2804	1639.2473	1635.2392
Zn	1777.8481	1777.8344	1777.8307	1778.1196	1778.0870	1773.9099
Ga	1923.2612	1923.2487	1923.2454	1923.4735	1923.4402	1919.0951
Ge	2075.3603	2075.3483	2075.3453	2075.5287	2075.4937	2070.9811
As	2234.2399	2234.2281	2234.2251	2234.3657	2234.3291	2229.6475
Se	2399.8691	2399.8573	2399.8543	2399.9654	2399.9289	2395.0759
Br	2572.4418	2572.4300	2572.4269	2572.5159	2572.4780	2567.4546
Kr	2752.0550	2752.0430	2752.0398	2752.1006	2752.0613	2746.8661

In Tables 1, 2 and 3 we show the total ground-state energies obtained with the x-only DFT methods mentioned above and the SUHF scheme for neutral atoms with nuclear charge Z from 1 to 54. The OEP ground-state configurations used for the transition elements are listed in Table 4.

Comparing the first two columns of Tables 1, 2 and 3, it is evident that the SUHF and OEP results are very close to each other [9]. For some transition elements marked with an asterisk in Tables 2 and 3 the SUHF results have been obtained as expectation value of the Hamiltonian with respect to a linear combination of *two or more* Slater determinants in order to ensure a wave function exhibiting the appropriate symmetry, whereas the OEP and the other DFT calculations have been performed with a *single-determinant wave function.* This is the reason why for these systems the difference between the two values is larger. OEP and KLI results using more than a single Slater determinant have been performed by Li et. al. [15].

Table 3: *Absolute total ground-state energies for Rb through Xe calculated self-consistently employing various x-only approximations. For elements denoted with an asterisk the SUHF results have been obtained by using wave functions consisting of more than one Slater determinant, whereas the DFT results are obtained with a single-determinant wave function. SUHF results have been taken from [15]. All numbers in atomic units.*

	SUHF	OEP	KLI	B88	xPW91	xLDA
Rb	2938.3576	2938.3455	2938.3421	2938.3909	2938.3517	2932.9835
Sr	3131.5457	3131.5334	3131.5299	3131.5791	3131.5422	3125.9981
Y	3331.6846	3331.6710	3331.6670	3331.7256	3331.6880	3325.9712
Zr*	3539.0117	3538.9700	3538.9656	3539.0327	3538.9941	3533.1035
Nb	3753.6006	3753.5855	3753.5807	3753.6672	3753.6223	3747.5635
Mo	3975.5530	3975.5371	3975.5320	3975.6140	3975.5678	3969.3323
Tc	4204.7949	4204.7793	4204.7741	4204.8362	4204.7943	4198.3724
Ru*	4441.5409	4441.5088	4441.5032	4441.5925	4441.5498	4434.9512
Rh*	4685.8822	4685.8485	4685.8429	4685.9394	4685.8959	4679.1175
Pd	4937.9210	4937.9060	4937.9016	4938.0157	4937.9712	4931.0100
Ag	5197.6989	5197.6815	5197.6758	5197.7652	5197.7196	5190.5783
Cd	5465.1331	5465.1144	5465.1084	5465.1907	5465.1463	5457.8218
In	5740.1694	5740.1514	5740.1455	5740.1999	5740.1550	5732.6492
Sn	6022.9325	6022.9149	6022.9091	6022.9450	6022.8985	6015.2128
Sb	6313.4870	6313.4697	6313.4639	6313.4799	6313.4318	6305.5658
Te	6611.7856	6611.7683	6611.7625	6611.7725	6611.7246	6603.6768
I	6917.9814	6917.9642	6917.9582	6917.9671	6917.9181	6909.6900
Xe	7232.1384	7232.1210	7232.1150	7232.1165	7232.0662	7223.6572

As, by construction, the SUHF scheme gives the variationally best, i.e. lowest total energy, the x-only OEP solutions are always somewhat higher in energy, with the exception of one and two electron systems where the two methods coincide. The largest difference occurs for Be where the ground-state energies differ by 41 ppm. The disagreement decreases with increasing atomic number to about 2 ppm for Xe. We point out that these differences are due to the different nature of the HF and DFT approach and are resolved by the correlation contributions, which are defined differently for each scheme [81]. Therefore, the quality of the approximate x-only DFT approaches has to be judged by comparison with OEP rather than HF results [9].

To assess the quality of the KLI approximation, we have plotted on the left-hand side of Figure 3 the deviation of the KLI from the exact OEP results as a function of atomic number. As both methods use the same total-energy functional but the KLI scheme yields an approximate one-particle potential, the inequality $E^{\mathrm{OEP}} \leq E^{\mathrm{KLI}}$

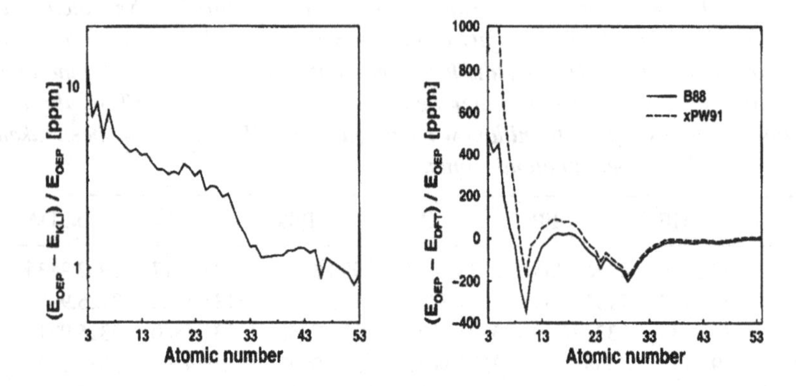

Figure 3: *Difference between the total energy from x-only OEP and KLI calculations (left) and between the total energy from x-only OEP and B88 and PW91 calculations (right) in ppm.*

[16] is always satisfied. The results differ only slightly [15, 16, 19], the difference being largest for Li with 13 ppm and dropping to 0.8 ppm for Xe. For one and two-particle systems the KLI approximation is exact and thus gives total energies identical to the exact OEP solutions. The mean absolute deviation from the exact DFT values provided by the exact OEP scheme for all atoms displayed in Tables 1, 2 and 3 is only 3.1 mH.

Turning to the conventional approximations, i.e. to explicit density functionals, the high accuracy of the KLI approximation becomes evident. In Figure 3, we have plotted the difference between the exact total x-only ground-state energies and the results of the two GGAs for the atoms listed in Tables 1, 2 and 3. Comparing the left with the right plot in Figure 3, the much larger error of the conventional density functionals is clearly visible. It is also apparent that the GGAs do not provide an upper bound for the total energy: For quite a few atoms the GGA results lie *below* the exact ones. For the B88 exchange functional, this is most pronounced for He, where the deviation is -594 ppm while the xPW91 functional underestimates the exact value for Ne by -183 ppm. As for the KLI approximation, the absolute magnitude of the error of the GGAs decreases with increasing atomic number. For the B88 functional, the error drops from 4200 ppm for the Hydrogen atom to 0.6 ppm for Xe, whereas the xPW91 functional yields errors of 9400 ppm and 7.6 ppm for the same atoms, respectively. The mean absolute deviation is 66 mH for the B88 and 53 mH for the xPW91 energy functional.

A glance at the xLDA results for the total energy displayed in Tables 1, 2 and 3 shows that these results differ considerably from the exact OEP ones. The error introduced by this approximation is clearly the largest of all the approximate x-only

Table 4: *OEP ground-state configurations of the transition elements.*

Atom	Configuration	Atom	Configuration
Sc	$3d^1 4s^2$	Y	$4d^1 5s^2$
Ti	$3d^2 4s^2$	Zr	$4d^3 5s^1$
V	$3d^3 4s^2$	Nb	$4d^4 5s^1$
Cr	$3d^5 4s^1$	Mo	$4d^5 5s^1$
Mn	$3d^5 4s^2$	Tc	$4d^5 5s^2$
Fe	$3d^6 4s^2$	Ru	$4d^7 5s^1$
Co	$3d^7 4s^2$	Rh	$4d^8 5s^1$
Ni	$3d^8 4s^2$	Pd	$4d^{10} 5s^0$
Cu	$3d^{10} 4s^1$	Ag	$4d^{10} 5s^1$
Zn	$3d^{10} 4s^2$	Cd	$4d^{10} 5s^2$

functionals considered here. It drops from 85800 ppm for Hydrogen to 1170 ppm for Xe, yielding a mean absolute deviation of 156.7 mH.

xLDA-SICKLI results for atoms H through Ar are given in Table 1. As for the KLI calculations employing the exact exchange-energy functional, the method yields exact results for one and spin-unpolarized two electron systems. For atoms with more electrons, the xLDA results are considerably improved by the SIC scheme but for heavier atoms the resulting total energies are still worse than the ones obtained from the two GGA approaches. The mean absolute deviation for the atoms H through Ar is 299 mH.

The trends found for the total ground-state energies remain valid for most quantities of interest: The HF and OEP results differ very little, the KLI scheme provides the best approximation to the exact DFT results, the GGAs are (sometimes only slighty) worse while the xLDA is by far the least accurate approximation. However, the xLDA-SICKLI results are often of higher quality than those obtained from any of the GGAs.

For example, in Tables 5 and 6 we have listed the eigenvalues $\varepsilon_{i\sigma}$ corresponding to the occupied orbitals of Ar and Cu, respectively, as obtained with the various methods. In DFT, the energy eigenvalues have no physical interpretation except for the highest occupied one which — in the exact theory including all correlation effects — is equal to the ionization potential of the system [47]. However, the eigenvalues of the inner orbitals may indicate the quality of the exchange potential $V_x(\mathbf{r})$. While the highest occupied eigenvalues obtained within the exact x-only OEP scheme are very close to the SUHF values, the inner eigenvalues differ more, illustrating their auxiliary nature within DFT calculations. The KLI approximation yields results very close to the exact OEP values, indicating the high quality of the KLI exchange potential. Most notable is the agreement of the highest occupied eigenvalues. As a typical example, we have listed the highest occupied

Table 5: *Absolute eigenvalues of Ar obtained from various self-consistent x-only calculations. SUHF values have been taken from [82]. All values in atomic units.*

	SUHF	OEP	KLI	B88	xPW91	xLDA	xLDA-SICKLI
1s	118.6104	114.4524	114.42789	114.1890	114.1887	113.7159	114.3650
2s	12.3222	11.1534	11.1820	10.7911	10.7932	10.7299	10.9804
2p	9.5715	8.7339	8.7911	8.4107	8.4141	8.3782	8.6191
3s	1.2774	1.0993	1.0942	0.8459	0.8481	0.8328	1.0497
3p	0.5910	0.5908	0.5893	0.3418	0.3441	0.3338	0.5493

Table 6: *Absolute eigenvalues of Cu obtained from various self-consistent x-only calculations. SUHF values have been taken from [37]. All values in atomic units.*

	SUHF	OEP	KLI	B88	xPW91	xLDA	xLDA-SICKLI
1s↑	328.7940	321.5109	321.3829	321.4929	321.4910	320.7080	321.4691
2s↑	40.8187	38.2779	38.2515	38.1898	38.1923	38.0830	38.2498
2p↑	35.6168	33.5336	33.5481	33.4859	33.5336	33.4214	33.5755
3s↑	5.0124	4.2328	4.1868	4.0360	4.0387	4.0054	4.1463
3p↑	3.3222	2.7535	2.7191	2.5751	2.5782	2.5577	2.6973
3d↑	0.4891	0.3046	0.2854	0.1616	0.1648	0.1575	0.2923
4s↑	0.2396	0.2405	0.2440	0.1625	0.1639	0.1588	0.2693
1s↓	328.7921	321.7166	321.5309	321.4919	321.4900	320.7069	321.6692
2s↓	40.8195	38.4734	38.4660	38.1925	38.1946	38.0860	38.4544
2p↓	35.6193	33.7275	33.7608	33.4877	33.4913	33.4235	33.7789
3s↓	5.0116	4.4212	4.4131	4.0399	4.0419	4.0093	4.3469
3p↓	3.3274	2.9410	2.9443	2.5782	2.5807	2.5609	2.8968
3d↓	0.4933	0.4893	0.4984	0.1558	0.1587	0.1512	0.4801

eigenvalues for the atoms of the nitrogen group in Table 7. Both the GGA and the xLDA calculations are seriously in error, yielding results which are too small by roughly a factor of two. For Cu, the two outermost majority-spin eigenvalues are incorrectly ordered if the B88 or xLDA functionals are used. These errors in the outermost orbitals are due to the incorrect asymptotic behavior of the approximate potentials. The inner orbitals are usually in better agreement with the exact results. The xLDA-SICKLI results are clearly much better than the results from any of the explicitly density-dependent functionals. In particular, the improvement over the xLDA results is very remarkable. This demonstrates the

Table 7: *Absolute eigenvalues of the highest occupied orbital for atoms of the nitrogen group obtained from various self-consistent x-only calculations. SUHF and OEP values have been taken from [15]. All values in atomic units.*

		SUHF	OEP	KLI	B88	xPW91	xLDA	xLDA-SICKLI
N	2p↑	0.5709	0.5712	0.5705	0.2846	0.2867	0.2763	0.5363
	2s↓	0.7258	0.7258	0.7245	0.5036	0.5059	0.4820	0.7247
P	3p↑	0.3921	0.3916	0.3905	0.2100	0.2113	0.2033	0.3582
	3s↓	0.5562	0.5562	0.5554	0.3921	0.3944	0.3840	0.5690
As	4p↑	0.3702	0.3691	0.3678	0.1975	0.1988	0.1929	0.3356
	4s↓	0.5561	0.5562	0.5559	0.4145	0.4174	0.4106	0.5896
Sb	5p↑	0.3358	0.3347	0.3336	0.1819	0.1832	0.1785	0.3032
	5s↓	0.4689	0.4693	0.4696	0.3491	0.3520	0.3478	0.5032

great importance of the self-interaction correction part of the exchange functional, which is responsible for the correct $-1/r$ decay of the xLDA-SICKLI exchange potential. The xLDA-SICKLI description even corrects the wrong ordering found in xLDA for the two outermost majority-spin eigenvalues in Cu.

In Figures 4 and 5 we have plotted $V_x(\mathbf{r})$ resulting from various x-only schemes for the Ar atom. Figure 5 shows the asymptotic region in greater detail. As expected, the KLI potential closely follows the exact OEP curve, the two being indistinguishable in the asymptotic region where they both decay like $-1/r$, as was shown in section 2.3.2. For the xLDA-SICKLI potential, the same holds true. However, the exact OEP and the xLDA-SICKLI potential agree less closely in the inner regions as compared to the KLI potential. The exchange potentials corresponding to all of the conventional functionals decay too rapidly, the one obtained from the xPW91 functional even introduces a spurious dip in this region. The B88 potential is known [37] to decay like $-1/r^2$. The so-called *intershell peaks*, i.e. the maxima of the potential occuring in the intershell regions, clearly visible in the OEP and KLI potentials, are not properly reproduced in the conventional DFT approximations. This is also the case for the xLDA-SICKLI potential, which exhibits no outer peak at all. Despite the rather close agreement of the innermost orbital energies, both of the GGA potentials exhibit an unphysical divergence at the position of the nucleus which may be traced back to the density gradients in the expressions for the exchange energy, cf. Eqs. (248) and (250). In Figure 6 we have plotted the difference between the spin-up and spin-down exchange potentials for Cu, which is a measure for the tendency of the atom to favour spin-polarization.

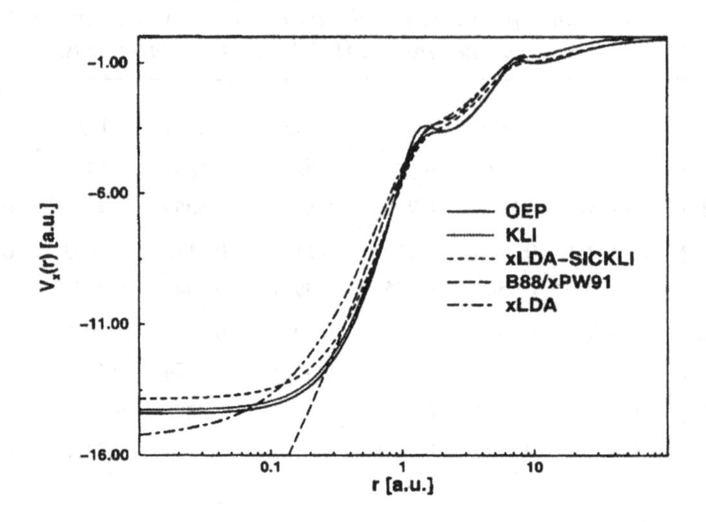

Figure 4: *Exchange potentials $V_x(\mathbf{r})$ for Ar from various self-consistent x-only calculations.*

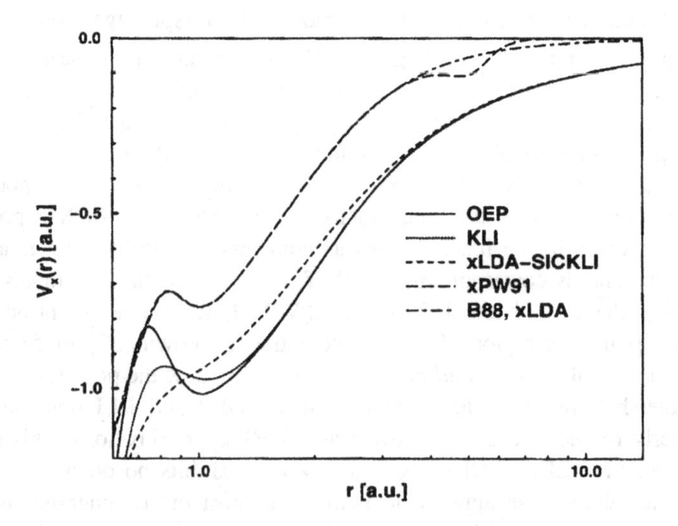

Figure 5: *Exchange potentials $V_x(\mathbf{r})$ for Ar from various self-consistent x-only calculations in the valence region. In this region, the xLDA potential is identical to the B88 one.*

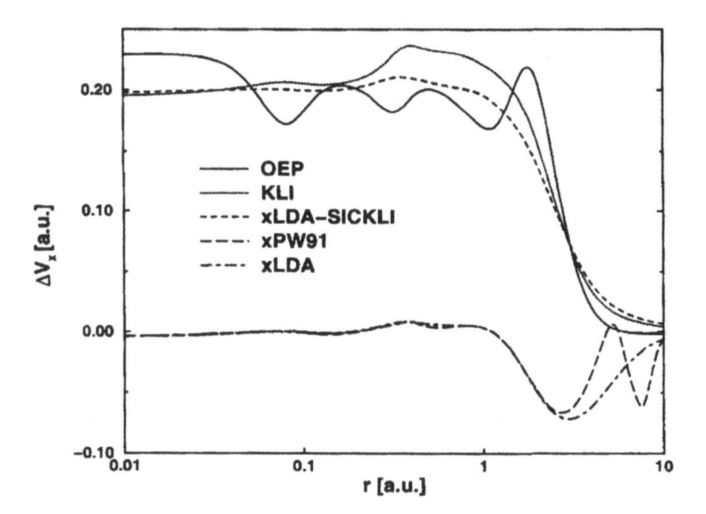

Figure 6: *Difference between spin-up and spin-down exchange potentials,* $\Delta V_{\mathrm{x}}(\mathbf{r})$, *for Cu from various self-consistent x-only calculations*

While the KLI and xLDA-SICKLI approximations roughly follow the overall shape of the exact result but miss most of the finer structure, both of the GGAs and the xLDA give very different results. Most strikingly, they fail to reproduce the constant shift between $V_{\mathrm{x}\uparrow}(\mathbf{r})$ and $V_{\mathrm{x}\downarrow}(\mathbf{r})$ in the inner region of the atom. This is, to a lesser extent, also true for other atoms [37].

The accuracy of the calculated electron densities may be assessed by comparing various r^n expectation values. In Table 8 we list the r^2 expectation values for the noble gas atoms He through Xe as calculated with the various schemes. This quantity puts a strong weight on the electron density of the outer shells and therefore represents a measure of its quality in that region. The SUHF and OEP values are almost identical, while the KLI approximation introduces only a small error. The errors introduced by the conventional DFT methods are about two orders of magnitude larger, while the xLDA-SICKLI results lie somewhere in between, which is largely due to the correct asymptotic form of the potential. Again, the remarkable improvement over the LDA due to the inclusion of the self-interaction correction part is noteworthy.

The density in the inner region of the atom heavily contributes to the $1/r$-expectation value, which we show for the four lightest alkaline earths in Table 9. While the SUHF and OEP values are identical to all given digits and the xLDA gives the worst results, the comparison of the KLI and GGA results shows an unexpected feature as the latter are always better than the former. This is somewhat surprising as the GGA exchange potentials show an unphysical singularity for $r \rightarrow 0$ as discussed above.

Table 8: *Expectation values of r^2 for some noble gases. SUHF and OEP values from [15]. All numbers in atomic units.*

	SUHF	OEP	KLI	B88	xPW91	xLDA	xLDA-SICKLI
He	1.1848	1.1848	1.1848	1.2644	1.2805	1.3275	1.1848
Ne	0.9372	0.9372	0.9367	0.9952	1.0015	1.0036	0.9524
Ar	1.4464	1.4465	1.4467	1.4791	1.4876	1.4889	1.4699
Kr	1.0981	1.0980	1.0985	1.1176	1.1232	1.1803	1.1124
Xe	1.1602	1.1600	1.1607	1.1716	1.1768	1.1716	1.1695

Table 9: *Expectation values of r^{-1} for some alkaline earth metals. SUHF and OEP values from [15]. All numbers in atomic units.*

	SUHF	OEP	KLI	B88	xPW91	xLDA	xLDA-SICKLI
Be	2.1022	2.1022	2.1039	2.1017	2.1008	2.0766	2.0934
Mg	3.3267	3.3267	3.3258	3.3261	3.3259	3.3147	3.3256
Ca	4.0080	4.0080	4.0086	4.0085	4.0084	4.0010	4.0050
Sr	5.1729	5.1729	5.1723	5.1729	5.1728	5.1685	5.1712

Table 10: *Magnetization density $\zeta(0) = (\rho_\uparrow(0) - \rho_\downarrow(0))/(\rho_\uparrow(0) + \rho_\downarrow(0))$ at the position of the nucleus for atoms of the nitrogen group from various self-consistent calculations. SUHF, OEP, and KLI values have been calculated from the results given in [15]. All values in units of 10^{-6}.*

	SUHF	OEP	KLI	B88	xPW91	xLDA	xLDA-SICKLI
N	910.8	921.5	−1623	125.0	195.7	−255.3	−98.2
P	−63.8	−72.1	−97.5	−116.9	−115.6	−129.4	−199.4
As	−24.2	−27.1	−26.3	−39.2	−36.4	−40.5	−46.8
Sb	−9.0	−10.4	−9.6	−15.1	−13.7	−16.0	−15.4

The magnetization density $\zeta(0) = (\rho_\uparrow(0) - \rho_\downarrow(0))/(\rho_\uparrow(0) + \rho_\downarrow(0))$ at the nucleus provides a very sensitive test for the quality of the spin densities. In Table 10 this quantity is shown for the four lightest atoms of the nitrogen group. The relatively large difference between the SUHF and OEP results underlines the sensitivity of this quantity. For the nitrogen atom the KLI approximation leads to a completely worthless result as the sign is reversed compared to the OEP value.

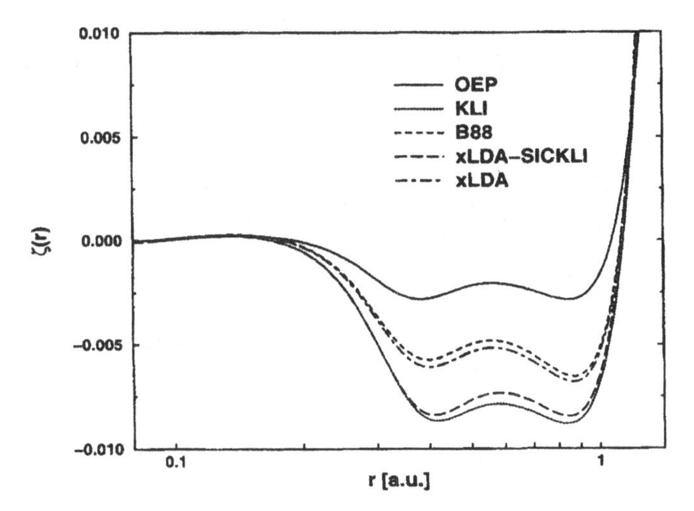

Figure 7: *Magnetization density* $\zeta = (\rho_\uparrow - \rho_\downarrow)/(\rho_\uparrow + \rho_\downarrow)$ *for Cu from various self-consistent x-only calculations. The result from the PW91 functional is very similar to the one from the B88 one.*

For this atom, the GGAs give the right sign but only one tenth of the correct number. For P, As and Sb the KLI scheme provides the best approximation, whereas the xLDA is worst. The pattern apparent from these few systems is by no means special but rather typical throughout the periodic table [15]. We conclude that no current approximation yields even moderate accuracy for the magnetization density at the nucleus.

Another failure of the KLI and xLDA-SICKLI approximations is apparent from Figure 7, where we have plotted the magnetization density for Cu. Clearly, the KLI results are twice as far off the exact curve as the GGA and xLDA approximations. For most systems, however, the differences are not as dramatic as for Cu.

4.1.2 Diatomic molecules

In order to demonstrate the validity of the KLI approach for more complex systems, we have performed KLI calculations for diatomic molecules employing the exact exchange-energy functional as defined by equation (27). Our calculations have been performed with a fully numerical basis-set-free code, developed from the Xα program written by Laaksonen, Sundholm and Pyykkö [83–85]. The code solves the one-particle Schrödinger equation for diatomic molecules

$$\left(-\frac{\nabla^2}{2} - \frac{Z_1}{|\mathbf{R}_1 - \mathbf{r}|} - \frac{Z_2}{|\mathbf{R}_2 - \mathbf{r}|} + V_H(\mathbf{r}) + V_{x\sigma}^{\mathrm{KLI}}(\mathbf{r})\right) \varphi_{j\sigma}(\mathbf{r}) = \varepsilon_{j\sigma}\varphi_{j\sigma}(\mathbf{r}), \tag{258}$$

where $\mathbf{R}_i$ denotes the location and Z_i the nuclear charge of the i-th nucleus in the molecule. This partial differential equation is solved in prolate spheroidal coordinates on a two-dimensional mesh by a relaxation method. The third variable, the azimuthal angle, is treated analytically. The Hartree potential

$$V_H(\mathbf{r}) = \int d^3r' \, \frac{\rho(\mathbf{r}')}{|\mathbf{r} - \mathbf{r}'|} \tag{259}$$

and the functions

$$u_{\mathrm{x}i\sigma}(\mathbf{r}) = -\frac{1}{\varphi^*_{i\sigma}(\mathbf{r})} \sum_{k=1}^{N_\sigma} \varphi^*_{k\sigma}(\mathbf{r}) \int d^3r' \, \frac{\varphi^*_{i\sigma}(\mathbf{r}')\varphi_{k\sigma}(\mathbf{r}')}{|\mathbf{r} - \mathbf{r}'|} \tag{260}$$

needed for the calculation of the exchange potential $V^{\mathrm{KLI}}_{\mathrm{x}\sigma}(\mathbf{r})$ (cf. Eq. (76)) are calculated as solutions of a Poisson and Poisson-like equation, respectively. In this step, the same relaxation technique as for the solution of the Schrödinger equation (258) is employed. Starting with an initial guess for the wave functions $\varphi_{i\sigma}(\mathbf{r})$, equations (258)–(260) together with (76) are iterated until self-consistency is achieved. A detailed description of the code is given in [86].

For comparison, we have performed additional x-only calculations with two other approximations of $V_{x\sigma}(\mathbf{r})$ and E_x, respectively. The first one of these, denoted by *Slater* in the following, employs – like the HF and x-only KLI methods – the exact representation (27) of E_x but uses the averaged exchange potential due to Slater [87] given by

$$V^S_{x\sigma}(\mathbf{r}) = -\frac{1}{\rho_\sigma(\mathbf{r})} \sum_{i,j=1}^{N_\sigma} \varphi^*_{j\sigma}(\mathbf{r})\varphi_{i\sigma}(\mathbf{r}) \int d^3r' \, \frac{\varphi^*_{i\sigma}(\mathbf{r}')\varphi_{j\sigma}(\mathbf{r}')}{|\mathbf{r} - \mathbf{r}'|}. \tag{261}$$

This expression can also be obtained from (76) by setting the constants $\bar{V}_{\mathrm{xc}i\sigma} - \bar{u}_{\mathrm{xc}i\sigma}$ equal to zero for all i. The other is the well known x-only local density approximation (xLDA) of conventional DFT, where the exchange energy is given by Eq. (247). As for the KLI calculations, we have successfully tested our implementations on atomic systems.

Results are given in Tables 11 through 18 for LiH, BH, FH, He_2, Li_2, Be_2, N_2 and OH^-. For each system we show the total ground-state energy E_{tot}, the various orbital energies ε and the nonzero electronic contributions to the dipole, quadrupole, octopole and hexadecapole moments denoted by Q^e_1, Q^e_2, Q^e_3 and Q^e_4, calculated from the geometrical center of the respective molecule. For FH, N_2 and OH^- we give the total moments Q^{tot}_L (including nuclear contributions), calculated from the center of mass of the respective molecule. For these three molecules we also present the expectation values of $1/r$, denoted by $< 1/r >$, calculated at the nuclei.

Table 11: *X-only results for LiH. HF values for bond length of 3.015 a.u. from [86]. Present calculations performed on a* 153 × 193 *grid with bond length of 3.015 a.u. All numbers in atomic units. Taken from [88].*

	HF	KLI	Slater	xLDA
E_{tot}	−7.9874	−7.9868	−7.9811	−7.7043
$\varepsilon_{1\sigma}$	−2.4452	−2.0786	−2.3977	−1.7786
$\varepsilon_{2\sigma}$	−0.3017	−0.3011	−0.3150	−0.1284
Q_1^e	0.6531	0.6440	0.8614	0.8679
Q_2^e	7.1282	7.1365	6.9657	6.7717
Q_3^e	2.9096	2.9293	3.0799	2.6924
Q_4^e	16.0276	16.1311	15.5881	15.0789

Table 12: *X-only results for BH. HF values for bond length of 2.336 a.u. from [86]. Present calculations performed on a* 193 × 265 *grid with bond length of 2.336 a.u. All numbers in atomic units. Taken from [88].*

	HF	KLI	Slater	xLDA
E_{tot}	−25.1316	−25.1290	−25.1072	−24.6299
$\varepsilon_{1\sigma}$	−7.6863	−6.8624	−7.4837	−6.4715
$\varepsilon_{2\sigma}$	−0.6482	−0.5856	−0.6358	−0.3956
$\varepsilon_{3\sigma}$	−0.3484	−0.3462	−0.3721	−0.1626
Q_1^e	5.3525	5.3498	5.2991	5.3154
Q_2^e	12.1862	12.1416	11.4720	11.9542
Q_3^e	15.6411	15.5618	14.3328	14.0904
Q_4^e	25.8492	25.4188	25.2152	21.9134

Table 13: *X-only results for FH. HF values for bond length of 1.7328 a.u. from [86]. Present calculations performed on a* 161 × 321 *grid with bond length of 1.7328 a.u. All numbers in atomic units. Taken from [88].*

	HF	KLI	Slater	xLDA
E_{tot}	−100.0708	−100.0675	−100.0225	−99.1512
$\varepsilon_{1\sigma}$	−26.2946	−24.5116	−25.6625	−24.0209
$\varepsilon_{2\sigma}$	−1.6010	−1.3994	−1.4327	−1.0448
$\varepsilon_{3\sigma}$	−0.7682	−0.7772	−0.8167	−0.4483
$\varepsilon_{1\pi}$	−0.6504	−0.6453	−0.6897	−0.3109
Q_1^{tot}	−0.7561	−0.8217	−0.8502	−0.6962
Q_2^{tot}	1.7321	1.8012	1.8472	1.7124
Q_3^{tot}	−2.5924	−2.7222	−2.8781	−2.4662
Q_4^{tot}	5.0188	5.1825	5.3720	4.7068
$< 1/r >_H$	6.1130	6.0878	6.0909	6.0901
$< 1/r >_F$	27.1682	27.1622	27.6049	27.0289

Table 14: *X-only results for He_2. HF values for bond length of 5.6 a.u. from [86]. Present calculations performed on a 209 × 225 grid with bond length of 5.6 a.u. All numbers in atomic units. Taken from [88].*

	HF	KLI	Slater	xLDA
E_{tot}	−5.72333	−5.72332	−5.72332	−5.44740
$\varepsilon_{1\sigma g}$	−0.92017	−0.91929	−0.91977	-0.51970
$\varepsilon_{1\sigma u}$	−0.91570	−0.91566	−0.91614	-0.51452
Q_2^e	31.36165	31.35931	31.35907	31.35507
Q_4^e	245.8779	245.8643	245.8615	245.8255

Table 15: *X-only results for Li_2. HF values for bond length of 5.051 a.u. from [86]. Present calculations performed on a 209 × 225 grid with bond length of 5.051 a.u. All numbers in atomic units. Taken from [88].*

	HF	KLI	Slater	xLDA
E_{tot}	−14.8716	−14.8706	−14.8544	−14.3970
$\varepsilon_{1\sigma g}$	−2.4531	−2.0276	−2.3875	−1.7869
$\varepsilon_{1\sigma u}$	−2.4528	−2.0272	−2.3873	−1.7864
$\varepsilon_{2\sigma g}$	−0.1820	−0.1813	−0.1989	−0.0922
Q_2^e	27.6362	27.4993	29.0014	29.4401
Q_4^e	159.9924	159.6809	169.1300	172.8505

Table 16: *X-only results for Be_2. HF values for bond length of 4.6 a.u. from [86]. Present calculations performed on a 209 × 225 grid with bond length of 4.6 a.u. All numbers in atomic units. Taken from [88].*

	HF	KLI	Slater	xLDA
E_{tot}	−29.1337	−29.1274	−29.0939	−28.4612
$\varepsilon_{1\sigma g}$	−4.73150	−4.09876	−4.60353	−3.78576
$\varepsilon_{1\sigma u}$	−4.73147	−4.09872	−4.60351	−3.87571
$\varepsilon_{2\sigma g}$	−0.39727	−0.33452	−0.37659	−0.23067
$\varepsilon_{2\sigma u}$	−0.24209	−0.23489	−0.26524	−0.13163
Q_2^e	46.0878	46.2475	43.4833	46.2501
Q_4^e	261.774	277.365	249.135	281.950

Table 17: *X-only results for N_2. HF values for bond length of 2.07 a.u. from [86]. Present calculations performed on a* 209 × 225 *grid with bond length of 2.07 a.u. All numbers in atomic units. Taken from [88].*

	HF	KLI	Slater	xLDA
E_{tot}	−108.9936	−108.9856	−108.9109	−107.7560
$\varepsilon_{1\sigma g}$	−15.6822	−14.3722	−15.2692	−13.8950
$\varepsilon_{1\sigma u}$	−15.6787	−14.3709	−15.2682	−13.8936
$\varepsilon_{2\sigma g}$	−1.4726	−1.3076	−1.3316	−0.9875
$\varepsilon_{2\sigma u}$	−0.7784	−0.7452	−0.7473	−0.4434
$\varepsilon_{3\sigma g}$	−0.6347	−0.6305	−0.6521	−0.3335
$\varepsilon_{1\pi u}$	−0.6152	−0.6818	−0.6960	−0.3887
Q_2^{tot}	−0.9372	−0.9489	−1.1757	−1.1643
Q_4^{tot}	−7.3978	−6.7481	−7.1272	−6.2553
$< 1/r >_N$	21.6543	21.6439	21.9749	21.5820

Table 18: *X-only results for OH^-. HF values for bond length of 1.835 a.u. from [86]. Present calculations performed on a* 105 × 145 *grid with bond length of 1.835 a.u. All numbers in atomic units.*

	HF	KLI	Slater
E_{tot}	−75.4180	−75.4145	−75.3681
$\varepsilon_{1\sigma}$	−20.1858	−18.5775	−19.6091
$\varepsilon_{2\sigma}$	−0.9009	−0.7062	−0.7316
$\varepsilon_{3\sigma}$	−0.2508	−0.2363	−0.2668
$\varepsilon_{1\pi}$	−0.1097	−0.1066	−0.1355
Q_1^{tot}	0.4855	0.5792	0.5941
Q_2^{tot}	−1.8314	−1.9808	−1.7975
Q_3^{tot}	1.4003	1.7646	1.9280
Q_4^{tot}	−4.0346	−4.5159	−4.6758
$< 1/r >_H$	5.7355	5.7061	5.7399
$< 1/r >_O$	23.2588	23.2524	23.6925

For the quantities of physical interest, i.e. for E_{tot}, the energies ε_{HOMO} of the highest occupied orbitals and the multipole moments, the x-only KLI and HF results differ only slightly, typically by a few hundredths of a percent for the total energies, a few tenths of a percent for ε_{HOMO} and a few percent for the multipole moments. The largest difference for ε_{HOMO} is found for Be_2, where the results differ by 3%. For N_2, the energetic order of the $1\pi_u$ and $3\sigma_g$ orbitals obtained with the HF method is reversed in all DFT approaches, which reproduce the experimentally observed order of the outer valence ionization potentials [89]. As far as the multipole moments are concerned, the largest discrepancy between the x-only KLI and HF approach occurs for the total hexadecapole moment of OH^-, where the results differ by 11.9%. The $1/r$-expectation values obtained with the HF and x-only KLI methods are almost identical, differing by only a few hundredths of a precent with the exception of the ones for the hydrogen nuclei in FH and OH^-, where the difference is an order of magnitude larger. In these cases, the Slater and for FH also the xLDA approximation give values closer to the HF results.

Except for the cases mentioned above, the Slater method gives values for E_{tot}, ε_{HOMO}, the multipole moments and $1/r$-expectation values which differ to a larger extent from both the KLI and HF results than the latter from each other. From the energy eigenvalues of the inner orbitals it is obvious that the Slater exchange potential $V_{x\sigma}^{S}(\mathbf{r})$ is deeper than the one obtained in the KLI method which therefore yields results closer to the HF ones.

Finally, the xLDA results differ strongly from the other methods, yielding much higher total energies. Especially prominent are the values for ε_{HOMO}, which are roughly twice as large as the ones from any of the other methods. Once again, this is due to the wrong exponential decay of $V_{x\sigma}^{LDA}(\mathbf{r})$ for large r. For the negatively charged molecule OH^- there is no convergence of the self-consistency cycle if the xLDA approximation is used.

We expect that the bulk part of the differences between the x-only KLI and the HF results is not caused by the KLI approximation, but is rooted in the different nature of the HF and the DFT approaches. This was found for atomic systems as discussed in the previous subsection and we see no reason why this should not be the case for molecular systems as well.

However, the x-only KLI approach is also prone to reproduce some failures of HF theory. For example, the dissociation energy of Be_2 is found to be -12.3 mH in the HF and -17.2mH in the x-only KLI scheme, i.e. the molecule is predicted to be unstable by both methods. The xLDA, on the other hand, gives 14.6 mH for the dissociation energy, which is still far from the exact value of 3.8mH (calculated from the results given in [90] and [91]) but at least leads to a stable molecule. This effect is also present for other systems and has to be corrected by properly chosen correlation functionals. Calculations on diatomic molecules using the LDA-SIC functional in KLI approximation have been reported by Krieger et. al. [92, 93].

4.2 Comparison of nonrelativistic with relativistic results

The discussions in the preceding section were solely based on nonrelativistic DFT. In order to investigate the influence of relativistic effects, we present, in this section, results of fully relativistic calculations on closed-shell atoms.

Since we are not concerned with any magnetic fields in our calculations, the following analysis is done in the framework of the "electrostatic case", discussed in section 3.3. First of all, a brief remark concerning the nature of relativistic interactions has to be made: In QED, the electron-electron interaction, mediated by the exchange of photons, is properly described by the (free) photon propagator $D^0_{\mu\nu}(x-y)$. It can be decomposed into an instantaneous (longitudinal) Coulomb part and a transversal contribution that contains all retardation effects [94]:

$$D^0_{\mu\nu}(x-y) = i g_{0\mu} g_{0\nu} \frac{\delta(x^0-y^0)}{|\mathbf{x}-\mathbf{y}|} + D^{0,T}_{\mu\nu}(x-y). \tag{262}$$

Based on this decomposition, we may split $E_{\mathrm{H}}[\rho]$ as well as $E_{\mathrm{xc}}[\rho]$ into their longitudinal and transverse contributions, i.e.

$$E_{\mathrm{H}}[\rho] = E^{\mathrm{L}}_{\mathrm{H}}[\rho] + E^{\mathrm{T}}_{\mathrm{H}}[\rho] \tag{263}$$

$$E_{\mathrm{xc}}[\rho] = E^{\mathrm{L}}_{\mathrm{xc}}[\rho] + E^{\mathrm{T}}_{\mathrm{xc}}[\rho]. \tag{264}$$

Aiming at calculations for atomic systems, we can expect the interaction to be dominated by the familiar Coulomb term. Therefore, as a starting point, it seems plausible to neglect all transverse contributions. This approach represents the RDFT analogue of the standard relativistic many-body treatment employing the Dirac-Coulomb Hamiltonian.

Restricting ourselves to the longitudinal (Coulomb) interactions, we have performed self-consistent calculations in the x-only limit of RDFT. In analogy to the nonrelativistic case, the x-only limit of the xc energy functional is defined by using the exact expression for the exchange-energy functional, i.e. the relativistic Fock term

$$E^{\mathrm{L,exact}}_{\mathrm{x}}[\rho] = -\frac{1}{2} \sum_{-c^2<\varepsilon_j,\varepsilon_k<e_F} \int d^3r \int d^3r' \, \frac{\varphi^\dagger_j(\mathbf{r})\varphi_k(\mathbf{r})\varphi^\dagger_k(\mathbf{r}')\varphi_j(\mathbf{r}')}{|\mathbf{r}-\mathbf{r}'|} \tag{265}$$

in the case of longitudinal interactions only. As discussed in section 3.3, the exact longitudinal exchange potential $V^{\mathrm{L}}_{\mathrm{x}}(\mathbf{r})$ can be obtained by solving the full ROEP integral equation (214) with E_{xc} replaced by $E^{\mathrm{L,exact}}_{\mathrm{x}}$. Simultaneous solution of the ROEP integral equation and the RKS equation (202) therefore represents the exact implementation of the longitudinal x-only limit of RDFT. It is compared to the RKLI method, which employs the same exact expression (265) for the exchange energy and only approximates the local exchange potential $V^{\mathrm{L}}_{\mathrm{x}}(\mathbf{r})$ by means of Eq. (246).

Besides, we list results of traditional x-only RKS schemes obtained from the longitudinal x-only LDA (xRLDA), given by

$$E^{\mathrm{L,RLDA}}_{\mathrm{x}}[\rho] = \int d^3r \, e^{\mathrm{NRLDA}}_{\mathrm{x}}(\rho)\Phi^{\mathrm{L}}_0(\beta), \tag{266}$$

with the nonrelativistic energy density

$$e_{\mathrm{x}}^{\mathrm{NRLDA}}(\rho) = -\frac{3}{4}\left(\frac{3}{\pi}\right)^{\frac{1}{3}} \rho^{\frac{4}{3}}(\mathbf{r}) \tag{267}$$

and the relativistic correction

$$\Phi_0^L(\beta) := \frac{5}{6} + \frac{1}{3\beta^2} + \frac{2\eta}{3\beta} \operatorname{arsinh}\beta - \frac{2\eta^4}{3\beta^4} \ln\eta - \frac{1}{2}\left(\frac{\eta}{\beta} - \frac{arsinh\,\beta}{\beta^2}\right)^2 \tag{268}$$

with

$$\beta(\mathbf{r}) := \frac{1}{c}\left(3\pi^2\rho(\mathbf{r})\right)^{\frac{1}{3}} \tag{269}$$

and

$$\eta(\mathbf{r}) := \left(1 + \beta(\mathbf{r})^2\right)^{\frac{1}{2}}. \tag{270}$$

Furthermore, we performed calculations with two recently introduced relativistic GGAs (RGGAs) [70, 95]. The first one is given by the functional

$$E_{\mathrm{x}}^{\mathrm{L,RGGA}}[\rho] = \int d^3r\, e_{\mathrm{x}}^{\mathrm{NRLDA}}(\rho) \left(\Phi_0^{\mathrm{L}}(\beta) + g(\xi)\Phi_2^{\mathrm{L}}(\beta)\right), \tag{271}$$

with the quantity $g(\xi)$ proposed by Becke (RB88)

$$g_{B88}(\xi) = \frac{d\xi}{1 + 9d\xi^{1/2}\operatorname{arcsinh}\left[2\left(6\pi^2\right)^{1/3}\xi^{1/2}\right]/(4\pi)} \tag{272}$$

(d=0.2743). The second one is a [2/2]-Padé approximant (RECMV92) [52]

$$g_{ECMV92}(\xi) = \frac{A_1\xi + A_2\xi^2}{1 + B_1\xi + B_2\xi^2} \tag{273}$$

($A_1 = 0.3402$, $A_2 = 5.9955$, $B_1 = 27.5026$, $B_2 = 5.7728$), where

$$\xi(\mathbf{r}) = \left(\frac{\nabla\rho(\mathbf{r})}{2(3\pi^2\rho(\mathbf{r}))^{1/3}\rho(\mathbf{r})}\right)^2. \tag{274}$$

The relativistic effects are accounted for via the function $\Phi_2^{\mathrm{L}}(\beta)$, which is also given by a [2/2]-Padé approximant

$$\Phi_2^{\mathrm{L}}(\beta) = \frac{1 + a_1\beta^2 + a_2\beta^4}{1 + b_1\beta^2 + b_2\beta^4} \tag{275}$$

with the constants given in Table 19.

Table 19: *Parameters for the longitudinal correction factor (275). Taken from [70].*

	RECMV92	RB88
a1	2.21259	2.20848
a2	0.669152	0.668684
b1	1.32998	1.33075
b2	0.794803	0.795105

These various approaches are analyzed for spherical (closed-shell) atoms. To this end, the spin-angular part of the wave function is treated analytically and the remaining radial part of the Dirac equation is solved numerically on a logarithmic mesh employing an Adams-Bashforth-Moulton predictor-corrector scheme [64]. We also note, that for these closed-shell atoms the transverse Hartree energy vanishes, i.e. $E_{\mathrm{H}}^{\mathrm{T}} = 0$. In all our calculations we use finite nuclei modeled by a homogeneously charged sphere with the radii given by

$$R_{\mathrm{nucl}} = 1.0793\ A^{1/3} + 0.73587\ \mathrm{fm} \tag{276}$$

and A being the atomic mass taken from [96]. We mention in passing that employing finite nuclei is not necessary to ensure convergent results as, for example, in the relativistic Thomas-Fermi model. We incorporate finite nuclei because they represent the physically correct approach.

In Table 20, we show the longitudinal ground-state energy $E_{\mathrm{tot}}^{\mathrm{L}}$ obtained from the various self-consistent x-only RDFT approaches and, in addition, from relativistic Hartree-Fock (RHF) calculations. Furthermore, since we are interested in the effects induced by relativity, we calculated the relativistic contribution to $E_{\mathrm{tot}}^{\mathrm{L}}$, defined by

$$\Delta E_{\mathrm{tot}} := E_{\mathrm{tot}}^{\mathrm{L}}[\rho^{\mathrm{R}}] - E_{\mathrm{tot}}^{\mathrm{NR}}[\rho^{\mathrm{NR}}], \tag{277}$$

which is listed in Table 21. We note that the nonrelativistic total ground-state energies $E_{\mathrm{tot}}^{\mathrm{NR}}[\rho^{\mathrm{NR}}]$ are *not* those of section 4.1, but have also been calculated employing finite nuclei. From Table 21, we realize that the inclusion of relativistic effects leads to drastic corrections especially for high-Z atoms. For example, Table 21 shows that the relativistic correction of Hg amounts for about 6.7% of the total energy thus demonstrating the need for a fully relativistic treatment. Comparing the different approaches, we basically find the same trends as in the nonrelativistic context: The RHF and ROEP data agree closely with each other, the small differences resulting again from the different nature of the two approaches which correspond to different definitions of the respective correlation energies. Since with increasing atomic number, the inner orbitals, contributing most to the total energy, become more and more localized, the differences between the non-local RHF potential and the local ROEP decrease. In fact, we see from

Table 20: *Longitudinal ground-state energy* $-E^{L}_{\mathrm{tot}}$ *from various self-consistent x-only and RHF calculations.* $\bar{\Delta}$ *denotes the mean absolute deviation and* $\bar{\delta}$ *the average relative deviation (in 0.1 percent) from the exact ROEP values. All numbers in atomic units. Taken from [68].*

	RHF	ROEP	RKLI	RB88	RECMV92	xRLDA
He	2.862	2.862	2.862	2.864	2.864	2.724
Be	14.576	14.575	14.575	14.569	14.577	14.226
Ne	128.692	128.690	128.690	128.735	128.747	127.628
Mg	199.935	199.932	199.931	199.952	199.970	198.556
Ar	528.684	528.678	528.677	528.666	528.678	526.337
Ca	679.710	679.704	679.702	679.704	679.719	677.047
Zn	1794.613	1794.598	1794.595	1794.892	1794.880	1790.458
Kr	2788.861	2788.848	2788.845	2788.907	2788.876	2783.282
Sr	3178.080	3178.067	3178.063	3178.111	3178.079	3172.071
Pd	5044.400	5044.384	5044.380	5044.494	5044.442	5036.677
Cd	5593.319	5593.299	5593.292	5593.375	5593.319	5585.086
Xe	7446.895	7446.876	7446.869	7446.838	7446.761	7437.076
Ba	8135.644	8135.625	8135.618	8135.612	8135.532	8125.336
Yb	14067.669	14067.621	14067.609	14068.569	14068.452	14054.349
Hg	19648.865	19648.826	19648.815	19649.141	19649.004	19631.622
Rn	23602.005	23601.969	23601.959	23602.038	23601.892	23582.293
Ra	25028.061	25028.027	25028.017	25028.105	25027.962	25007.568
No	36740.682	36740.625	36740.609	36741.900	36741.783	36714.839
$\bar{\Delta}$			0.006	0.189	0.168	8.668
$\bar{\delta}$			0.002	0.103	0.108	6.20

Table 20 that the smallest deviations are found for No. Comparing the second and third columns of Table 20, it is obvious that the RKLI method yields results in very close agreement with the exact ROEP ones. The mean absolute deviation from the exact ROEP data of the 18 neutral atoms listed in Table 20 is only 5 mH. Hence it is of the same order of magnitude as the corresponding mean absolute deviation in the nonrelativistic case as discussed in the preceding section. Moreover, when looking at Table 21, we realize that the relativistic contributions to $E^{\mathrm{L}}_{\mathrm{tot}}$ are reproduced almost perfectly within the RKLI scheme. In other words, almost no additional deviations are introduced by the relativistic treatment of the KLI scheme, so that the high quality of the nonrelativistic KLI approximation is maintained in the relativistic domain. Turning towards the conventional x-only schemes, the conclusions drawn in section 4.1 can be repeated: The RGGAs, although clearly improving over the xRLDA, are worse by more than one order of magnitude when compared to the RKLI, whereas the xRLDA yields the least accurate results.

Table 21: *Relativistic Contribution* $-\Delta E^L_{\text{tot}}$ *from various self-consistent x-only and RHF calculations.* $\bar{\Delta}$ *denotes the mean absolute deviation and* $\bar{\delta}$ *the average relative deviation (in 0.1 percent) from the exact ROEP values. All numbers in atomic units. Taken from [68].*

	RHF	ROEP	RKLI	RB88	RECMV92	xRLDA
He	0.000	0.000	0.000	0.000	0.000	0.000
Be	0.003	0.003	0.003	0.003	0.003	0.002
Ne	0.145	0.145	0.145	0.145	0.145	0.138
Mg	0.320	0.320	0.320	0.321	0.321	0.308
Ar	1.867	1.867	1.867	1.867	1.867	1.821
Ca	2.953	2.953	2.953	2.952	2.953	2.888
Zn	16.771	16.770	16.770	16.779	16.779	16.555
Kr	36.821	36.820	36.820	36.822	36.821	36.432
Sr	46.554	46.553	46.553	46.552	46.551	46.092
Pd	106.527	106.526	106.526	106.526	106.525	105.715
Cd	128.245	128.243	128.243	128.243	128.241	127.323
Xe	214.860	214.858	214.858	214.825	214.822	213.522
Ba	252.223	252.222	252.221	252.176	252.173	250.725
Yb	676.559	676.551	676.549	676.590	676.588	673.785
Hg	1240.521	1240.513	1240.511	1240.543	1240.538	1236.349
Rn	1736.153	1736.144	1736.142	1736.151	1736.151	1730.890
Ra	1934.777	1934.770	1934.768	1934.781	1934.783	1929.116
No	3953.172	3953.155	3953.151	3953.979	3954.015	3944.569
$\bar{\Delta}$			0.001	0.056	0.058	1.788
$\bar{\delta}$			0.009	1.14	1.35	33.7

These trends also remain valid when other quantities of interest are considered. For example, in Table 22 we have listed the relativistic contributions to the longitudinal exchange energy, defined analogously to Eq. (277). Again, only small deviations between the ROEP and RKLI results are found. From the third and fourth columns of Table 22 we notice that the results of the RGGA functionals — with the exception of No — are also in excellent agreement with the exact data. However, since the RGGAs are optimized for exactly these quantities [70], this might not come as a surprise. Again, the RLDA is the least accurate approximation. It is worthwhile noting that the exchange energy $E^{\mathrm{L}}_{\mathrm{x}}$ is influenced quite substantially by relativistic effects, too. Taking again Hg as an example, we realize that the 5.8%-contribution to $E^{\mathrm{L}}_{\mathrm{x}}$ is of the same order as for the total energy. Furthermore, even for lighter atoms such as Mg, the relativistic corrections to $E^{\mathrm{L}}_{\mathrm{x}}$ are comparable or even larger than the differences between the currently best nonrelativistic

Table 22: *Relativistic Contribution $-\Delta E_x$ to the exchange energy from various self-consistent x-only calculations. $\bar{\Delta}$ denotes the mean absolute deviation and $\bar{\delta}$ the average relative deviation (in percent) from the exact ROEP values. All energies in atomic units. Taken from [68].*

	ROEP	RKLI	RB88	RECMV92	xRLDA
He	0.000	0.000	0.000	0.000	0.000
Be	0.001	0.001	0.001	0.001	0.000
Ne	0.015	0.015	0.015	0.015	0.007
Mg	0.029	0.029	0.029	0.029	0.015
Ar	0.118	0.118	0.117	0.118	0.069
Ca	0.172	0.172	0.171	0.171	0.104
Zn	0.627	0.626	0.632	0.632	0.402
Kr	1.215	1.214	1.212	1.211	0.814
Sr	1.478	1.477	1.473	1.472	1.005
Pd	2.785	2.787	2.782	2.780	1.958
Cd	3.264	3.264	3.255	3.252	2.322
Xe	5.021	5.020	4.977	4.974	3.657
Ba	5.739	5.736	5.684	5.680	4.215
Yb	12.043	12.024	12.027	12.024	9.194
Hg	19.963	19.956	19.965	19.957	15.734
Rn	26.637	26.620	26.612	26.610	21.307
Ra	29.241	29.218	29.225	29.224	23.513
No	52.403	52.402	53.168	53.205	43.683
$\bar{\Delta}$		0.004	0.053	0.056	1.819
$\bar{\delta}$		0.079	1.03	0.857	35.9

exchange-energy functionals as might be seen by comparing the results to the ones of section 4.1. As a consequence, a relativistic treatment is indispensable for the ultimate comparison with experiments [64].

Next, we turn our attention to local properties such as the exchange potential $V_x^L(r)$. In Fig. 8, the exchange potential is plotted for the case of Hg. As expected, the RKLI potential follows the exact curve most closely, although the strong intershell peaks of the ROEP curve are not fully reproduced. However, it again improves significantly over the conventional RDFT results, where this structure is smeared out or even absent. In addition, large errors are introduced within the conventional RDFT schemes in the asymptotic regions near as well as far off the nucleus. Since these observations closely resemble the ones made in the analysis of nonrelativistic systems in section 4.1, we consider the relativistic contribution

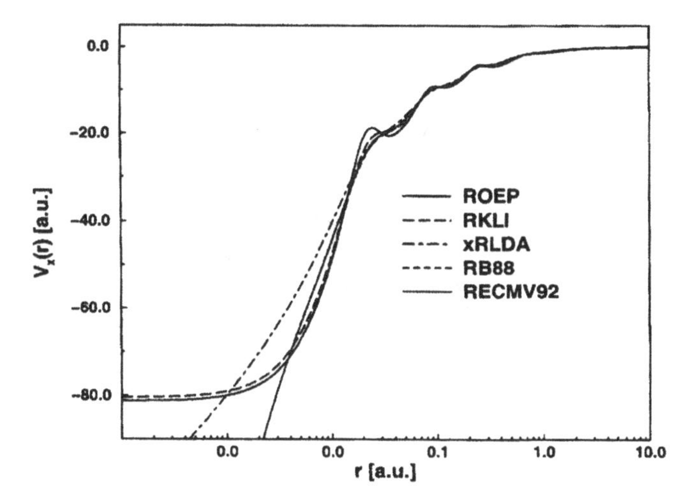

Figure 8: *Longitudinal exchange potential $V_x^L(r)$ for Hg from various self-consistent x-only calculations. Taken from [68].*

separately. The relativistic contribution to the exchange potential, given by

$$\Delta V_x(r) := \frac{V_x^L[\rho^R](r) - V_x^{NR}[\rho^{NR}](r)}{V_x^{NROEP}[\rho^{NR}](r)}, \tag{278}$$

is plotted in Fig. 9. We first observe strong oscillations between 0.1 a.u. and 5 a.u. These oscillations are introduced by the displacement of the density due to relativistic effects and thus represent a direct consequence of the atomic shell structure. As the shell structure of the exchange potential is not fully reproduced within the RKLI approach, the amplitudes of the oscillations are somewhat smaller compared to $\Delta V_x^{OEP}(r)$. While these deviations are clearly visible, the RKLI curve is still closest to the exact one, especially in the region near the nucleus and in the valence region, where large deviations occur for the conventional RDFT methods.

As a consequence of the failures of the conventional RDFT schemes in the asymptotic region, large deviations are found for the energies of the outermost orbitals. This observation – well-known from nonrelativistic calculations – is also found in fully relativistic treatments as shown in Table 23, where the single-particle energies of Hg are listed. In contrast, the RKLI approximation, yielding the correct $-1/r$-behavior of the potential as $r \to \infty$, produces results very similar to the exact ROEP data. Apart from the fact that the single-particle energies are only auxiliary quantities and do not possess any actual physical meaning, the influence

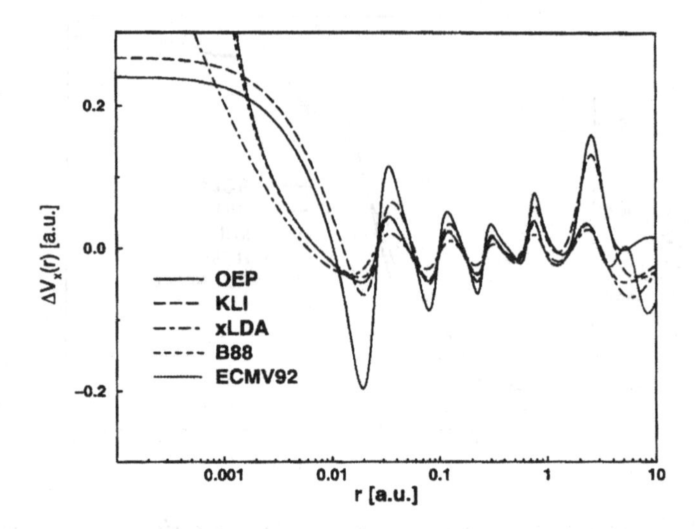

Figure 9: *Relativistic contribution to the exchange potential $\Delta V_{\mathrm{x}}(r)$, Eq. (278), for Hg from various self-consistent x-only calculations. Taken from [68].*

of relativity is seen very clearly in Table 23. As could be expected, the effects are strongest for the innermost orbitals whose energies are considerably lowered. However, due to the requirements of orthogonality this also largely influences the outer orbitals which, for example, causes the energy of the $6s\,1/2$ state of Hg to decrease by about 25%. Furthermore, due to the contraction of the s- and p-orbitals, the nucleus is screened more efficiently, leading to more weakly bound d- and f-levels as can be seen in Table 23. Finally, we clearly recognize the influence of spin-orbit coupling, removing the degeneracy in the angular momentum quantum number l when relativistic effects come into play.

4.3 Inclusion of correlation contributions for nonrelativistic systems

The inclusion of correlation effects into the OEP scheme is straightforward, as indicated by the subscripts xc in section 2. As we will demonstrate in this section, the correlation-energy functional developed by Colle and Salvetti [97, 98] is well suited for atomic systems.

Table 23: *Single particle energies ($-\epsilon_{nlj}$) for Hg from various self-consistent x-only calculations. All numbers in atomic units.*

	ROEP	NROEP	RKLI	RB88	RECMV92	xRLDA
1S1/2	3047.430	2756.925	3047.764	3047.706	3047.644	3044.410
2S1/2	540.056	461.647	539.821	539.782	539.788	539.250
2P1/2	518.061	444.015	517.964	518.080	518.049	517.746
2P3/2	446.682	444.015	446.543	446.616	446.599	446.399
3S1/2	128.272	108.762	128.187	128.054	128.058	127.905
3P1/2	118.350	100.430	118.333	118.238	118.233	118.148
3P3/2	102.537	100.430	102.507	102.407	102.404	102.346
3D3/2	86.201	84.914	86.193	86.093	86.095	86.060
3D5/2	82.807	84.914	82.802	82.699	82.700	82.668
4S1/2	28.427	23.522	28.360	28.112	28.114	28.046
4P1/2	24.161	19.895	24.125	23.896	23.895	23.854
4P3/2	20.363	19.895	20.312	20.072	20.072	20.030
4D3/2	13.411	13.222	13.386	13.166	13.166	13.146
4D5/2	12.700	13.222	12.674	12.452	12.453	12.432
4F5/2	3.756	4.250	3.755	3.561	3.561	3.559
4F7/2	3.602	4.250	3.601	3.407	3.407	3.404
5S1/2	4.403	3.501	4.407	4.289	4.289	4.286
5P1/2	3.012	2.344	3.015	2.896	2.896	2.896
5P3/2	2.363	2.344	2.353	2.224	2.225	2.218
5D3/2	0.505	0.538	0.496	0.366	0.367	0.363
5D5/2	0.439	0.538	0.429	0.299	0.300	0.296
6S1/2	0.329	0.262	0.332	0.223	0.222	0.222

This functional has been obtained via a Jastrow-type ansatz for the correlated total wavefunction through a series of approximations. It may be written in spinpolarized form as [99]

$$E_c^{\mathrm{CS}}[\{\varphi_{j\tau}\}] = -ab\int d^3r\,\gamma(\mathbf{r})\xi(\mathbf{r})\Bigg[\sum_\sigma \rho_\sigma(\mathbf{r})\sum_i |\nabla\varphi_{i\sigma}(\mathbf{r})|^2 - \frac{1}{4}|\nabla\rho(\mathbf{r})|^2 - \frac{1}{4}\sum_\sigma \rho_\sigma(\mathbf{r})\Delta\rho_\sigma(\mathbf{r}) + \frac{1}{4}\rho(\mathbf{r})\Delta\rho(\mathbf{r})\Bigg] - a\int d^3r\,\gamma(\mathbf{r})\frac{\rho(\mathbf{r})}{\eta(\mathbf{r})}, \tag{279}$$

where

$$\gamma(\mathbf{r}) = 4\,\frac{\rho_\uparrow(\mathbf{r})\rho_\downarrow(\mathbf{r})}{\rho(\mathbf{r})^2}, \tag{280}$$

$$\eta(\mathbf{r}) = 1 + d\rho(\mathbf{r})^{-\frac{1}{3}}, \tag{281}$$

$$\xi(\mathbf{r}) = \frac{\rho(\mathbf{r})^{-\frac{5}{3}} e^{-c\rho(\mathbf{r})^{-\frac{1}{3}}}}{\eta(\mathbf{r})}. \tag{282}$$

The constants a, b, c and d are given by

$$a = 0.04918, \qquad b = 0.132,$$
$$c = 0.2533, \qquad d = 0.349.$$

Performing the functional derivative with respect to the one-particle orbitals, one obtains for $u_{cj\sigma}(\mathbf{r})$ [34]

$$\begin{aligned}
u_{cj\sigma}(\mathbf{r}) = & -\frac{a}{\eta(\mathbf{r})}\left(\gamma(\mathbf{r}) + \rho(\mathbf{r})\frac{1}{\varphi^*_{j\sigma}(\mathbf{r})}\frac{\partial\gamma(\mathbf{r})}{\partial\varphi_{j\sigma}(\mathbf{r})}\right) - \frac{ad}{3}\gamma(\mathbf{r})\frac{\rho(\mathbf{r})^{-\frac{1}{3}}}{\eta(\mathbf{r})^2} \\
& - \frac{ab}{4}\frac{1}{\varphi^*_{j\sigma}(\mathbf{r})}\left[\frac{\partial}{\partial\varphi_{j\sigma}(\mathbf{r})}\Big(\gamma(\mathbf{r})\xi(\mathbf{r})\Big)\right]\left[4\sum_{\sigma'}\rho_{\sigma'}(\mathbf{r})\sum_i |\nabla\varphi_{i\sigma'}(\mathbf{r})|^2\right. \\
& \left. - \Big(\nabla\rho(\mathbf{r})\Big)^2 + \Big(\rho_\uparrow(\mathbf{r})\Delta\rho_\downarrow(\mathbf{r}) + \rho_\downarrow(\mathbf{r})\Delta\rho_\uparrow(\mathbf{r})\Big)\right] \\
& - \frac{ab}{2}\nabla\Big(\gamma(\mathbf{r})\xi(\mathbf{r})\Big)\Big(\nabla\rho(\mathbf{r}) + \nabla\rho_{\tilde\sigma}(\mathbf{r})\Big) \\
& - \frac{ab}{4}\Delta\Big(\gamma(\mathbf{r})\xi(\mathbf{r})\Big)\rho_{\tilde\sigma}(\mathbf{r}) \\
& - ab\,\gamma(\mathbf{r})\xi(\mathbf{r})\left[\sum_i |\nabla\varphi_{i\sigma}(\mathbf{r})|^2 + \frac{1}{2}\Big(\Delta\rho(\mathbf{r}) + \Delta\rho_{\tilde\sigma}(\mathbf{r})\Big)\right] \\
& + ab\,\frac{\nabla\varphi^*_{j\sigma}(\mathbf{r})}{\varphi^*_{j\sigma}(\mathbf{r})}\nabla\Big(\gamma(\mathbf{r})\xi(\mathbf{r})\rho_\sigma(\mathbf{r})\Big) \\
& + ab\,\frac{\Delta\varphi^*_{j\sigma}(\mathbf{r})}{\varphi^*_{j\sigma}(\mathbf{r})}\rho_\sigma(\mathbf{r})\gamma(\mathbf{r})\xi(\mathbf{r})
\end{aligned} \tag{283}$$

where $\tilde\sigma$ denotes the spin projection opposite to σ, i.e. $\tilde\sigma = \uparrow$ if $\sigma = \downarrow$ and vice versa. The KLI scheme obtained by combining the correlation-energy functional of Colle and Salvetti with the exact exchange-energy expression (27) will be denoted by KLICS.

For comparison, we have also performed calculations on atoms using the conventional Kohn-Sham method with two standard exchange-correlation energy functionals. The first one of these is the exchange-energy functional by Becke, c.f. Eq. (248), combined with the correlation-energy functional by Lee, Yang and Parr [99],

$$
\begin{aligned}
E_{\mathrm{c}}^{\mathrm{LYP}}[\rho_\uparrow,\rho_\downarrow] = & -ab\int d^3r\,\gamma(\mathbf{r})\xi(\mathbf{r})\Bigg[2^{\frac{5}{3}}C_F\left(\sum_\sigma \rho_\sigma(\mathbf{r})^{\frac{8}{3}}\right) \\
& +\frac{1}{4}\,|\nabla\rho(\mathbf{r})|^2 - \frac{1}{4}\rho(\mathbf{r})\Delta\rho(\mathbf{r}) \\
& -\frac{1}{36}\left(\sum_\sigma |\nabla\rho_\sigma(\mathbf{r})|^2\right) - \frac{1}{12}\left(\sum_\sigma \rho_\sigma(\mathbf{r})\Delta\rho_\sigma(\mathbf{r})\right)\Bigg] \\
& -a\int d^3r\,\gamma(\mathbf{r})\frac{\rho(\mathbf{r})}{\eta(\mathbf{r})}
\end{aligned}
\tag{284}
$$

where

$$C_F = \frac{3}{10}\left(3\pi^2\right)^{2/3} \tag{285}$$

and $\gamma(\mathbf{r})$, $\eta(\mathbf{r})$ and $\xi(\mathbf{r})$ are given by Eqs. (280), (281) and (282), respectively. In the following, this procedure is referred to as BLYP. The other is the generalized gradient approximation by Perdew and Wang [77] referred to as PW91. Its exchange part is given by Eq. (250) and the correlation part by

$$E_{\mathrm{c}}^{PW91}[\rho_\uparrow,\rho_\downarrow] = \int d^3r\rho(\mathbf{r})\left(\varepsilon_{\mathrm{c}}^{\mathrm{LDA}}[\rho_\uparrow,\rho_\downarrow](\mathbf{r}) + H_0(\mathbf{r}) + H_1(\mathbf{r})\right) \tag{286}$$

where

$$H_0(\mathbf{r}) = \phi^3(\mathbf{r})\frac{\beta^2}{2\alpha}\ln\left(1+\frac{2\alpha}{\beta}\frac{t^2(\mathbf{r})+A(\mathbf{r})t^4(\mathbf{r})}{1+A(\mathbf{r})t^2(\mathbf{r})+A^2(\mathbf{r})t^4(\mathbf{r})}\right), \tag{287}$$

$$H_1(\mathbf{r}) = 15.7559\,(C_c - 0.003521)\,\phi^3(\mathbf{r})t^2(\mathbf{r})\exp\left(-100\phi^4(\mathbf{r})\left(\frac{k_s^2(\mathbf{r})}{k_F^2(\mathbf{r})t^2(\mathbf{r})}\right)\right), \tag{288}$$

$$k_F(\mathbf{r}) = (3\pi\rho(\mathbf{r}))^{1/3}, \tag{289}$$

$$k_s(\mathbf{r}) = \left(\frac{4k_F}{\pi}\right)^{1/2}, \tag{290}$$

$$r_s(\mathbf{r}) = \left(\frac{3}{4\pi\rho(\mathbf{r})}\right)^{1/3}, \tag{291}$$

$$\zeta(\mathbf{r}) = \frac{\rho_\uparrow(\mathbf{r}) - \rho_\downarrow(\mathbf{r})}{\rho(\mathbf{r})}, \tag{292}$$

$$\phi(\mathbf{r}) = \frac{1}{2}\left[(1+\zeta(\mathbf{r}))^{2/3}(1-\zeta(\mathbf{r}))^{2/3}\right], \tag{293}$$

$$t^2(\mathbf{r}) = \frac{(\nabla\rho(\mathbf{r}))^2}{4\phi(\mathbf{r})k_s^2(\mathbf{r})\rho^2(\mathbf{r})}, \tag{294}$$

$$A(\mathbf{r}) = \frac{2\alpha}{\beta}\left[\exp\left(\frac{2\alpha\varepsilon_c^{LDA}[\rho_\uparrow,\rho_\downarrow](\mathbf{r})}{\phi^3(\mathbf{r})\beta^2}\right) - 1\right]^{-1}, \tag{295}$$

$$C_c(\mathbf{r}) = \frac{c_1 + c_2 r_s(\mathbf{r}) + c_3 r_s^2(\mathbf{r})}{1 + c_4 r_s(\mathbf{r}) + c_5 r_s^2(\mathbf{r}) + c_6 r_s^3(\mathbf{r})} - c_x \tag{296}$$

and $\alpha = 0.09, \beta = 15.7559 \times 4.235 \times 10^{-3}, c_x = -1.667212 \times 10^{-3}$, $c_1 = 2.568 \times 10^{-3}, c_2 = 2.3266 \times 10^{-2}, c_3 = 7.389 \times 10^{-6}, c_4 = 8.723$, $c_5 = 0.472, c_6 = 7.389 \times 10^{-2}$. $\varepsilon_c^{\mathrm{LDA}}$ is the correlation energy per particle of a homogeneous electron gas in the parameterization by Perdew and Wang [100] given by

$$\begin{aligned}\varepsilon_c^{\mathrm{LDA}}[\rho_\uparrow,\rho_\downarrow](\mathbf{r}) = {} & \varepsilon^U(\mathbf{r})\left(1 - f(\mathbf{r})\zeta^4(\mathbf{r})\right) + \varepsilon^P(\mathbf{r})f(\mathbf{r})\zeta^4(\mathbf{r}) \\ & + \alpha_c(\mathbf{r})f(\mathbf{r})\left(1 - f\zeta^4(\mathbf{r})\right)/d\end{aligned} \tag{297}$$

where

$$f(\mathbf{r}) = \frac{1}{\gamma}\left[(1+\zeta(\mathbf{r}))^{(4/3)} + (1-\zeta(\mathbf{r}))^{(4/3)} - 2\right], \tag{298}$$

$$\begin{aligned}\varepsilon^{\mathrm{U}}(\mathbf{r}) = & -2a_{01}\left(1 + a_{11}r_s(\mathbf{r})\right) \times \\ & \ln\left(1 + \frac{1}{2a_{01}\left(b_{11}x(\mathbf{r}) + b_{21}x^2(\mathbf{r}) + b_{31}x^3(\mathbf{r}) + b_{41}x^4(\mathbf{r})\right)}\right),\end{aligned} \tag{299}$$

$$\begin{aligned}\varepsilon^{\mathrm{P}}(\mathbf{r}) = & -2a_{02}\left(1 + a_{12}r_s(\mathbf{r})\right) \times \\ & \ln\left(1 + \frac{1}{2a_{02}\left(b_{12}x(\mathbf{r}) + b_{22}x^2(\mathbf{r}) + b_{32}x^3(\mathbf{r}) + b_{42}x^4(\mathbf{r})\right)}\right),\end{aligned} \tag{300}$$

$$\begin{aligned}\alpha_c(\mathbf{r}) = & -2a_{03}\left(1 + a_{13}r_s(\mathbf{r})\right) \times \\ & \ln\left(1 + \frac{1}{2a_{03}\left(b_{13}x(\mathbf{r}) + b_{23}x^2(\mathbf{r}) + b_{33}x^3(\mathbf{r}) + b_{43}x^4(\mathbf{r})\right)}\right),\end{aligned} \tag{301}$$

and

$$x(\mathbf{r}) = r_s^{1/2}(\mathbf{r}). \tag{302}$$

$r_s(\mathbf{r})$ and $\zeta(\mathbf{r})$ are given by Eqs. (291) and (292), respectively and the constants are $\gamma = 0.5198421$, $d = 1.709921$, $a_{01} = 0.031097$, $a_{02} = 0.01554535$, $a_{03} = 0.0168869$, $a_{11} = 0.21370$, $a_{12} = 0.20548$, $a_{13} = 0.11125$, $b_{11} = 7.5957$, $b_{12} = 14.1189$, $b_{13} = 10.357$, $b_{21} = 3.5876$, $b_{22} = 6.1977$, $b_{23} = 3.6231$, $b_{31} = 1.6382$, $b_{32} = 3.3662$, $b_{33} = 0.88026$, $b_{41} = 0.49294$, $b_{42} = 0.62517$, $b_{43} = 0.49671$.

As far as available, we will also list SIC results using the exchange-energy functional given by Eq. (253) and the correlation energy

$$E_{\mathrm{c}}^{\mathrm{LDA}-SIC}[\{\varphi_{i\sigma}\}] = \int d^3r\rho(\mathbf{r})\varepsilon_{\mathrm{c}}^{\mathrm{LDA}}[\rho_\uparrow, \rho_\downarrow](\mathbf{r}) - \sum_{\sigma=\uparrow,\downarrow}\sum_{i=1}^{N_\sigma}\int d^3r\rho(\mathbf{r})\varepsilon_{\mathrm{c}}^{\mathrm{LDA}}[|\varphi_{i\sigma}|^2, 0](\mathbf{r}) \tag{303}$$

where $\varepsilon_{\mathrm{c}}^{\mathrm{LDA}}$ is given by Eq. (297). If the corresponding potential is calculated using the KLI approximation (76), we will denote it by LDA-SICKLI.

4.3.1 Atomic systems

Table 24 shows the total absolute ground-state energies of the first-row atoms. For these systems, there exist accurate estimates of the exact non-relativistic values obtained from experimental ionization energies and improved *ab initio* calculations by Davidson et al. [102]. It is evident from this table, that the density functional methods perform quite well. The mean absolute errors, denoted by $\bar{\Delta}$ and given in the last row of Table 24, clearly show that the KLICS approach is significantly more accurate than the LDA-SICKLI and the conventional Kohn-Sham methods and nearly as accurate as recent CI based quantum chemical results by Montgomery et al. [101]. The situation is similar for second-row atoms, as can be seen from Table 25. As relativistic effects for these atoms are more important and experiments increasingly difficult, the comparison of the calculated values with the Lamb-shift corrected experimental ones from [2] has to be done cautiously and is by no means as rigorous as for first-row atoms. Nevertheless, the values calculated with the KLICS approach mirror these experimental values more closely than the other approximations. Despite the somewhat larger errors the SIC results constitute a significant improvement over the conventional LDA results which are not shown here. For the latter the mean absolute deviation from the exact data is 0.3813 Hartrees for the first-row atoms and 1.225 Hartrees for the second-row atoms.

For further analysis, we list, in Tables 26 and 27, the values of E_x and E_c separately. Ignoring the LDA-SICKLI results for a moment, the data show two main features: First, the results for E_x are lowest for the KLICS and highest for the PW91 method, while the BLYP-values lie somewhere in between. And second,

Table 24: *Total absolute ground-state energies for first-row atoms from various self-consistent calculations. Quantum chemistry (QC) values from [101].* $\bar{\Delta}$ *denotes the mean absolute deviation from the exact nonrelativistic values [102]. All numbers in atomic units.*

	KLICS	LDA-SICKLI	BLYP	PW91	QC	exact
He	2.9033	2.9198	2.9071	2.9000	2.9049	2.9037
Li	7.4829	7.5058	7.4827	7.4742	7.4743	7.4781
Be	14.6651	14.6953	14.6615	14.6479	14.6657	14.6674
B	24.6564	24.7022	24.6458	24.6299	24.6515	24.6539
C	37.8490	37.9335	37.8430	37.8265	37.8421	37.8450
N	54.5905	54.7295	54.5932	54.5787	54.5854	54.5893
O	75.0717	75.2590	75.0786	75.0543	75.0613	75.067
F	99.7302	99.9995	99.7581	99.7316	99.7268	99.734
Ne	128.9202	129.2868	128.9730	128.9466	128.9277	128.939
$\bar{\Delta}$	0.0047	0.1282	0.0108	0.0114	0.0045	

Table 25: *Total absolute ground-state energies for second-row atoms from various self-consistent calculations.* $\bar{\Delta}$ *denotes the mean absolute deviation from Lamb-shift corrected experimental values, taken from [2]. All numbers in atomic units.*

	KLICS	LDA-SICKLI	BLYP	PW91	experiment
Na	162.256	162.672	162.293	162.265	162.257
Mg	200.062	200.536	200.093	200.060	200.059
Al	242.362	242.891	242.380	242.350	242.356
Si	289.375	289.969	289.388	289.363	289.374
P	341.272	341.930	341.278	341.261	341.272
S	398.128	398.852	398.128	398.107	398.139
Cl	460.164	460.967	460.165	460.147	460.196
Ar	527.553	528.432	527.551	527.539	527.604
$\bar{\Delta}$	0.013	0.624	0.026	0.023	

for E_c, this trend is reversed, as now the KLICS results are highest and the ones from BLYP and PW91 are lower in nearly all cases. In Table 28 we show results of various *x-only* calculations performed with only the exchange-energy parts of the respective functionals. For the spherical atoms listed, there exist exact x-only OEP values [37, 52]. It is evident, that the KLI-approximation gives values much

Table 26: *Absolute exchange energies from various approximations. All values in atomic units.*

	KLICS	LDA-SICKLI	BLYP	PW91
He	1.028	1.031	1.018	1.009
Li	1.784	1.781	1.771	1.758
Be	2.674	2.665	2.658	2.644
B	3.760	3.758	3.727	3.711
C	5.064	5.099	5.028	5.010
N	6.610	6.701	6.578	6.558
O	8.200	8.327	8.154	8.136
F	10.025	10.228	9.989	9.972
Ne	12.110	12.416	12.099	12.082
Na	14.017	14.371	14.006	13.985
Mg	15.997	16.401	15.986	15.967
Al	18.081	18.523	18.053	18.033
Si	20.295	20.787	20.260	20.238
P	22.649	23.196	22.609	22.587
S	25.021	25.618	24.967	24.944
Cl	27.530	28.195	27.476	27.453
Ar	30.192	30.928	30.139	30.116

Table 27: *Absolute correlation energies from various approximations. All values in atomic units.*

	KLICS	LDA-SICKLI	BLYP	PW91
He	0.0416	0.0582	0.0437	0.0450
Li	0.0509	0.0715	0.0541	0.0571
Be	0.0934	0.1169	0.0954	0.0942
B	0.1289	0.1532	0.1287	0.1270
C	0.1608	0.1886	0.1614	0.1614
N	0.1879	0.2232	0.1925	0.1968
O	0.2605	0.2967	0.2640	0.2587
F	0.3218	0.3645	0.3256	0.3193
Ne	0.3757	0.4283	0.3831	0.3784
Na	0.4005	0.4555	0.4097	0.4040
Mg	0.4523	0.5089	0.4611	0.4486
Al	0.4905	0.5502	0.4979	0.4891
Si	0.5265	0.5910	0.5334	0.5322
P	0.5594	0.6314	0.5676	0.5762
S	0.6287	0.7037	0.6358	0.6413
Cl	0.6890	0.7700	0.6955	0.7055
Ar	0.7435	0.8330	0.7515	0.7687

Table 28: *Total absolute exchange energies of spherical first and second-row atoms for various self-consistent x-only calculations. The exact OEP data are from [37, 52]. All values in atomic units.*

	KLI	xLDA-SICKLI	B88	PW91	OEP
He	1.026	1.026	1.016	1.005	1.026
Li	1.781	1.777	1.768	1.754	1.781
Be	2.667	2.658	2.652	2.638	2.666
N	6.603	6.691	6.569	6.547	6.604
Ne	12.099	12.398	12.086	12.061	12.105
Na	14.006	14.355	13.993	13.968	14.013
Mg	15.983	16.383	15.972	15.950	15.988
P	22.633	23.177	22.593	22.565	22.634
Ar	30.174	30.905	30.122	30.089	30.175

Table 29: *Ionization potentials calculated from ground-state-energy differences of first-row atoms. QC values are from [101].* $\bar{\Delta}$ *denotes the mean absolute deviation from the experimental values, taken from [103]. All values in atomic units.*

	KLICS	LDA-SICKLI	BLYP	PW91	QC	exp
He	0.903	0.920	0.912		0.905	0.903
Li	0.203	0.200	0.203	0.207	0.198	0.198
Be	0.330	0.335	0.330	0.333	0.344	0.343
B	0.314	0.327	0.309	0.314	0.304	0.305
C	0.414	0.445	0.425	0.432	0.413	0.414
N	0.527	0.565	0.542	0.551	0.534	0.534
O	0.495	0.520	0.508	0.505	0.499	0.500
F	0.621	0.673	0.656	0.660	0.639	0.640
Ne	0.767	0.825	0.808	0.812	0.792	0.792
$\bar{\Delta}$	0.009	0.022	0.010	0.014	0.001	

closer to the exact ones than the generalized gradient approximations. From this and from Tables 26 and 27 one may conclude that an error cancellation between exchange and correlation energies occurs in the BLYP and PW91 schemes which leads to rather good total energies. Exchange and correlation energies *separately*, however, are reproduced less accurately in the BLYP and PW91 approaches. In the KLICS scheme, both exchange and correlation energies are of high quality. As far as the LDA-SICKLI approach is concerned, both exchange and correlation

Table 30: *Ionization potentials calculated from ground-state-energy differences of second-row atoms. $\bar{\Delta}$ denotes the mean absolute deviation from the experimental values, taken from [103]. All values in atomic units.*

	KLICS	LDA-SICKLI	BLYP	PW91	experiment
Na	0.191	0.195	0.197	0.198	0.189
Mg	0.275	0.284	0.280	0.281	0.281
Al	0.218	0.222	0.212	0.221	0.220
Si	0.294	0.307	0.294	0.305	0.300
P	0.379	0.392	0.376	0.389	0.385
S	0.380	0.394	0.379	0.379	0.381
Cl	0.471	0.495	0.476	0.482	0.477
Ar	0.575	0.595	0.576	0.583	0.579
$\bar{\Delta}$	0.004	0.009	0.005	0.004	

energies are too low. Therefore, no error cancellation occurs leading to rather large errors in the total energies.

Limitations of the DFT approaches become evident for ionization potentials and electron affinities. In Tables 29 and 30 we show ionization potentials calculated from ground-state-energy differences and QC values from [101] as well as experimental ones from [103]. The performance of the KLICS, B88 and PW91 methods is similar, while the LDA-SICKLI scheme leads to results showing a mean absolute deviation from the experimental values which is roughly twice as large as for any of the other DFT methods. On the whole, QC calculations lead to clearly better results. Somewhat surprisingly, the DFT methods work better for the second-row than for the first-row atoms.

In *exact* DFT, the highest occupied orbital energy of the neutral atom is identical with the ionization potential, while for negative ions the highest occupied energy level coincides with the electron affinity of the neutral atom [47]. How well ionization potentials and electron affinities are reproduced by the highest occupied energy eigenvalues resulting from an *approximate* xc functional is therefore a measure of the quality of the xc potential. Table 31 shows the ionization energies obtained from the highest occupied single-particle-energy eigenvalue of the neutral atoms. For the BLYP and PW91 approaches the resulting values are much worse than the ones in Tables 29 and 30. The deviation from experiment is around 100 percent for all atoms. This is due to the incorrect asymptotic behavior of the BLYP and PW91 potentials. The KLICS and LDA-SICKLI potentials, on the other hand, have the correct $-1/r$ behavior for large r and the resulting highest occupied orbital energies are much closer to the experimental ionization potentials. Nevertheless, the KLICS values obtained from ground-state-energy differences (see Tables 29 and 30) are considerably more accurate.

Table 31: *Ionization potentials from the highest occupied orbital energy of neutral atoms. $\bar{\Delta}$ denotes the mean absolute deviation from the experimental (exp) values, taken from [103]. All values in atomic units.*

	KLICS	LDA-SICKLI	BLYP	PW91	exp
He	0.945	0.948	0.585	0.583	0.903
Li	0.200	0.197	0.111	0.119	0.198
Be	0.329	0.329	0.201	0.207	0.343
B	0.328	0.306	0.143	0.149	0.305
C	0.448	0.427	0.218	0.226	0.414
N	0.579	0.550	0.297	0.308	0.534
O	0.559	0.527	0.266	0.267	0.500
F	0.714	0.686	0.376	0.379	0.640
Ne	0.884	0.843	0.491	0.494	0.792
Na	0.189	0.190	0.106	0.113	0.189
Mg	0.273	0.275	0.168	0.174	0.281
Al	0.222	0.205	0.102	0.112	0.220
Si	0.306	0.287	0.160	0.171	0.300
P	0.399	0.371	0.219	0.233	0.385
S	0.404	0.383	0.219	0.222	0.381
Cl	0.506	0.481	0.295	0.301	0.477
Ar	0.619	0.580	0.373	0.380	0.579
$\bar{\Delta}$	0.030	0.016	0.183	0.177	

Table 32: *Self-consistent electron affinities for first-row atoms. QC values are from [101] and experimental (exp) values from [103]. $\bar{\delta}$ denotes the mean value of $|A_{DFT} - A_{exp}|/|A_{exp}|$. All values in atomic units.*

	KLICS^a	KLICS^b	LDA-SICKLI^a	LDA-SICKLI^b	QC	exp
Li	0.016	0.024	0.021	0.025	0.023	0.023
B	−0.002	0.033	0.025	0.028	0.008	0.010
C	0.028	0.083	0.062	0.073	0.045	0.046
O	0.017	0.110	0.065	0.100	0.052	0.054
F	0.082	0.208	0.138	0.189	0.125	0.125
$\bar{\delta}$	50.5	97.0	44.8	76.8	5.2	

^{a} from ground-state-energy differences.
^{b} from the highest occupied orbital energies of the negative ions.

Table 33: *Self-consistent electron affinities for second-row atoms. Experimental (exp) values are from [103].* $\bar{\delta}$ *denotes the mean value of* $|A_{\mathrm{DFT}} - A_{\mathrm{exp}}|/|A_{\mathrm{exp}}|$. *All values in atomic units.*

	KLICS[a]	KLICS[b]	LDA-SICKLI[a]	LDA-SICKLI[b]	exp
Na	0.015	0.022	0.021	0.024	0.020
Al	0.007	0.024	0.023	0.020	0.016
Si	0.040	0.065	0.058	0.054	0.051
P	0.022	0.048	0.038	0.041	0.027
S	0.065	0.106	0.092	0.095	0.076
Cl	0.122	0.174	0.147	0.151	0.133
$\bar{\delta}$	24.0	39.3	22.5	23.5	

[a] from ground-state-energy differences
[b] from the highest occupied orbital energies of the negative ions.

For electron affinities, the situation is much worse, as may be seen from Tables 32 and 33. First of all, because of the wrong asymptotic behavior of the xc potential for large r, there is no convergence for negative ions within the self-consistent BLYP and PW91 schemes. This is not the case for the KLICS and LDA-SICKLI approaches. However, the resulting electron affinities obtained either from ground-state-energy differences or from the highest orbital energies of negative ions are far less accurate than the ionization energies. The KLICS method even gives the wrong sign for the Boron atom if the electron affinity is calculated from the ground-state-energy differences. On average, the values obtained from ground-state-energy differences are more accurate than the results obtained from the highest occupied orbital energies. Comparing the KLICS with the LDA-SICKLI results, the latter are slightly better on average. The fact that the KLICS and LDA-SICKLI approaches allow for a fully self-consistent calculation of electron affinities is encouraging, but the poor accuracy of the results clearly shows that the xc potentials need further improvement. Here, quantum-chemical approaches are definitely superior.

Two-electron systems The CS and LDA-SIC correlation functionals may be studied more thoroughly in two-electron atoms. There are two reasons for this: First of all, as pointed out above, the solution of the full OEP integral equation for these systems is identical to the one obtained from the KLI-scheme. Furthermore, the exact exchange-energy functional (27) is identical with $E_x^{\text{LDA-SIC}}$ for spin-saturated two-electron systems, so that the only error made is due to the approximation for E_c. Secondly there exist practically exact solutions [7] of the two-particle Schrödinger equation. Hence the various DFT-related quantities of interest can be compared with exact results.

Table 34: *Total absolute ground-state energies for the Helium-isoelectronic series from various self-consistent calculations.* $\bar{\Delta}$ *denotes the mean absolute deviation from the exact values from [102]. All numbers in atomic units.*

	KLICS	LDA-SICKLI	BLYP	PW91	exact
H^-	0.5189	0.5263			0.5278
He	2.9033	2.9198	2.9071	2.9000	2.9037
Li^+	7.2803	7.3057	7.2794	7.2676	7.2799
Be^{2+}	13.6556	13.6886	13.6500	13.6340	13.6556
B^{3+}	22.0301	22.0698	22.0200	21.9996	22.0310
C^{4+}	32.4045	32.4499	32.3896	32.3649	32.4062
N^{5+}	44.7788	44.8293	44.7592	44.7299	44.7814
O^{6+}	59.1531	59.2081	59.1286	59.0948	59.1566
F^{7+}	75.5274	75.5864	75.4981	75.4595	75.5317
Ne^{8+}	93.9017	93.9644	93.8675	93.8241	93.9068
Na^{9+}	114.2761	114.3422	114.2369	114.1886	114.2819
Mg^{10+}	136.6505	136.7197	136.6064	136.5531	136.6569
Al^{11+}	161.0250	161.0970	160.9758	160.9175	161.0320
Si^{12+}	187.3995	187.4742	187.3453	187.2819	187.4070
P^{13+}	215.7740	215.8512	215.7147	215.6462	215.7821
S^{14+}	246.1485	246.2281	246.0842	246.0105	246.1571
Cl^{15+}	278.5231	278.6049	278.4536	278.3748	278.5322
Ar^{16+}	312.8977	312.9816	312.8231	312.7390	312.9072
K^{17+}	349.2723	349.3582	349.1926	349.1032	349.2822
Ca^{18+}	387.6470	387.7347	387.5620	387.4674	387.6572
$\bar{\Delta}$	0.0053	0.0533	0.0450	0.0943	

In Table 34 we show total absolute ground-state energies of the atoms isoelectronic with helium. The exact nonrelativistic results in the last column are taken from [102]. Note that there is no convergence for negative ions in the conventional Kohn Sham method. In Figure 10, we have plotted the errors $E_{tot}^{DFT} - E_{tot}^{exact}$ corresponding to the numbers in Table 34. It is obvious that the KLICS scheme gives superior results, the mean absolute error $\bar{\Delta}$ being smaller by an order of magnitude compared to the LDA-SICKLI and the two conventional Kohn-Sham approaches.

In Table 35 we have listed the highest occupied orbital energies for the two-electron series as obtained from various *self-consistent* calculations. Comparing the results with the exact values it is obvious that the KLICS and LDA-SICKLI schemes perform much better than the conventional Kohn-Sham approaches. The difference is less pronounced for the highly charged ions as the nuclear potential

Table 35: *Absolute highest occupied orbital energies from various self consistent calculations. Exact values calculated from [102]. All values in atomic units.*

	KLI x-only	KLICS xc	LDA-SICKLI xc	BLYP xc	PW91	exact
He	0.9180	0.9446	0.9481	0.5849	0.5833	0.9037
Li^{+}	2.7924	2.8227	2.8293	2.2312	2.2269	2.7799
Be^{2+}	5.6671	5.6992	5.6556	4.8760	4.8701	5.6556
B^{3+}	9.5420	9.5751	9.5871	8.5201	8.5129	9.5310
C^{4+}	14.4169	14.4507	14.4648	13.1638	13.1554	14.4062
N^{5+}	20.2918	20.3261	20.3421	18.8072	18.7978	20.2814
O^{6+}	27.1668	27.2014	27.2191	25.4504	25.4401	27.1566
F^{7+}	35.0418	35.0766	35.0959	33.0935	33.0823	35.0317
Ne^{8+}	43.9167	43.9517	43.9725	41.7366	41.7245	43.9068
Na^{9+}	53.7917	53.8269	53.8489	51.3796	51.3666	53.7819
Mg^{10+}	64.6667	64.7020	64.7252	62.0225	62.0086	64.6569
Al^{11+}	76.5417	76.5770	76.6015	73.6654	73.6506	76.5320
Si^{12+}	89.4167	89.4521	89.4776	86.3083	86.2926	89.4071
P^{13+}	103.2917	103.3272	103.3536	99.9511	99.9345	103.2821
S^{14+}	118.1666	118.2022	118.2296	114.5939	114.5764	118.1571
Cl^{15+}	134.0416	134.0773	134.1055	130.2367	130.2183	134.0322
Ar^{16+}	150.9166	150.9523	150.9814	146.8795	146.8602	150.9072
K^{17+}	168.7916	168.8273	168.8572	164.5223	164.5021	168.7822
Ca^{18+}	187.6666	187.7024	187.7330	183.1650	183.1439	187.6572

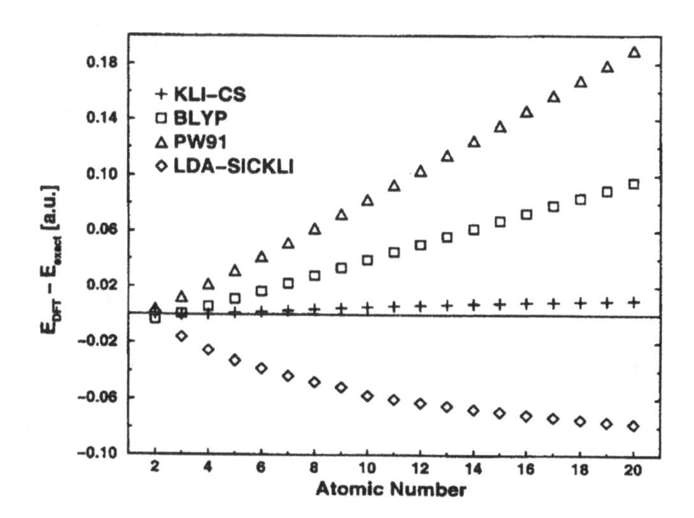

Figure 10: *Energy differences corresponding to Table 34.*

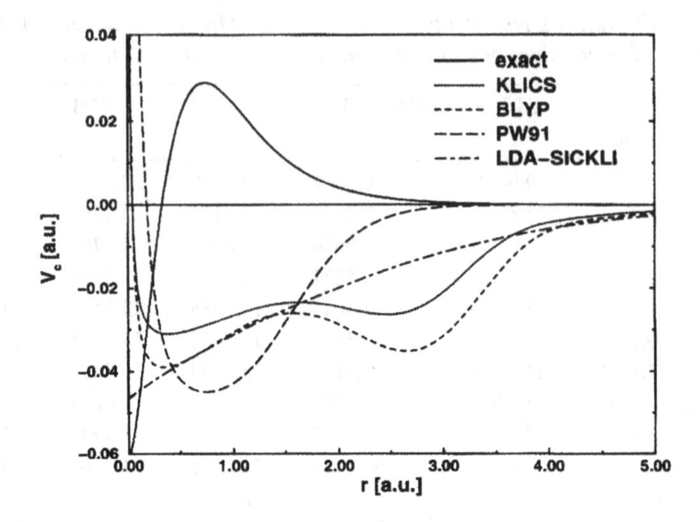

Figure 11: *Comparison of the exact and self consistently calculated correlation potentials of helium. Exact potential from [7].*

becomes more and more dominant as Z increases. A glance at the second column, in which we give the corresponding values from an *x-only* KLI calculation employing the exact functional (27), shows that the superior quality is due to the inclusion of the exact exchange in the KLI scheme leading to the correct asymptotic behavior of the KS potential. In fact, adding the CS correlation potential worsens the results, as may be seen by comparing the second and third columns: The correlation contribution lowers the already too small values from the x-only calculations for the highest occupied orbital energy even more. This indicates that the CS correlation potential has the wrong sign in the physically relevant regions of space. As the x-only LDA-SICKLI is identical with the x-only KLI scheme for these systems, the same conclusion can be drawn for the LDA-SICKLI correlation potential.

These findings are confirmed by Figure 11 where we plot the exact [7] and various self-consistent correlation potentials. It is evident that the approximate potentials show very large deviations from the exact one. In the region where most of the charge density is located, the approximate correlation potentials have the wrong sign. Furthermore, the potentials obtained with the CS functional within the KLICS scheme and the two conventional Kohn-Sham approximations exhibit spurious divergences at the origin. These may be traced back to gradients of the density and of the one-particle orbitals occurring in these correlation energy functionals. As the LDA-SICKLI functional does not contain any density gradients, these divergences are not found for the LDA-SICKLI potential. The need for further improvement of the correlation energy functionals in these respects is obvious.

Table 36: *Total absolute ground-state energies for the Beryllium-isoelectronic series from various self-consistent calculations.* $\bar{\Delta}$ *denotes the mean absolute deviation from the exact values from [91] and* $\bar{\delta}$ *denotes* $|E_{tot}^{DFT} - E_{tot}^{exact}|/|E_{tot}^{exact}|$. *All numbers in atomic units.*

	KLICS	BLYP	PW91	exact
Be	14.6651	14.6615	14.6479	14.6674
B^{+}	24.3427	24.3366	24.3160	24.3489
C^{2+}	36.5224	36.5143	36.4881	36.5349
N^{3+}	51.2025	51.1927	51.1618	51.2228
O^{4+}	68.3825	68.3713	68.3362	68.4117
F^{5+}	88.0624	88.0499	88.0110	88.1011
Ne^{6+}	110.2420	110.2285	110.1859	110.2909
Na^{7+}	134.9216	134.9071	134.8610	134.9809
Mg^{8+}	162.1010	162.0857	162.0361	162.1710
Al^{9+}	191.7803	191.7642	191.7113	191.8613
Si^{10+}	223.9595	223.9427	223.8864	224.0516
P^{11+}	258.6387	258.6212	258.5616	258.7420
S^{12+}	295.8178	295.7996	295.7367	295.9324
Cl^{13+}	335.4968	335.4781	335.4119	335.6229
Ar^{14+}	377.6758	377.6566	377.5870	377.8134
K^{15+}	422.3548	422.3350	422.2621	422.5040
Ca^{16+}	469.5338	469.5134	469.4372	469.6946
Sc^{17+}	519.2127	519.1919	519.1122	519.3851
Ti^{18+}	571.3917	571.3703	571.2873	571.5757
V^{19+}	626.0708	626.0487	625.9623	626.2663
Cr^{20+}	683.2497	683.2271	683.1373	683.4570
Mn^{21+}	742.9286	742.9056	742.8123	743.1476
Fe^{22+}	805.1074	805.0840	804.9873	805.3382
Co^{23+}	869.7863	869.7624	869.6623	870.0289
Ni^{24+}	936.9652	936.9408	936.8373	937.2195
$\bar{\Delta}$	0.1183	0.1352	0.1973	
$\bar{\delta}$	0.0351	0.0443	0.0755	

Beryllium and Neon isoelectronic series For further analysis we have calculated the total ground state energies of positive ions isoelectronic with Beryllium (shown in Table 36) and Neon (shown in Table 37). Again, we compare various self-consistent DFT methods with exact data from Ref. [91] and plot the errors in Figures 12 and 13, respectively. The data for both series show the same trends: The overall best results are obtained with the KLICS scheme, where the absolute total and relative mean deviations from the exact values are smallest. The BLYP

Table 37: *Total absolute ground-state energies for the Neon-isoelectronic series from various self-consistent calculations.* $\bar{\Delta}$ *denotes the mean absolute deviation from the exact values from [91] and* $\bar{\delta}$ *denotes* $|E_{tot}^{DFT} - E_{tot}^{exact}|/|E_{tot}^{exact}|$. *All numbers in atomic units.*

	KLICS	BLYP	PW91	exact
Ne	128.9202	128.9730	128.9466	128.9376
Na^{+}	162.0645	162.0956	162.0668	162.0659
Mg^{2+}	199.2291	199.2448	199.2136	199.2204
Al^{3+}	240.4071	240.4102	240.3768	240.3914
Si^{4+}	285.5945	285.5867	285.5509	285.5738
P^{5+}	334.7888	334.7712	334.7331	334.7642
S^{6+}	387.9885	387.9616	387.9212	387.9608
Cl^{7+}	445.1922	445.1567	445.1138	445.1622
Ar^{8+}	506.3993	506.3554	506.3101	506.3673
K^{9+}	571.6091	571.5570	571.5092	571.5754
Ca^{10+}	640.8211	640.7610	640.7107	640.7861
Sc^{11+}	714.0350	713.9671	713.9141	713.9988
Ti^{12+}	791.2504	791.1748	791.1191	791.2132
V^{13+}	872.4671	872.3839	872.3255	872.4291
Cr^{14+}	957.6850	957.5942	957.5331	957.6463
Mn^{15+}	1046.9039	1046.8056	1046.7417	1046.8646
Fe^{16+}	1140.1237	1140.0179	1139.9511	1140.0838
Co^{17+}	1237.3440	1237.2309	1237.1613	1237.3039
Ni^{18+}	1338.5652	1338.4447	1338.3722	1338.5247
$\bar{\Delta}$	0.0293	0.0334	0.0694	
$\bar{\delta}$	0.0054	0.0067	0.0097	

scheme is only slightly worse, but the PW91 functional gives errors almost twice as large as the other DFT approaches. From the plots in Figures 12 and 13 it is obvious that these statements hold for most ions individually.

There are two other trends worth noting: (a) While the absolute errors rise within the isoelectronic series as the atomic number increases, the percentage errors remain almost constant. (b) The mean absolute error is smaller by almost an order of magnitude for the ten-electron series compared to the four-electron series.

The ionization potentials from the various approaches as calculated from the highest occupied orbital energies are shown in Tables 38 and 39 for the four- and ten-electron series, respectively. The exact nonrelativistic values have been calculated from the data given in [91]. Owing to the correct asymptotic behavior of the KLICS potential it comes as no surprise that the KLICS data are superior to the ones obtained from the conventional Kohn-Sham approaches.

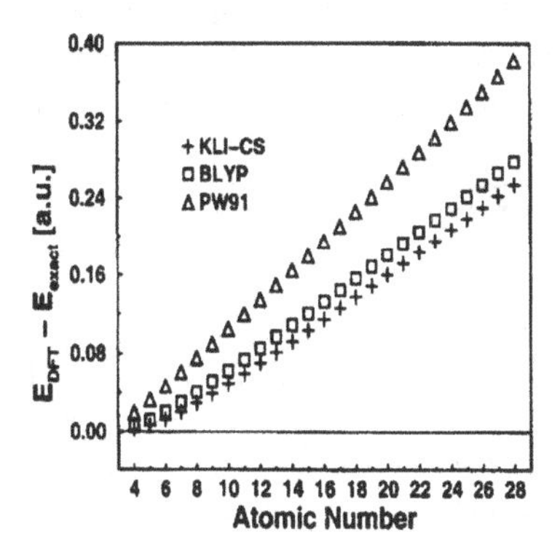

Figure 12: *Energy differences corresponding to Table 36.*

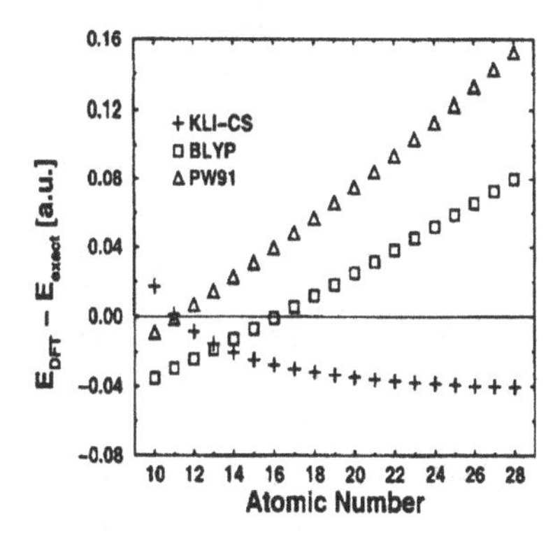

Figure 13: *Energy differences corresponding to Table 37.*

The effect of the correlation potential within the OEP scheme is — like in the two-electron case — a lowering of the energy eigenvalue of the highest occupied orbital. This is seen by comparing the second and third columns showing the OEP results in x-only approximation and with inclusion of CS correlation in the KLI scheme, respectively. In contrast to the Helium and Neon isoelectronic series, this effect improves the quality of the results in the Beryllium isoelectronic series. We mention that the ionization potentials are in much better agreement with the exact results when calculated as ground-state energy differences.

4.3.2 Diatomic molecules

To examine the effect of correlation contributions, we have implemented the CS correlation-energy functional (279) in our fully numerical basis-set-free code for

Table 38: *Ionization potentials from highest occupied Kohn-Sham orbital energies for the Beryllium-isoelectronic series from various self-consistent calculations. Exact nonrelativistic values calculated from [91]. All numbers in atomic units.*

	KLI x-only	KLICS xc	BLYP xc	PW91 xc	exact
Be	0.3089	0.3294	0.2009	0.2072	0.3426
B^{+}	0.8732	0.8992	0.7129	0.7185	0.9243
C^{2+}	1.6933	1.7226	1.4804	1.4856	1.7594
N^{3+}	2.7659	2.7975	2.5000	2.5049	2.8459
O^{4+}	4.0898	4.1231	3.7706	3.7754	4.1832
F^{5+}	5.6644	5.6991	5.2918	5.2964	5.7708
Ne^{6+}	7.4896	7.5253	7.0633	7.0678	7.6087
Na^{7+}	9.5652	9.6017	9.0850	9.0894	9.6967
Mg^{8+}	11.8910	11.9281	11.3569	11.3611	12.0348
Al^{9+}	14.4669	14.5047	13.8788	13.8829	14.6230
Si^{10+}	17.2930	17.3312	16.6508	16.6548	17.4613
P^{11+}	20.3692	20.4078	19.6729	19.6768	20.5496
S^{12+}	23.6955	23.7345	22.9450	22.9487	23.8880
Cl^{13+}	27.2718	27.3111	26.4671	26.4707	27.4763
Ar^{14+}	31.0982	31.1378	30.2393	30.2427	31.3147
K^{15+}	35.1747	35.2145	34.2615	34.2647	35.4031
Ca^{16+}	39.5012	39.5412	38.5336	38.5367	39.7416
Sc^{17+}	44.0777	44.1179	43.0558	43.0587	44.3300
Ti^{18+}	48.9043	48.9447	47.8280	47.8307	49.1684
V^{19+}	53.9808	54.0214	52.8502	52.8527	54.2569
Cr^{20+}	59.3074	59.3481	58.1224	58.1247	59.5954
Mn^{21+}	64.8840	64.9249	63.6447	63.6467	65.1838
Fe^{22+}	70.7107	70.7516	69.4169	69.4187	71.0223
Co^{23+}	76.7873	76.8284	75.4391	75.4408	77.1108
Ni^{24+}	83.1140	83.1551	81.7113	81.7128	83.4493

diatomic molecules [104]. For comparison, we have also performed calculations employing the conventional LDA [100] and the PW91 functional for E_{xc}.

To demonstrate the accuracy of our implementation, we compare results for the Neon atom obtained with our molecular code with the ones from our one-dimensional atomic structure program in Table 40. The deviation from the exact results obtained with the atomic code is a few μHartrees at most.

Ground-state properties of the closed-shell-first-row dimers and hydrides were calculated using the approximations mentioned above in a fully self-consistent fashion. As our program uses no basis functions, the results are free of

Table 39: *Ionization potentials from highest occupied Kohn-Sham orbital energies for the Neon-isoelectronic series from various self-consistent calculations. Exact nonrelativistic values calculated from [91]. All numbers in atomic units.*

	KLI x-only	KLICS xc	BLYP xc	PW91 xc	exact
Ne	0.8494	0.8841	0.4914	0.4942	0.7945
Na^{+}	1.7959	1.8340	1.3377	1.3416	1.7410
Mg^{2+}	3.0047	3.0450	2.4531	2.4579	2.9499
Al^{3+}	4.4706	4.5125	3.8285	3.8339	4.4161
Si^{4+}	6.1912	6.2343	5.4601	5.4661	6.1371
P^{5+}	8.1651	8.2091	7.3458	7.3524	8.1112
S^{6+}	10.3914	10.4362	9.4846	9.4917	10.3378
Cl^{7+}	12.8696	12.9150	11.8757	11.8832	12.8162
Ar^{8+}	15.5992	15.6451	14.5185	14.5264	15.5460
K^{9+}	18.5800	18.6263	17.4126	17.4209	18.5270
Ca^{10+}	21.8118	21.8585	20.5579	20.5665	21.7589
Sc^{11+}	25.2943	25.3413	23.9541	23.9630	25.2416
Ti^{12+}	29.0274	29.0747	27.6010	27.6102	28.9748
V^{13+}	33.0112	33.0587	31.4986	31.5080	32.9586
Cr^{14+}	37.2453	37.2931	35.6467	35.6563	37.1929
Mn^{15+}	41.7299	41.7779	40.0453	40.0551	41.6776
Fe^{16+}	46.4649	46.5130	44.6943	44.7042	46.4126
Co^{17+}	51.4501	51.4984	49.5936	49.6037	51.3979
Ni^{18+}	56.6856	56.7341	54.7432	54.7535	56.6335

Table 40: *Results for the Neon atom using various DFT approaches. 1D denotes the exact values obtained with our one-dimensional code, 2D the results from our 2D code with a bond distance of 1 a.u. and one nuclear charge set to zero. 1D calculations performed with a grid size of 800 points, 2D calculations with a grid size of 145 × 241 points. All numbers in atomic units.*

	KLICS		PW91		xcLDA	
	1D	2D	1D	2D	1D	2D
E_{tot}	−128.920235	−128.920234	−128.946580	−128.946580	−128.229914	−128.229915
ε_{1s}	−30.841442	−30.841447	−30.507920	−30.507920	−30.305770	−30.305770
ε_{2s}	−1.741044	−1.741044	−1.334970	−1.334970	−1.322601	−1.322601
$\varepsilon_{2p\sigma}$	−0.884057	−0.884057	−0.494228	−0.494228	−0.497847	−0.497847
$\varepsilon_{2p\pi}$	−0.884057	−0.884057	−0.494228	−0.494228	−0.497847	−0.497847
$< \frac{1}{r} >$	3.111736	3.111751	3.110075	3.110091	3.099824	3.099840
$< r >$	0.787235	0.787235	0.799878	0.799878	0.801547	0.801547
$< r^2 >$	0.931987	0.931987	0.982608	0.982608	0.985199	0.985200

Table 41: *Calculated bond lengths of the closed-shell-first-row dimers and hydrides. HF values taken from [105]. Experimental values from [106] except where noted. All values in atomic units. Taken from [104].*

	KLICS	PW91	xcLDA	HF	experiment
H_2	1.378	1.414	1.446	1.379	1.401[a]
Li_2	5.086	5.153	5.120	5.304	5.051
Be_2	–[b]	4.588	4.522	–	4.63[c]
C_2	2.306	2.367	2.354	–	2.3481
N_2	1.998	2.079	2.068	2.037	2.074
F_2	2.465	2.669	2.615	2.542	2.6682
LiH	2.971	3.030	3.030	3.092	3.0154
BH	2.274	2.356	2.373	–	2.3289
FH	1.684	1.756	1.761	1.722	1.7325

[a] Exact value from [107]
[b] There is no local minimum in the electronic potential curve.
[c] From [90].

Table 42: *Absolute total ground-state energies of the closed-shell-first-row dimers and hydrides calculated at the bond lengths given in Table 41. Estimates for exact values calculated using dissociation energies from Table 43 and nonrelativistic, infinite nuclear mass atomic ground-state energies from [91]. All numbers in atomic units. Taken from [104].*

	KLICS	PW91	xcLDA	exact
H_2	1.171444	1.170693	1.137692	1.174448[a]
Li_2	14.9982	14.9819	14.7245	14.9954
Be_2	29.3197[b]	29.3118	28.9136	29.3385
C_2	75.7736	75.8922	75.2041	75.922
N_2	109.4683	109.5449	108.6959	109.5424
F_2	199.4377	199.5699	198.3486	199.5299
LiH	8.0723	8.0625	7.9189	8.0705
BH	25.2857	25.2688	24.9770	25.29
FH	100.4241	100.4715	99.8490	100.4596

[a] Exact value from [107]
[b] Calculated at the experimental bond length of 4.63 a.u.

basis-set truncation and basis-set superposition errors. Where available, we have also included HF results obtained with conventional codes using basis-set expansions. Therefore, the comparisons between DFT and HF results in Tables 41, 43 and 44 have to be interpreted with due care.

In Table 41 we display results for the bond lengths. It is apparent that the KLICS scheme leads to equilibrium distances which are generally too short, an effect present in the HF approximation as well. Most notable is the fact that the potential energy curve of Be_2 displays no local minimum in this approximation. Good agreement with experiment is found only for Li_2. Except for this molecule, the PW91 values are clearly superior. In most cases even the LDA results are better than the KLICS values. With the exception of Li_2 and C_2, the GGA reduces the error of the LDA significantly.

Total absolute ground-state energies calculated at the bond lengths given in Table 41 are shown in Table 42. The exact values for the dimers are from [90], for the hydrides they are calculated by the same method using the exact nonrelativistic atomic ground-state energies in [91] and the experimental dissociation energies in [106]. For the lighter molecules H_2, Li_2, Be_2, LiH and BH the KLICS and PW91 results are of the same good quality, yielding errors of a few mHartrees. For the heavier molecules, however, the KLICS results are worse. Being the simplest approximation, it is not surprising that the LDA gives values for the total energies which show the largest errors.

Apart from H_2 and LiH, the dissociation energies obtained within the KLICS approach are disappointing, as may be seen in Table 43. In most cases, the magnitude is underestimated considerably and for Be_2 and F_2 even the wrong sign is obtained. Since the corresponding *atomic* ground-state energies given in the previous section are of excellent quality, the error must be due to correlation effects present in molecules only. In particular, the left-right correlation error [110] well-known in HF theory also occurs in DFT when the exact Fock expression (27) for E_x is employed. Apparently, the error is not sufficiently corrected for by the CS functional. The LDA and PW91 results are clearly much better, the latter reducing the over-binding tendencies of the former.

Despite these shortcomings, the KLICS xc potential is of better quality than the conventional xc potentials. In Table 44 we list the absolute values of the highest occupied molecular orbital energies. In an *exact* implementation their values should be equal to the ionization potentials of the systems under consideration. It is evident that the conventional KS approaches represented by the LDA and PW91 functionals yield results which are typically 40 percent too high, while the KLICS values are much closer to the experimental results. As for atomic systems, this fact may be traced back to the correct asymptotic behavior of the KLICS xc potential for large r. In order to examine the quality of the KLICS potential in the region closer to the nuclei, we compare the orbital energies of FH obtained using the various DFT approximations with results from the Zhao-Parr (ZP) method [5] in Table 45. This scheme allows for the calculation of the Kohn-Sham potential

Table 43: *Dissociation energies of the closed-shell-first-row dimers and hydrides calculated at the bond lengths given in Table 41. HF values taken from [105]. All numbers in mHartrees. Taken from [104].*

	KLICS	PW91	xcLDA	HF	experiment
H_2	171.444	167.665	180.270	121.0	174.475[b]
Li_2	32.4	33.5	37.9	3.5	39.3[c]
Be_2	−10.5[a]	15.9	20.6	–	3.8[c]
C_2	75.6	239.2	267.5	–	232[d]
N_2	287.3	387.5	427.1	167.5	364.0[d]
F_2	−22.7	106.7	126.2	−54.7	62.1[d]
LiH	89.4	86.8	96.9	48.4	92.4[d]
BH	129.3	137.4	145.8	–	135[e]
FH	193.9	238.4	259.1	130.8	225.7[e]

[a] Calculated at the experimental bond length 4.63 a.u.
[b] Exact value from [107]
[c] From [90]
[d] From [106]
[e] From [108].

Table 44: *Absolute values for the highest occupied orbital energies of the closed-shell-first-row dimers and hydrides calculated at the bond lengths given in Table 41. Experimental values are the ionization potentials taken from [106]. All numbers in atomic units. Taken from [104].*

	KLICS	PW91	xcLDA	experiment
H_2	0.621563	0.382656	0.373092	0.5669
Li_2	0.1974	0.1187	0.1187	0.18
Be_2	0.2560[a]	0.1678	0.1660	—
C_2	0.4844	0.2942	0.2987	0.4465
N_2	0.6643	0.3804	0.3826	0.5726
F_2	0.6790	0.3512	0.3497	0.5764
LiH	0.3237	0.1621	0.1612	0.283[b]
BH	0.3692	0.2058	0.2041	0.359
FH	0.6803	0.3567	0.3594	0.5894

[a] Calculated at the experimental bond length 4.63 a.u.
[b] From [109].

Table 45: *Kohn-Sham orbital energies for FH using a bond length of* 1.6373 *a.u. ZP denotes results from the Zhao-Parr method using accurate densities from coupled cluster calculations, taken from [10]. These numbers are close to the exact ones. All values in atomic units.*

	KLICS	PW91	xcLDA	ZP
$\varepsilon_{1\sigma}$	−24.5479	−24.2671	−24.0863	−24.6209
$\varepsilon_{2\sigma}$	−1.4492	−1.1234	−1.1114	−1.3732
$\varepsilon_{3\sigma}$	−0.8220	−0.5080	−0.5085	−0.7418
$\varepsilon_{1\pi}$	−0.6835	−0.3622	−0.3655	−0.5996

which uniquely corresponds to a given density. Ingamells et. al. [10] have used accurate densities obtained within the coupled cluster approach and determined the orbital energies of FH shown in the last column of Table 45. Although these values are expected to be close to the exact ones, they still contain some errors. For example, the energy of the highest molecular orbital, −0.5996 Hartrees, is not equal to the experimentally observed ionization potential of −0.5894 Hartrees. Nevertheless, it is evident from the table that the KLICS results are much closer to the ZP values than the ones from the conventional DFT approaches. We conclude that the KLICS potential is a better approximation to the exact KS potential than any of the conventional approximations.

4.4 Solids

Few calculations applying OEP or KLI methods to solids have been reported in the literature [39, 40, 111–118]. In this section we concentrate on materials for which more than one calculation has been reported, i.e. on Si, Ge and diamond. In Table 46 we show the energy gaps for Si at various high-symmetry points in the Brillouin zone. The results have been obtained with different computational techniques for the band structure calculation but they all make use of the exact, orbital-dependent exchange energy functional combined with the correlation energy functional in LDA. KKR-ASA denotes the Korringa-Kohn-Rostoker method in the atomic sphere approximation (ASA), LMTO-ASA is a linear-muffin-tin-orbital calculation in the same (ASA) approximation. Furthermore we report several calculations using pseudopotentials (PP). The abbreviations EXX and KLI stand for exact exchange and KLI exchange, both combined with LDA correlation.

For comparison, we show in Table 47 the LDA gaps obtained with the same computational methods. This comparison serves two purposes: first it can be seen that the LDA band gaps from the different methods agree pretty well. In contrast, the corresponding EXX (and KLI) band gaps show somewhat larger discrepancies among each other, although the agreement is still reasonable.

Table 46: *Kohn-Sham energy gaps for Si (in eV) from various calculations using the exact exchange functional.*

	KKR-ASA EXX [a]	LMTO-ASA EXX [a]	local PP EXX [b]	nonlocal PP EXX [b]	KLI-PP [c] KLI [c]	expt.
E_g	1.12	1.25	1.43	1.44	—	1.17 [d]
L	1.98	2.09	2.36	2.30	1.82	2.4 [e]
Γ	2.87	2.95	3.46	3.29	2.87	3.05 [f]
X	1.24	1.38	—	1.58	0.94	

[a] from reference [111]
[b] from reference [118]
[c] from reference [117]
[d] from reference [119]
[e] from reference [120]
[f] from reference [121].

Table 47: *Kohn-Sham energy gaps for Si (in eV) from various LDA calculations.*

	KKR-ASA LDA [a]	LMTO-ASA LDA [a]	local PP LDA [b]	nonlocal PP LDA [b]	PP LDA [c]	expt.
E_g	0.54	0.55	0.52	0.49	—	1.17 [d]
L	1.43	1.43	1.54	1.45	1.43	2.4 [e]
Γ	2.57	2.57	2.79	2.55	2.57	3.05 [f]
X	0.66	0.66	—	—	0.60	

[a] from reference [111]
[b] from reference [118]
[c] from reference [116]
[d] from reference [119]
[e] from reference [120]
[f] from reference [121].

Second we recognize the general trend of the EXX and KLI calculations to enhance the gaps significantly over the LDA values, thus achieving considerably better agreement with the experimental results. This can be attributed to the fact that EXX and KLI calculations have no self-interaction errors in the exchange potentials. Comparison of the results of the KLI calculation (column 5 of Table 46) with the exact exchange calculations obtained by also employing a local pseudopotential (column 3 of Table 46) suggests that the KLI approximation yields somewhat smaller band gaps than exact exchange.

Table 48: *Kohn-Sham energy gaps for Ge (in eV) from various calculations.*

	KKR-ASA EXX [a]	LMTO-ASA EXX [a]	KKR- or LMTO-ASA LDA [a]	PP KLI [b]	PP LDA [b]	expt. [c]
L (E_g)	1.03	1.12	0.40	0.77	0.42	0.84
Γ	1.57	1.67	0.60	1.26	0.82	1.00
X	1.24	1.34	0.78	0.87	0.57	1.3 ± 0.2

[a] from reference [111]
[b] from reference [117]
[c] from reference [119].

Table 49: *Kohn-Sham energy gaps for diamond (in eV) from various calculations.*

	KKR-ASA EXX [a]	LMTO-ASA EXX [a]	PP EXX [b]	PP LDA [b]	expt. [c]
E_g	4.58	4.65	5.06	4.16	5.47
L	8.63	8.68	9.19	8.42	—
Γ	5.87	5.92	6.28	5.57	7.3

[a] from reference [111]
[b] from reference [118]
[b] from reference [122].

In Tables 48 and 49 we report band gaps for Ge and diamond, respectively. The general findings from the calculations on Si are preserved in both cases: compared to LDA the band gaps obtained with exact exchange are significantly enhanced, but not as much as for Si.

Städele et al. [118] also studied the x-only derivative discontinuity which should be taken into account in a rigorous calculation of the band gap (see Eq. (179)). They found that this discontinuity is roughly twice as large as the Kohn-Sham band gap obtained by simply taking the difference of the KS eigenvalues. This leads to total gaps close to the Hartree-Fock band gaps. The discontinuity in the correlation potential must therefore cancel a large part of the exchange discontinuity to reproduce the experimentally observed gaps. This demonstrates the importance of finding a good, orbital-dependent correlation energy functional. LDA correlation has no discontinuity and is therefore not able to do this job.

All the calculations discussed so far employed the exact exchange energy functional combined with LDA correlation. Städele et al. [118] also performed a self-consistent exchange-only calculation, i.e. employing the exact exchange potential without adding any correlation potential. The resulting exchange potential, being the exact exchange potential of x-only DFT, was compared for silicon

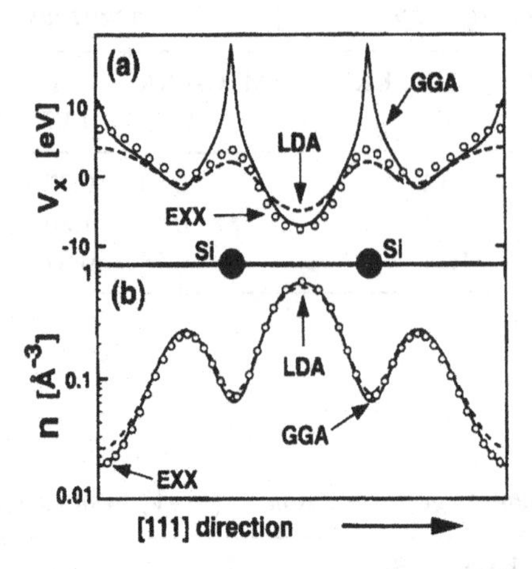

Figure 14: *(a) Comparison of the calculated exact exchange potential (open circles), in eV, along the [111] direction in Si with the approximate LDA (dashed line) and GGA (solid line) [111] exchange potentials. The filled black circles correspond to the positions of the Si atoms. The LDA and GGA potentials were evaluated at the exchange-only EXX density and the local ionic pseudopotential was employed. The mean values of all exchange potentials have been set equal to zero. (b) Self-consistent charge densities computed with the indicated exchange-only functionals; taken from [118].*

with approximate exchange functionals. Their results are shown in Fig. 14. In the physically important bonding region between the Si atoms the LDA potential significantly underestimates the spatial variations of the exact exchange. In this region the GGA exchange potential of Becke [75] is clearly superior to the LDA, while in the low-density regions the GGA potential is far too large.

In this self-consistent x-only calculation for Si, Städele et al. [118] determined the total energy per atom as well as the exchange energy per atom to be −104.75 eV and −29.40 eV, respectively. Similar to the results reported for atoms and molecules in the previous sections, these total-energy results agree very closely with the corresponding Hartree-Fock values [118] of −104.87 eV and 29.61 eV.

For metals the exact exchange functional is not expected to give good results. Kotani and Akai [112] found for transition metals such as Fe that the exact exchange plus LDA correlation potential yields the occupied bands too deep relative to the *s* bands and leads to magnetizations too large compared with experiment. In order to rectify this problem Kotani [113] has recently combined the exact exchange potential with a full (inhomogeneous) RPA for the correlation potential. It was found for the transition metals that the RPA correlation potential has a large contribution opposite in sign to the exchange potential. The resulting band structures and magnetizations are rather close to LDA results and to the experimental numbers. The magnetization of Fe, for example, is found to be 2.02 μ_B in the EXX+RPA calculation in contrast to 3.27 μ_B in the EXX+LDA calculation. The corresponding LDA and experimental values are 2.13 μ_B and 2.12 μ_B, respectively.

5 Beyond the OEP — a connection with Many–Body Perturbation Theory

We have discussed in detail the connection between the OEP method and the conventional density functional approach. The OEP method, where the xc energy is expressed as a functional of the orbitals, allows for more flexibility in approximating the xc energy, while preserving important properties of the exact KS potential such as the derivative discontinuities and the absence of spurious self-interactions. Apart from the use of the Colle-Salvetti functional in the OEP procedure as described in previous chapters, several approaches have been reported in the literature to derive approximations to the correlation-energy functional suitable for the OEP method. Görling and Levy [32, 123] proposed a scheme to derive a correlation-energy functional from a coupling-constant perturbation expansion. Casida [124] on the other hand, used an approximate perturbative expression for the ground-state energy to include correlation in the OEP scheme.

The purpose of the present chapter is to establish a connection between the OEP method and many-body perturbation theory which may serve to construct correlation energy functionals in a systematic way. We first derive exact equations for the xc potential and the xc energy in terms of the Green function and self-energy of the system by using a particular perturbative expansion of the interacting Hamiltonian. Then we demonstrate that the x-only OEP method is equivalent to the first-order approximation of this expansion. On the basis of this observation we then propose an iterative scheme to include correlation which relies on the Kohn-Sham Green function rather than on the Kohn-Sham orbitals.

The starting point of our discussion is the Hamiltonian of the system of interacting electrons (see Eq. (1)) written in second quantized notation:

$$\hat{H} = \hat{T} + \hat{V}_0 + \hat{W}_{\mathrm{Clb}} \tag{304}$$

where

$$\hat{T} = \sum_{\sigma} \int \mathrm{d}^3 r \; \hat{\psi}_{\sigma}^{\dagger}(\mathbf{r}) \left(-\frac{\nabla^2}{2} \right) \hat{\psi}_{\sigma}(\mathbf{r}) \tag{305}$$

is the kinetic energy operator,

$$\hat{V}_0 = \sum_{\sigma} \int \mathrm{d}^3 r \; \hat{\psi}_{\sigma}^{\dagger}(\mathbf{r}) v_0(\mathbf{r}) \hat{\psi}_{\sigma}(\mathbf{r}) \tag{306}$$

is the operator of the external (nuclear) potential and

$$\hat{W}_{\mathrm{Clb}} = \frac{1}{2} \sum_{\sigma,\sigma'} \int d^3r \int d^3r' \, \hat{\psi}^{\dagger}_{\sigma}(\mathbf{r}) \hat{\psi}^{\dagger}_{\sigma'}(\mathbf{r}') \frac{1}{|\mathbf{r}-\mathbf{r}'|} \hat{\psi}_{\sigma'}(\mathbf{r}') \hat{\psi}_{\sigma}(\mathbf{r}) \tag{307}$$

is the Coulomb interaction of the electrons.

The central idea is now to formally rewrite Eq. (304) by addition and subtraction of the exact Hartree and xc potentials in the following way:

$$\hat{H} = \hat{T} + \hat{V}_0 + \hat{V}_{\mathrm{H}} + \hat{V}_{\mathrm{xc}} + (\hat{W}_{\mathrm{Clb}} - \hat{V}_{\mathrm{H}} - \hat{V}_{\mathrm{xc}}) =: \hat{H}_{KS} + \hat{H}_1 \ . \tag{308}$$

Here $\hat{V}_{\mathrm{H}}$ and $\hat{V}_{\mathrm{xc}}$ are the operators of the Hartree and xc potentials defined in analogy to Eq. (306) with $v_0(\mathbf{r})$ replaced by

$$v_{\mathrm{H}}(\mathbf{r}) = \int d^3r \, \frac{\rho(\mathbf{r}')}{|\mathbf{r}-\mathbf{r}'|} \tag{309}$$

and $V_{\mathrm{xc}\sigma}(\mathbf{r})$, respectively.

$$\hat{H}_{KS} = \hat{T} + \hat{V}_0 + \hat{V}_{\mathrm{H}} + \hat{V}_{\mathrm{xc}} \tag{310}$$

is the Hamiltonian of non-interacting electrons moving in the spin-dependent KS potential

$$V_{S\sigma}(\mathbf{r}) = v_0(\mathbf{r}) + v_{\mathrm{H}}(\mathbf{r}) + V_{\mathrm{xc}\sigma}(\mathbf{r}) \tag{311}$$

and will serve as the unperturbed reference Hamiltonian while

$$\hat{H}_1 = \hat{W}_{\mathrm{Clb}} - \hat{V}_{\mathrm{H}} - \hat{V}_{\mathrm{xc}} \tag{312}$$

will be treated as a perturbation to $\hat{H}_{KS}$. It has to be emphasized that the xc potential $V_{\mathrm{xc}\sigma}(\mathbf{r})$ – and hence the unperturbed Hamiltonian $\hat{H}_{KS}$ – is not known at this stage and has to be determined by further arguments. Its uniqueness, on the other hand, is guaranteed by the Hohenberg-Kohn theorem. We also note that the perturbative approach based on the proposed splitting (308) of the interacting Hamiltonian is *not* equivalent to a Taylor expansion in powers of e^2 with e being the elementary charge. This is due to the fact that the unperturbed Hamiltonian $\hat{H}_{KS}$ through the exact KS potential $V_{\mathrm{xc}\sigma}$ already contains e^2 up to infinite order.

Application of the techniques of many-body perturbation theory to Hamiltonian (308) in the described way, i.e. treating $\hat{H}_1$ as a perturbation to the KS Hamiltonian $\hat{H}_{KS}$, leads to the Dyson equation

$$\begin{aligned} G_{\sigma}(\mathbf{r},\mathbf{r}',\omega) = {} & G^{KS}_{\sigma}(\mathbf{r},\mathbf{r}',\omega) \\ & + \int d^3y \int d^3y' \, G^{KS}_{\sigma}(\mathbf{r},\mathbf{y},\omega) \Sigma_{\sigma}(\mathbf{y},\mathbf{y}',\omega) G_{\sigma}(\mathbf{y}',\mathbf{r}',\omega) \end{aligned} \tag{313}$$

which relates the Green function G_σ of the interacting system to the KS Green function

$$G_\sigma^{KS}(\mathbf{r},\mathbf{r}',\omega) = \lim_{\eta\to 0^+} \sum_j \varphi_{j\sigma}(\mathbf{r})\varphi_{j\sigma}^*(\mathbf{r}') \left(\frac{\Theta(\varepsilon_{j\sigma}-\varepsilon_F)}{\omega-\varepsilon_{j\sigma}+i\eta} + \frac{\Theta(\varepsilon_F-\varepsilon_{j\sigma})}{\omega-\varepsilon_{j\sigma}-i\eta} \right) . \tag{314}$$

$\Sigma_\sigma(\mathbf{r},\mathbf{r}',\omega)$ is the irreducible self-energy corresponding to the perturbation $\hat{H}_1$. Since this perturbation contains the unknown xc potential one can write the self-energy as a functional of $V_{xc\sigma}(\mathbf{r})$:

$$\Sigma_\sigma(\mathbf{r},\mathbf{r}',\omega) = \Sigma_\sigma[V_{xc\sigma}](\mathbf{r},\mathbf{r}',\omega) . \tag{315}$$

Explicit approximations of this functional can be found, e.g., by evaluating all self-energy diagrams up to a given order in the perturbation. An example for this strategy will be discussed below.

In order to determine $V_{xc\sigma}(\mathbf{r})$ we finally make use of the fact that – due to the KS theorem – the spin density of the interacting system can be obtained both via the Green function G_σ and the KS Green function G_σ^{KS} by the expression

$$\rho_\sigma(\mathbf{r}) = -i\int \frac{d\omega}{2\pi}\, G_\sigma(\mathbf{r},\mathbf{r},\omega) = -i\int \frac{d\omega}{2\pi}\, G_\sigma^{KS}(\mathbf{r},\mathbf{r},\omega) . \tag{316}$$

Using Dyson's equation (313) one is thus led to the integral equation

$$\int \frac{d\omega}{2\pi} \int d^3y \int d^3y'\, G_\sigma^{KS}(\mathbf{r},\mathbf{y},\omega)\Sigma_\sigma(\mathbf{y},\mathbf{y}',\omega)G_\sigma(\mathbf{y}',\mathbf{r},\omega) = 0 . \tag{317}$$

Writing the self-energy as

$$\Sigma_\sigma(\mathbf{r},\mathbf{r}',\omega) = \Sigma_{xc\sigma}(\mathbf{r},\mathbf{r}',\omega) - \delta(\mathbf{r}-\mathbf{r}')V_{xc\sigma}(\mathbf{r}) \tag{318}$$

one obtains

$$\begin{aligned}
&\int d^3y\, V_{xc\sigma}(\mathbf{y}) \int \frac{d\omega}{2\pi}\, G_\sigma^{KS}(\mathbf{r},\mathbf{y},\omega)G_\sigma(\mathbf{y},\mathbf{r},\omega) \\
&\quad = \int d^3y \int d^3y' \int \frac{d\omega}{2\pi}\, G_\sigma^{KS}(\mathbf{r},\mathbf{y},\omega)\Sigma_{xc\sigma}(\mathbf{y},\mathbf{y}',\omega)G_\sigma(\mathbf{y}',\mathbf{r},\omega) .
\end{aligned} \tag{319}$$

This exact integral equation relating the xc potential to the xc part $\Sigma_{xc\sigma}(\mathbf{r},\mathbf{r}',\omega)$ of the irreducible self-energy was first derived by Sham and Schlüter [50, 54]

Sham also derived an exact expression for the xc energy in terms of the Green function and the self-energy by using the techniques of Luttinger and Ward [125]. This expression reads

$$\begin{aligned}
E_{xc}[\rho_\uparrow,\rho_\downarrow] &= i\sum_\sigma \int \frac{d\omega}{2\pi} \int d^3r\, \log\left(1 - G_\sigma^{KS}\Sigma_\sigma\right)(\mathbf{r},\mathbf{r},\omega) \\
&\quad + i\sum_\sigma \int \frac{d\omega}{2\pi} \int d^3r \int d^3r'\, \Sigma_\sigma(\mathbf{r},\mathbf{r}',\omega)G_\sigma(\mathbf{r}',\mathbf{r},\omega) \\
&\quad + \sum_{n=1}^{\infty} Y^{(n)}
\end{aligned} \tag{320}$$

where the logarithm of some integral operator $C(\mathbf{r}, \mathbf{r}')$ has to be read as

$$\log(1 - C)(\mathbf{r}, \mathbf{r}') = -\sum_{n=1}^{\infty} \frac{1}{n} \left(C(\mathbf{r}, \mathbf{r}') \right)^n \tag{321}$$

with

$$\left(C(\mathbf{r}, \mathbf{r}') \right)^n = \int d^3x_1 \ldots \int d^3x_{n-1} C(\mathbf{r}, \mathbf{x}_1) C(\mathbf{x}_1, \mathbf{x}_2) \ldots C(\mathbf{x}_{n-1}, \mathbf{r}') \,. \tag{322}$$

The quantities $Y^{(n)}$ are defined by

$$Y^{(n)} = -\frac{i}{2n} \sum_{\sigma} \int \frac{d\omega}{2\pi} \int d^3r \int d^3r' \; \Sigma^{(n)}_{\sigma,\mathrm{dressed}}(\mathbf{r}, \mathbf{r}', \omega) G_\sigma(\mathbf{r}', \mathbf{r}, \omega) \tag{323}$$

where, for $n \geq 2$, $\Sigma^{(n)}_{\sigma,\mathrm{dressed}}(\mathbf{r}, \mathbf{r}', \omega)$ is the sum of all dressed skeleton diagrams of order n, while for $n = 1$ it is given solely by the nonlocal first-order dressed skeleton diagram, i.e.,

$$\Sigma^{(1)}_{\sigma,\mathrm{dressed}}(\mathbf{r}, \mathbf{r}') = i \int \frac{d\omega}{2\pi} \frac{G_\sigma(\mathbf{r}, \mathbf{r}', \omega)}{|\mathbf{r} - \mathbf{r}'|} \,. \tag{324}$$

As a consequence of the logarithm of integral operators, Eq. (320) is rather difficult to handle. In the following we derive a much simpler, though still exact, expression for E_{xc}. We begin with the standard expression [126] for the total ground state energy of the interacting system:

$$\begin{aligned}
E_{GS} &= -\frac{i}{2} \sum_{\sigma} \int \frac{d\omega}{2\pi} \int d^3r \lim_{\mathbf{r}' \to \mathbf{r}} \left(\omega - \frac{\nabla^2_{\mathbf{r}}}{2} + v_0(\mathbf{r}) \right) G_\sigma(\mathbf{r}, \mathbf{r}', \omega) \\
&= -\frac{i}{2} \sum_{\sigma} \int \frac{d\omega}{2\pi} \int d^3r \lim_{\mathbf{r}' \to \mathbf{r}} \left(\omega - h^{KS}_\sigma(\mathbf{r}) + 2h^{KS}_\sigma(\mathbf{r}) \right) G_\sigma(\mathbf{r}, \mathbf{r}', \omega) \\
&\quad -\frac{1}{2} \sum_{\sigma} \int d^3r \left(v_H(\mathbf{r}) + v_{xc\sigma}(\mathbf{r}) \right) \rho_\sigma(\mathbf{r}) \\
&= -i \sum_{\sigma} \int \frac{d\omega}{2\pi} \int d^3r \lim_{\mathbf{r}' \to \mathbf{r}} \left(-\frac{\nabla^2_{\mathbf{r}}}{2} \right) G_\sigma(\mathbf{r}, \mathbf{r}', \omega) + \int d^3r \, v_0(\mathbf{r}) \rho(\mathbf{r}) \\
&\quad + \frac{1}{2} \int d^3r \int d^3r' \frac{\rho(\mathbf{r}) \rho(\mathbf{r}')}{|\mathbf{r} - \mathbf{r}'|} \\
&\quad - \frac{i}{2} \sum_{\sigma} \int \frac{d\omega}{2\pi} \int d^3r \int d^3r' \Sigma_{xc\sigma}(\mathbf{r}, \mathbf{r}', \omega) G_\sigma(\mathbf{r}', \mathbf{r}, \omega)
\end{aligned} \tag{325}$$

where we used Dyson's equation (313) and the equation of motion for the KS Green function

$$(\omega - h_\sigma^{KS}(\mathbf{r}))G_\sigma^{KS}(\mathbf{r},\mathbf{r}',\omega) = \delta(\mathbf{r}-\mathbf{r}'), \tag{326}$$

with the KS single-particle Hamiltonian

$$h_\sigma^{KS}(\mathbf{r}) = -\frac{\nabla^2}{2} + V_{S\sigma}(\mathbf{r}) \,. \tag{327}$$

Comparison of Eq. (325) with the defining equation for E_{xc},

$$E_{GS}[\rho_\uparrow,\rho_\downarrow] = T_s[\rho_\uparrow,\rho_\downarrow] + \int \mathrm{d}^3r\, v_0(\mathbf{r})\rho(\mathbf{r}) + \frac{1}{2}\int \mathrm{d}^3r \int \mathrm{d}^3r' \frac{\rho(\mathbf{r})\rho(\mathbf{r}')}{|\mathbf{r}-\mathbf{r}'|} + E_{xc}[\rho_\uparrow,\rho_\downarrow] \;, \tag{328}$$

leads to the desired result

$$E_{xc}[\rho_\uparrow,\rho_\downarrow] = -\frac{i}{2}\sum_\sigma \int \frac{\mathrm{d}\omega}{2\pi}\int \mathrm{d}^3r \int \mathrm{d}^3r'\,\Sigma_{xc\sigma}(\mathbf{r},\mathbf{r}',\omega)G_\sigma(\mathbf{r}',\mathbf{r},\omega) + T[\rho_\uparrow,\rho_\downarrow] - T_s[\rho_\uparrow,\rho_\downarrow] \;, \tag{329}$$

where

$$T[\rho_\uparrow,\rho_\downarrow] = -i\sum_\sigma \int \frac{\mathrm{d}\omega}{2\pi}\int \mathrm{d}^3r \lim_{\mathbf{r}'\to\mathbf{r}}\left(-\frac{\nabla_\mathbf{r}^2}{2}\right) G_\sigma(\mathbf{r},\mathbf{r}',\omega) \tag{330}$$

is the kinetic energy of the interacting system while

$$T_s[\rho_\uparrow,\rho_\downarrow] = -i\sum_\sigma \int \frac{\mathrm{d}\omega}{2\pi}\int \mathrm{d}^3r \lim_{\mathbf{r}'\to\mathbf{r}}\left(-\frac{\nabla_\mathbf{r}^2}{2}\right) G_\sigma^{KS}(\mathbf{r},\mathbf{r}',\omega) \tag{331}$$

is the kinetic energy of the non-interacting KS system.

Sham noted that Eq. (319) reduces to the integral equation of the exchange-only OEP method if one approximates $\Sigma_{\mathrm{xc}\sigma}$ by the Fock self-energy

$$\Sigma_{\mathrm{xc}\sigma}(\mathbf{r},\mathbf{r}',\omega) \approx \Sigma_\sigma^F(\mathbf{r},\mathbf{r}') = -\sum_{j,\varepsilon_{j\sigma}\le\varepsilon_F} \frac{\varphi_{j\sigma}(\mathbf{r})\varphi_{j\sigma}^*(\mathbf{r}')}{|\mathbf{r}-\mathbf{r}'|} \tag{332}$$

and simultaneously replaces the Green function G_σ of the interacting system by the KS Green function G_σ^{KS}. This can easily be understood within the above framework by performing a perturbative analysis of the ground-state energy of the interacting system in terms of the proposed splitting (308) of the original Hamiltonian. The energy of the unperturbed KS reference system is

$$E^{(0)} = T_s[\rho] + \sum_\sigma \int \mathrm{d}^3r\; V_{S\sigma}(\mathbf{r})\rho_\sigma(\mathbf{r}) = T_s[\rho] + \int \mathrm{d}^3r\; v_0(\mathbf{r})\rho(\mathbf{r}) + \int \mathrm{d}^3r\; v_\mathrm{H}(\mathbf{r})\rho(\mathbf{r}) + \sum_\sigma \int \mathrm{d}^3r\; V_{\mathrm{xc}\sigma}(\mathbf{r})\rho_\sigma(\mathbf{r}) \;. \tag{333}$$

Due to this fact the resulting integral equation for $V_{xc\sigma}$ will not only contain the unknown xc potential but also the equally unknown xc kernel $f_{xc}^{\sigma\sigma'}$ defined as functional derivative of the xc potential:

$$f_{xc}^{\sigma\sigma'}(\mathbf{r},\mathbf{r}') = \frac{\delta V_{xc\sigma}(\mathbf{r})}{\delta\rho_{\sigma'}(\mathbf{r}')} = \frac{\delta^2 E_{xc}}{\delta\rho_\sigma(\mathbf{r})\delta\rho_{\sigma'}(\mathbf{r}')} . \tag{338}$$

Although it would be very interesting to obtain information on the xc kernel, the equation determining $f_{xc}^{\sigma\sigma'}$ and $V_{xc\sigma}$ can be expected to be *extremely* difficult to solve in practice. One should note that in the x-only method (i.e. for $n = 1$) the xc potential as well as the KS energies enter the energy functional only implicitly via the occupied KS orbitals and the resulting integral equation does not contain the xc kernel.

In the following we propose a third approach which might be more practical. The central idea of this scheme is to approximate the xc part $\Sigma_{xc\sigma}$ of the self-energy as a functional of the full Green function and then solve Dyson's equation (313) for G_σ together with Sham's integral equation (319) for $V_{xc\sigma}$ in a self-consistent fashion. To achieve this we first derive an *exact* diagrammatic representation of $\Sigma_{xc\sigma}$ in terms of the full Green function G_σ. To this end we return to the standard formulation of many-body perturbation theory and treat the interaction $\hat{W}_{Clb}$ as a perturbation to the non-interacting Hamiltonian $\hat{H}_0 = \hat{T} + \hat{V}_0$. In this case Dyson's equation of course reads

$$G_\sigma(\mathbf{r},\mathbf{r}',\omega) = G_\sigma^{(0)}(\mathbf{r},\mathbf{r}',\omega) + \int d^3y \int d^3y'\, G_\sigma^{(0)}(\mathbf{r},\mathbf{y},\omega)\tilde{\Sigma}_\sigma(\mathbf{y},\mathbf{y}',\omega)G_\sigma(\mathbf{y}',\mathbf{r}',\omega) \tag{339}$$

with the Green function $G_\sigma^{(0)}$ corresponding to $\hat{H}_0$ and the irreducible self-energy $\tilde{\Sigma}_\sigma$ corresponding to the perturbation $\hat{W}_{Clb}$. $\tilde{\Sigma}_\sigma$ can exactly be expressed as the sum of the two diagrams shown in Fig. 17 where Γ_σ is the irreducible vertex function and W_σ is the effective screened interaction [126].

The non-interacting Green function $G_\sigma^{(0)}$ is the solution of the equation of motion

$$(\omega - h^{(0)}(\mathbf{r}))G_\sigma^{(0)}(\mathbf{r},\mathbf{r}',\omega) = \delta(\mathbf{r}-\mathbf{r}') \tag{340}$$

with

$$h^{(0)}(\mathbf{r}) = -\frac{\nabla^2}{2} + v_0(\mathbf{r}) . \tag{341}$$

Operating with $(\omega - h^{(0)})$ on Eq. (339) yields

$$(\omega - h^{(0)}(\mathbf{r}))G_\sigma(\mathbf{r},\mathbf{r}',\omega) = \delta(\mathbf{r}-\mathbf{r}') + \int d^3y'\, \tilde{\Sigma}_\sigma(\mathbf{r},\mathbf{y}',\omega)G_\sigma(\mathbf{y}',\mathbf{r}',\omega) . \tag{342}$$

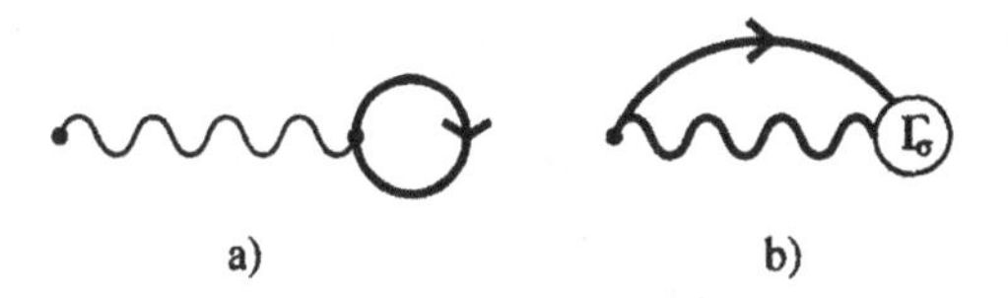

Figure 17: *The two self-consistent diagrams which constitute the self-energy* $\tilde{\Sigma}_\sigma$ *of Eq. (339). The thick straight lines represent the Green function of the interacting system, the thick wavy line is the screened interaction and* Γ_σ *is the irreducible vertex function. The xc self-energy* $\Sigma_{\text{xc}\sigma}$ *can exactly be represented solely by diagram b).*

In a similar manner, by using the equation of motion (326) of the KS Green function one obtains from Eq. (313)

$$(\omega - h_\sigma^{KS}(\mathbf{r}))G_\sigma(\mathbf{r},\mathbf{r}',\omega) = \delta(\mathbf{r}-\mathbf{r}') + \int d^3y'\ \Sigma_\sigma(\mathbf{r},\mathbf{y}',\omega)G_\sigma(\mathbf{y}',\mathbf{r}',\omega)$$
$$= \delta(\mathbf{r}-\mathbf{r}') - V_{\text{xc}\sigma}(\mathbf{r})G_\sigma(\mathbf{r},\mathbf{r}',\omega) + \int d^3y'\ \Sigma_{\text{xc}\sigma}(\mathbf{r},\mathbf{y}',\omega)G_\sigma(\mathbf{y}',\mathbf{r}',\omega) \tag{343}$$

where we used Eq. (318) in the last step. Eq. (343) can be rewritten as

$$(\omega - h^{(0)}(\mathbf{r}))G_\sigma(\mathbf{r},\mathbf{r}',\omega)$$
$$= \delta(\mathbf{r}-\mathbf{r}') + v_{\text{H}}(\mathbf{r})G_\sigma(\mathbf{r},\mathbf{r}',\omega) + \int d^3y'\ \Sigma_{\text{xc}\sigma}(\mathbf{r},\mathbf{y}',\omega)G_\sigma(\mathbf{y}',\mathbf{r}',\omega)\ . \tag{344}$$

Comparison of Eqs. (344) and (342) relates the self-energy $\tilde{\Sigma}_\sigma$ to $\Sigma_{\text{xc}\sigma}$ via

$$\tilde{\Sigma}_\sigma(\mathbf{r},\mathbf{r}',\omega) = v_{\text{H}}(\mathbf{r})\delta(\mathbf{r}-\mathbf{r}') + \Sigma_{\text{xc}\sigma}(\mathbf{r},\mathbf{r}',\omega)\ . \tag{345}$$

As mentioned before, $\tilde{\Sigma}_\sigma$ can be represented by the sum of the two diagrams of Fig. 17. Since the diagram of Fig. 17 a) yields $v_{\text{H}}(\mathbf{r})\delta(\mathbf{r}-\mathbf{r}')$, the xc-part of the self-energy $\Sigma_{\text{xc}\sigma}$ can be represented solely by diagram 17 b). Since one can express the self-energy $\tilde{\Sigma}_\sigma$ also in terms of skeleton diagrams [126], $\Sigma_{\text{xc}\sigma}$ can alternatively be represented as the sum of all fully dressed skeleton diagrams except the diagram of Fig. 17 a). This diagrammatic representation of the xc-part of the self-energy can then easily be used to approximate $\Sigma_{\text{xc}\sigma}$ as a functional of the full Green function G_σ. One possible approximation would be to use only the dressed skeleton diagrams up to a certain order, e.g., by the first-order skeleton diagram given by Eq. (324).

In the following we assume that some approximation to $\Sigma_{\text{xc}\sigma}$ as a functional of G_σ has been specified:

$$\Sigma_{\text{xc}\sigma}(\mathbf{r},\mathbf{r}',\omega) \approx \Sigma_{\text{xc}\sigma}^{\text{approx}}[G_\sigma](\mathbf{r},\mathbf{r}',\omega) \tag{346}$$

Using this approximation we propose the following iterative scheme:

1. Start with an approximation to $V_{xc\sigma}$ (e.g. the x-only V_{xc}^{OEP}) and calculate the corresponding KS Green function G_σ^{KS}.
2. Solve Dyson's equation (313) for the Green function G_σ with Σ_σ approximated by

$$\Sigma_\sigma(\mathbf{r}, \mathbf{r}', \omega) \approx \Sigma_{xc\sigma}^{approx}[G_\sigma](\mathbf{r}, \mathbf{r}', \omega) - \delta(\mathbf{r} - \mathbf{r}')V_{xc\sigma}(\mathbf{r}) . \qquad (347)$$

3. With the solution G_σ of Dyson's equation and the corresponding self-energy solve Sham's integral equation (319) for $V_{xc\sigma}$. Use that xc-potential to calculate the corresponding KS Green function G_σ^{KS}.
4. Return to point 2. and iterate until self-consistency is achieved.
5. Calculate the ground state energy of the interacting system via Eq. (325).

We emphasize that the only approximation in the above scheme is the approximation to the functional $\Sigma_{xc\sigma}[G_\sigma]$. Even for the simplest approximation to the xc self-energy, namely using only the first-order dressed skeleton diagram, the converged result of the above scheme goes far beyond the x-only OEP solution since in this case, as discussed at the beginning of this chapter, the xc self-energy is expressed as the first-order skeleton diagram using the *bare* KS Green function instead of using the dressed Green function. Therefore we hope that the above iterative scheme might have the merit of including the important part of electron correlation already through low-order approximations of the xc self-energy. Of course one could in principle also use more sophisticated approximations of the xc self-energy such as the GW approximation [127], but due to the necessity of solving the integral equation for $V_{xc\sigma}$ in each iteration cycle this would result in a very high computational effort.

Since in the x-only case of the OEP method the KLI approximation to the integral equation for the xc potential has proven to be extremely useful, an analogous approximation for Sham's integral equation seems highly desirable.

The proposed iterative scheme has another desirable feature: a converged solution of that scheme not only gives an approximate xc potential, it also yields an approximation of the Green function of the interacting system. This Green function yields the ground-state energy of the interacting system via Eq. (325), but it also allows for the calculation of photoemission and inverse photoemission spectra of the interacting system.

Acknowledgments

We thank Eberhard Engel for providing us with his KS and RKS computer codes, and John Perdew for his PW91 xc subroutine, Dage Sundholm and Pekka Pyykkö for their two-dimensional Xα code for molecules and for the warm hospitality during a stay of one of us (T.G.) in Helsinki. The entire manuscript was read by Kieron Burke, who deserves special thanks for numerous comments. T.K. gratefully acknowledges a fellowship of the Studienstiftung des deutschen Volkes. Numerous discussions with Dage Sundholm, Eberhard Engel, Klaus Capelle, Martin Petersilka and Martin Lüders are gratefully acknowledged. This work was supported in part by the Deutsche Forschungsgemeinschaft.

References

[1] *Density Functional Theory*, Vol. 337 of *NATO ASI Series B*, edited by E.K.U. Gross and R.M. Dreizler (Plenum Press, New York, 1995).
[2] R.M. Dreizler and E.K.U. Gross, *Density Functional Theory* (Springer, Berlin, 1990).
[3] R.G. Parr and W. Yang, *Density-Functional Theory of Atoms and Molecules* (Oxford University Press, New York, 1989).
[4] P. Hohenberg and W. Kohn, *Phys. Rev.* **136**, B864 (1964).
[5] Q. Zhao and R.G. Parr, *Phys. Rev. A* **46**, 2337 (1992).
[6] C.J. Umrigar and X. Gonze, in *High Performance Computing and its Application to the Physical Sciences*, edited by D.A. Browne et al. (World Scientific, Singapore, 1993).
[7] C.J. Umrigar and X. Gonze, *Phys. Rev. A* **50**, 3827 (1994).
[8] R. van Leeuwen and E.J. Baerends, *Phys. Rev. A* **49**, 2421 (1994).
[9] A. Görling and M. Ernzerhof, *Phys. Rev. A* **51**, 4501 (1995).
[10] V.E. Ingamells and N.C. Handy, *Chem. Phys. Lett.* **248**, 373 (1996).
[11] W. Kohn and L.J. Sham, *Phys. Rev.* **140**, A1133 (1965).
[12] R.T. Sharp and G.K. Horton, *Phys. Rev.* **90**, 317 (1953).
[13] J.D. Talman and W.F. Shadwick, *Phys. Rev. A* **14**, 36 (1976).
[14] J.B. Krieger, Y. Li, and G.J. Iafrate, *Phys. Rev. A* **46**, 5453 (1992).
[15] Y. Li, J.B. Krieger, and G.J. Iafrate, *Phys. Rev. A* **47**, 165 (1993).
[16] J.B. Krieger, Y. Li, and G.J. Iafrate, in *Density Functional Theory*, edited by R.M. Dreizler and E.K.U. Gross (Plenum Press, New York, 1995), p. 191.
[17] J.B. Krieger, Y. Li, and G.J. Iafrate, *Phys. Lett. A* **146**, 256 (1990).
[18] J.B. Krieger, Y. Li, M.R. Norman, and G.J. Iafrate, *Phys. Rev. B* **44**, 10437 (1991).
[19] J.B. Krieger, Y. Li, and G.J. Iafrate, *Phys. Rev. A* **45**, 101 (1992).
[20] J.B. Krieger, Y. Li, and G.J. Iafrate, *Int. J. Quantum Chem.* **41**, 489 (1992).
[21] Y. Li, J.B. Krieger, and G.J. Iafrate, *Chem. Phys. Lett.* **191**, 38 (1992).
[22] J.B. Krieger, J. Chen, Y. Li, and G.J. Iafrate, *Int. J. Quantum Chem. Symp.* **29**, 79 (1995).
[23] C.A. Ullrich, U.J. Gossmann, and E.K.U. Gross, *Phys. Rev. Lett.* **74**, 872 (1995).

[24] C.A. Ullrich, S. Erhard, and E.K.U. Gross, in *Super Intense Laser Atom Physics IV*, edited by H.G. Muller and M.V. Fedorov (Kluwer, Dordrecht, 1996), p. 267.
[25] M. Petersilka, U.J. Gossmann, and E.K.U Gross, *Phys. Rev. Lett.* **76**, 1212 (1996).
[26] E.K.U. Gross, J.F. Dobson, and M. Petersilka, in *Density Funktional Theory II*, Vol. 181 of *Topics in Current Chemistry*, edited by R.F. Nalewajski (Springer, Berlin, 1996), p. 81.
[27] K. Burke and E.K.U. Gross, in *Density Functionals: Theory and Applications*, Vol. 500 of *Lecture Notes in Physics*, edited by D. Joubert (Springer, Berlin, 1998), p. 116.
[28] V.A. Khodel, V.R. Shaginyan, and V.V. Khodel, *Phys. Rep.* **249**, 1 (1994).
[29] U. von Barth and L. Hedin, *J. Phys. C* **5**, 1629 (1972).
[30] M.M. Pant and A.K. Rajagopal, *Sol. State Commun.* **10**, 1157 (1972).
[31] V.R. Shaginyan, *Phys. Rev. A* **47**, 1507 (1993).
[32] A. Görling and M. Levy, *Phys. Rev. A* **50**, 196 (1994).
[33] M.R. Norman and D.D. Koelling, *Phys. Rev. B* **30**, 5530 (1984).
[34] T. Grabo and E.K.U. Gross, *Chem. Phys. Lett.* **240**, 141 (1995).
[35] V. Sahni, J. Gruenebaum, and J.P. Perdew, *Phys. Rev. B* **26**, 4371 (1982).
[36] J.P. Perdew and M.R. Norman, *Phys. Rev. B* **26**, 5445 (1982).
[37] E. Engel and S.H. Vosko, *Phys. Rev. A* **47**, 2800 (1993).
[38] K. Aashamar, T.M. Luke, and J.D. Talman, *At. Data Nucl. Data Tables* **22**, 443 (1978).
[39] T. Kotani, *Phys. Rev. B* **50**, 14816 (1994), and *Phys. Rev. B* **51**, 13903 (1995)(E).
[40] T. Kotani, *Phys. Rev. Lett.* **74**, 2989 (1995).
[41] T. Kotani and H. Akai, *Phys. Rev. B* **52**, 17153 (1995).
[42] J.B. Krieger, Y. Li, and G.J. Iafrate, *Phys. Lett. A* **148**, 470 (1990).
[43] J. Chen, J.B. Krieger, Y. Li, and G.J. Iafrate, *Phys. Rev. A* **54**, 3939 (1996).
[44] J.B. Krieger, Y. Li, Y. Liu, and G.J. Iafrate, *Int. J. Quantum Chem.* **61**, 273 (1997).
[45] T. Kreibich, S. Kurth, T. Grabo, and E.K.U. Gross, *Adv. Quantum Chem.* **33**, 31 (1998).
[46] C. Filippi, C.J. Umrigar, and X. Gonze, *Phys. Rev. A* **54**, 4810 (1996).
[47] C.O. Almbladh and U. von Barth, *Phys. Rev. B* **31**, 3231 (1985).
[48] N.C. Handy, M.T. Marron, and H.J. Silverstone, *Phys. Rev.* **180**, 45 (1969).
[49] M. Levy and A. Görling, *Phys. Rev. A* **53**, 3140 (1996).
[50] L. Sham, *Phys. Rev. B* **32**, 3876 (1985).
[51] M.K. Harbola and V. Sahni, *Phys. Rev. Lett.* **62**, 489 (1989).
[52] E. Engel, J.A. Chevary, L.D. MacDonald, and S.H. Vosko, *Z. Phys. D* **23**, 7 (1992).
[53] J.P. Perdew, R.G. Parr, M. Levy, and J.L. Balduz, *Phys. Rev. Lett.* **49**, 1691 (1982).

[54] M. Schlüter L.J. Sham, *Phys. Rev. Lett.* **51**, 1888 (1983).
[55] M. Schlüter L.J. Sham, *Phys. Rev. B* **32**, 3883 (1985).
[56] R. van Leeuwen, O.V. Gritsenko, and E.J. Baerends, *Z. Phys. D* **33**, 229 (1995).
[57] M. Petersilka and E.K.U Gross, *Int. J. Quantum Chem.* **60**, 181 (1996).
[58] M. Petersilka, U.J. Gossmann, and E.K.U Gross, in *Electronic Density Functional Theory: Recent Progress and New Directions*, edited by J.F. Dobson, G. Vignale, and M.P. Das (Plenum, New York, 1997), p. 177.
[59] P. Pyykkö, *Chem. Rev.* **88**, 563 (1988).
[60] O.K. Andersen, *Phys. Rev. B* **2**, 883 (1970).
[61] A.K. Rajagopal and J. Callaway, *Phys. Rev. B* **7**, 1912 (1973).
[62] E. Engel, H. Müller, C. Speicher, and R.M. Dreizler, in *Density Functional Theory*, edited by E.K.U. Gross and R.M. Dreizler (Plenum Press, New York, 1995), p. 65.
[63] E. Engel and R.M. Dreizler, in *Density Functional Theory II*, Vol. 181 of *Topics in Current Chemistry*, edited by R.F. Nalewajski (Springer, Berlin, 1996), p. 1.
[64] E. Engel, S. Keller, A. Facco Bonetti, H. Müller, and R.M. Dreizler, *Phys. Rev. A* **52**, 2750 (1995).
[65] A.K. Rajagopal, *J. Phys. C* **11**, L943 (1978).
[66] A.H. MacDonald and S.H. Vosko, *J. Phys. C* **12**, 2977 (1979).
[67] C. Itzykson and J.-B. Zuber, *Quantum Field Theory* (McGraw-Hill, New York, 1980).
[68] T. Kreibich, E.K.U. Gross, and E. Engel, *Phys. Rev. A* **57**, 138 (1998).
[69] B.A. Shadwick, J.D. Talman, and M.R. Norman, *Comput. Phys. Commun.* **54**, 95 (1989).
[70] E. Engel, S. Keller, and R.M. Dreizler, *Phys. Rev. A* **53**, 1367 (1996).
[71] E. Engel and S.H. Vosko, *Phys. Rev. B* **47**, 13164 (1993).
[72] J.D. Talman, *Comput. Phys. Commun.* **54**, 85 (1989).
[73] Y. Wang, J.P. Perdew, J.A. Chevary, L.D. Macdonald, and S.H. Vosko, *Phys. Rev. A* **41**, 78 (1990).
[74] Y. Li, J.B. Krieger, J.A. Chevary, and S.H. Vosko, *Phys. Rev. A* **43**, 5121 (1991).
[75] A.D. Becke, *Phys. Rev. A* **38**, 3098 (1988).
[76] J.P. Perdew, in *Electronic Structure of Solids '91*, edited by P. Ziesche and H. Eschrig (Akademie Verlag, Berlin, 1991), p. 11.
[77] J.P. Perdew, K. Burke, and Y. Wang, *Phys. Rev. B* **54**, 16533 (1996).
[78] K. Burke, J.P. Perdew, and Y. Wang, in *Electronic Density Functional Theory: Recent Progress and New Directions*, edited by J.F. Dobson, G. Vignale, and M.P. Das (Plenum, New York, 1997), p. 81.
[79] C. Froese Fischer, *The Hartree-Fock method for atoms* (Wiley, New York, 1977).

[80] J.C. Slater, *Structure of Molecules and Solids* (McGraw-Hill, New York, 1960), Vol. 4.
[81] E.K.U. Gross, M. Petersilka, and T. Grabo, in *Chemical Applications of Density Functional Theory, ACS Symposium Series 629*, edited by B.B Laird, R.B. Ross, and T. Ziegler (American Chemical Society, Washington, DC, 1996), p. 42.
[82] D. Heinemann, A. Rosén, and B. Fricke, Phys. Scr. **42**, 692 (1990).
[83] L. Laaksonen, P. Pyykkö, and D. Sundholm, *Int. J. Quantum Chem.* **23**, 309 (1983).
[84] L. Laaksonen, P. Pyykkö, and D. Sundholm, *Int. J. Quantum Chem.* **23**, 319 (1983).
[85] L. Laaksonen, D. Sundholm, and P. Pyykkö, *Int. J. Quantum Chem.* **28**, 601 (1985).
[86] L. Laaksonen, P. Pyykkö, and D. Sundholm, *Comp. Phys. Reports* **4**, 313 (1986).
[87] J.C. Slater, *Phys. Rev.* **81**, 385 (1951).
[88] T. Grabo and E.K.U. Gross, *Int. J. Quantum Chem.* **64**, 95 (1997).
[89] C. Jamorski, M.E. Casida, and D.R. Salahub, *J. Chem. Phys.* **104**, 5134 (1996).
[90] C. Filippi and C.J. Umrigar, *J. Chem. Phys.* **105**, 213 (1996).
[91] S.J. Chakravorty, S.R. Gwaltney, E.R. Davidson, F.A. Parpia, and C. Froese Fischer, *Phys. Rev. A* **47**, 3649 (1993).
[92] J.B. Krieger, J. Chen, and G.J. Iafrate, *Bull. Am. Phys. Soc.* **41**, 748 (1996).
[93] J. Chen, J.B. Krieger, and G.J. Iafrate, *Bull. Am. Phys. Soc.* **41**, 748 (1996).
[94] J.D. Bjorken and S.D. Drell, *Relativistic Quantum Fields* (McGraw-Hill, New York, 1965).
[95] E. Engel, S. Keller, and R.M. Dreizler, in *Electronic Density Functional Theory: Recent Progress and New Directions*, edited by J.F. Dobson, G. Vignale, and M.P.Das (Plenum, New York, 1997), p. 149.
[96] K. Hikasa et al. (Particle Data Group), Phys. Rev. D **45**, S1 (1992).
[97] R. Colle and D. Salvetti, *Theoret. Chim. Acta* **37**, 329 (1975).
[98] R. Colle and D. Salvetti, *Theoret. Chim. Acta* **53**, 55 (1979).
[99] C. Lee, W. Yang, and R.G. Parr, *Phys. Rev. B* **37**, 785 (1988).
[100] J.P. Perdew and Y. Wang, *Phys. Rev. B* **45**, 13244 (1992).
[101] J.A. Montgomery, J.W. Ochterski, and G.A. Petersson, *J. Chem. Phys.* **101**, 5900 (1994).
[102] E.R. Davidson, S.A. Hagstrom, S.J. Chakravorty, V.M. Umar, and C. Froese Fischer, *Phys. Rev. A* **44**, 7071 (1991).
[103] A.A. Radzig and B.M. Smirnov, *Reference Data on Atoms and Molecules* (Springer, Berlin, 1985).
[104] T. Grabo, T. Kreibich, and E.K.U. Gross, *Mol. Engineering* **7**, 27 (1997).
[105] B.G. Johnson, P.M.W. Gill, and J.A. Pople, *J. Chem. Phys.* **98**, 5612 (1993).

[106] K.P. Huber and G. Herzberg, *Molecular Spectra and Molecular Structure: IV. Constants of Diatomic Molecules* (Van Nostrand Reinhold, New York, 1979).

[107] W. Kolos and L. Wolniewicz, *J. Chem. Phys.* **49**, 404 (1968).

[108] K.A. Peterson, R.A. Kendall, and T.H. Dunning, *J. Chem. Phys.* **99**, 1930 (1993).

[109] O.V. Gritsenko, R. van Leeuwen, and E.J. Baerends, *J. Chem. Phys.* **104**, 8535 (1996).

[110] M.A. Buijse and E.J. Baerends, in *Density Functional Theory of Molecules, Clusters and Solids*, edited by D.E. Ellis (Kluwer Academic Publishers, Amsterdam, 1995), p. 1.

[111] T. Kotani and H. Akai, *Phys. Rev. B* **54**, 16502 (1996).

[112] T. Kotani and H. Akai, *Physica* **B237–238**, 332 (1997).

[113] T. Kotani, *J. Phys.: Condens. Matter* **10**, 9241 (1998).

[114] D.M. Bylander and L. Kleinman, *Phys. Rev. Lett.* **74**, 3660 (1994).

[115] D.M. Bylander and L. Kleinman, *Phys. Rev. B* **52**, 14566 (1995).

[116] D.M. Bylander and L. Kleinman, *Phys. Rev. B* **54**, 7891 (1996).

[117] D.M. Bylander and L. Kleinman, *Phys. Rev. B* **55**, 9432 (1997).

[118] M. Städele, J. A. Majewski, P. Vogl, and A. Görling, *Phys. Rev. Lett.* **79**, 2089 (1997).

[119] *Numerical Data and Functional Relationships in Science and Technology*, edited by K.-H. Hellwege, Landolt-Börnstein, *New Series, Group III, Vol 17*, edited by O. Madelung amd M. Schulz and H. Weiss (Springer, Berlin, 1982).

[120] D. Straub, L. Ley, and F.J. Himpsel, *Phys. Rev. Lett.* **54**, 142 (1985).

[121] J.E. Ortega and F.J. Himpsel, *Phys. Rev. B* **47**, 2130 (1993).

[122] A. Mainwood, in *Properties and Growth of Diamond*, edited by G. Davies (Electronic Materials Information Service, London, 1994), p. 3.

[123] A. Görling and M. Levy, *Phys. Rev. B* **47**, 13105 (1993).

[124] M.E. Casida, *Phys. Rev. A* **51**, 2005 (1995).

[125] J. Luttinger and J. Ward, *Phys. Rev.* **118**, 1417 (1960).

[126] E.K.U. Gross, E. Runge, and O. Heinonen, *Many-Particle Theory* (Adam Hilger, Bristol, 1991).

[127] L. Hedin and S. Lundqvist, *Solid State Physics* **23**, 1 (1969).

Index

J

K

L

M

N

T

V

W

Y

Z

Printed in the United States
by Baker & Taylor Publisher Services